THE
INTERNATIONAL SERIES
OF
MONOGRAPHS ON PHYSICS

Undulators and Free-electron Lasers

P. Luchini

Institute of Gas Dynamics, School of Engineering, University of Naples, Italy
and
National Institute for Nuclear Physics (INFN), Naples, Italy

and

H. Motz†

University of Oxford, England

CLARENDON PRESS · OXFORD
1990

Oxford University Press, Walton Street, Oxford OX2 6DP
Oxford New York Toronto
Delhi Bombay Calcutta Madras Karachi
Petaling Jaya Singapore Hong Kong Tokyo
Nairobi Dar es Salaam Cape Town
Melbourne Auckland
and associated companies in
Berlin Ibadan

Published in the United States
by Oxford University Press, New York

British Library Cataloguing in Publication Data
Luchini, P. (Paolo)
Undulators and free-electron lasers.
1. Free-electron lasers
I. Title II. Motz, H. (Hans), 1909–1987 III. Series
621.36'6
ISBN 0-19-852019-0

Library of Congress Cataloging in Publication Data
Luchini, P. (Paolo)
Undulators and free-electron lasers / P. Luchini and H. Motz.
p. cm. — (The International series of monographs on physics)
Includes bibliographical references.
ISBN 0-19-852019-0
1. Free electron lasers I. Motz, H. (Hans), d. 1987
II. Title III. Series: International series of monographs on physics (Oxford, England)
TA1695.L82 1990
621.36'6—dc20 89-26542

Set by APS Ltd., Salisbury, Wiltshire and
Printed in Great Britain by
Courier International Ltd
Tiptree, Essex

PREFACE

This is not the first book written on free-electron lasers, coming after T. C. Marshall's *Free-electron Lasers* (Macmillan, New York, 1985) and other less widely known reviews such as the chapter by G. Dattoli and A. Renieri in Volume 4 of the *Laser Handbook* edited by M. L. Stitch and M. Bass (North Holland, Amsterdam, 1985), and the report of Tel Aviv University (IAS 840-1987) on 'Spontaneous and Stimulated Emission from Quasi-free Electrons' by A. Gover, A. Friedman, G. Kurizki, S. Ruschin, and A. Yariv.

However, there are substantial differences between these and the present work. Marshall's book is intended, according to its own preface, as an introduction for beginners which could be read in three evenings, something to fill in the gap between mass-media news and professional papers and, as such, does a good job in describing the basic mechanisms and presenting the main results, while avoiding complication and referring the details of theory to the literature. The chapter by Dattoli and Renieri and the report by Gover *et al.* emphasize the very basic quantum-mechanical foundations of FEL operation. The present book tries to give a self-contained and reasonably thorough theory of undulators and free-electron lasers, self-contained in the sense that everything is derived from the first principles of classical relativistic mechanics and electromagnetism, and thorough in the sense that, although a major effort has been put into selecting the simplest and most explicative derivation of any single issue, more elaborate mathematical formulations are not avoided where they are justified by the practical relevance of the result.

In addition, Marshall's longer experience with long-wavelength (millimetre- and submillimetre-wave) FELs shows up through a greater emphasis put on these devices in his book, whereas here the emphasis is primarily on short-wavelength (infrared, visible, ultraviolet and, possibly in the future, X-ray) FELs.

The concept of this book originated in 1984, when I was invited as a Merrill Visiting Associate by St. Catherine's College, Oxford, from the encounter of two seemingly very different personalities: an emeritus Oxford professor with a long teaching and research career, already author of several monographs (Motz), and a young Italian scientist with much enthusiasm and a number of publications in scientific journals but no previous book writing experience (Luchini). We shared an interest in undulators and free-electron lasers and an interdisciplinary approach to science which had led each one of us to study, in addition to the physics of particle beams and lasers, topics as different as nervous pulse conduction (Motz) and fluid dynamics (Luchini). Professor Motz had already been, for some time, considering the possibility of writing a book on the undulator which, unknown to the majority of physicists for many years after he invented it in 1951 (although important experimental developments were being carried out in a few synchrotron-radiation laboratories), had finally become a widespread topic of research in the late 1970s after the development of the free-electron laser, and when we met he thought that we were the right combination of people to do it. Our international collaboration was made possible, in subsequent years, by the financial assistance of INFN and of the Leverhulme Trust. We spent almost two years planning and re-thinking the structure of the book, which we wanted to be as self-contained and unified as possible. Our aim was a systematic organization of the subject, which could be useful both as a learning tool for scientists or graduate students wishing to enter the field of undulator and free-electron laser research, or just feeling curious about it, and as a reference text for those already working in this field. In order to make the book accessible to as large an audience as possible we decided that only a basic understanding of relativistic mechanics and electromagnetic theory would be required of the reader, with no previous, other than incidental, knowledge of related fields such as particle accelerators or conventional lasers. Also, since some readers might not be totally familiar with some of the elements of Hamiltonian relativistic mechanics and classical electromagnetic radiation theory which underlie the deductive structure of the presentation, we deemed that a brief introduction should have been provided to both, and did so for the first in Appendix A and for the second in the first three sections of Chapter 4. Although according to the initial division of tasks the historical and technical parts of the book were to be handled mainly by Motz and the theoretical part mainly by me, we have actually both contributed to both parts, with suggestions and heavy revisions of the original draft. In particular, my own tendency to abstract mathematical deductions (my 'latinity of thought', as he used to say) was moderated and complemented on many occasions by his (Anglo-Saxon?) pragmatic approach and by his long-time knowledge of which concepts would be more familiar to the typical reader and which needed further explanation.

Unfortunately, my good friend could not see the end of our work, as he died in August 1987. The memory of his earnest and passionate contribution to science will remain vivid in all those who knew him and, perhaps, in some of the readers of this book.

Naples P. L.
September 1988

CONTENTS

BIBLIOGRAPHICAL NOTE

As undulator and free-electron laser literature is recent and today's bibliographical reference systems are very efficient (the *Physics Abstracts Journal*, for one, has an entire section devoted to free-electron lasers) we decided against competing with such media and intentionally did not try to compile a complete bibliography, giving instead a reference only where we actually followed the development of a single paper. In many sections no references are given, not because the topic has not been dealt with earlier but because, in order to give a unified character to the exposition, we have mostly followed our own approach; the concepts involved in these sections have evolved in the scientific literature from the contributions of many writers in the whole world, and quoting just one of them would not do justice to the others. On the other hand, the main purpose of this book is to give a systematic exposition of undulator and FEL theory, not to assess priorities. There are bound to be many valuable scientists whose contributions are never mentioned, to whom we apologize.

For the reader who wishes to extend his study of the matter further, we may suggest the proceedings of the initially bi-yearly and now yearly Free-electron Laser Conferences, listed below, and the special issues of the *IEEE Journal of Quantum Electronics* devoted every other year since 1981 to free-electron lasers. A number of relevant papers are also published every year by such journals as *Physical Review A*, *Physics of Fluids*, *Journal of Applied Physics*, and *Applied Physics B*, or appear in the proceedings of conferences on related topics such as accelerator technology.

The following abbreviations will be used for the proceedings of the Free-electron Laser Conferences wherever references are given to them:

*FEL*77: ONR Workshop on Novel Sources of Coherent Radiation, Telluride, Colorado, U.S.A., August 1977; *Novel sources of coherent radiation* (ed. S. F. Jacobs, M. Sargent III, and M. O. Scully) Physics of Quantum Electronics, Vol. 5. Addison-Wesley, Reading, Massachusetts, 1978.

*FEL*79: 2nd ONR Workshop on Free-Electron Generators of Coherent Radiation, Telluride, Colorado, U.S.A., August 1979; *Free-electron generators of coherent radiation* (ed. S. F. Jacobs, H. S. Pilloff, M. Sargent III, M. O. Scully, and R. Spitzer) Physics of Quantum Electronics, Vol. 7. Addison-Wesley, Reading, Massachusetts, 1980.

*FEL*81: 3rd ONR Workshop on Free-Electron Laser Devices, Sun Valley, Idaho, U.S.A., June 1981; *Free-electron generators of coherent radiation* (ed. S. F. Jacobs, G. T. Moore, H. S. Pilloff, M. Sargent III, M. O. Scully, and R. Spitzer) Physics of Quantum Electronics, Vols 8 & 9. Addison-Wesley, Reading, Massachusetts, 1982.

*FEL*82: Free-Electron Laser Conference, Bendor, France, September 1982; *Free-electron generators of coherent radiation. J. Phys. Colloq.* **44-C1** (1983).

*FEL*83: Free-Electron Laser Conference, Orcas Island, Washington, U.S.A., July 1983; *Free-electron generators of coherent radiation* (ed. C. A. Brau, S. F. Jacobs, and M. O. Scully) Proc. SPIE, Vol. 453. Society of Photo-optical Instrumentation Engineers, Bellingham, Washington, 1984.

*FEL*84: 1984 Free-Electron Laser Conference, Castelgandolfo (Rome), Italy, September 1984; *Free-electron lasers* (ed. J. M. J. Madey and A. Renieri) *Nucl. Instr. Meth. Phys. Res.* **A237** (1985).

*FEL*85: Seventh International Conference on Free-Electron Lasers, Tahoe City, California, U.S.A., September 1985; *Free-electron lasers* (ed. E. T. Scharlemann and D. Prosnitz) *Nucl. Instr. Meth. Phys. Res.* **A250** (1986).

*FEL*86: VIII International FEL Conference, Glasgow, U.K., September 1986; *Free-electron lasers*, (ed. M. W. Poole) *Nucl. Instr. Meth. Phys. Res.* **A259** (1987).

*FEL*87: IX International FEL Conference, Williamsburg, Virginia, U.S.A., September 1987; *Free-electron lasers*, (ed. P. Sprangle, C. M. Tang and J. Walsh). *Nucl. Instr. Meth. Phys. Res.* **A272** (1988).

1

EARLY DEVELOPMENTS

An undulator is a device for generating electromagnetic radiation in a narrow spectral interval at frequencies ranging from radio or millimetre waves to X-rays. This is achieved by letting an electron beam with energy ranging from 1 MeV, say, to 1000 MeV pass through an array of magnets. If the device is placed into an optical resonator with mirrors at both ends, coherent radiation may be produced. By analogy with conventional lasers, it is then called a Free-electron Laser (FEL) (this term was introduced by Madey 1971a, b), although its manner of functioning is rather different from that of atomic or molecular lasing media.

In this chapter a general overview of the subject is presented through a description of the arguments and results contained in some of the early papers. The themes are systematically taken up and treated in detail in later chapters.

1.1 The undulator

Whenever an electron beam changes direction it shakes off radiation in a continuous frequency band. The most conspicuous example is the intense radiation produced by electrons in a synchrotron orbit. In many laboratories this synchrotron radiation, channelled tangentially out of the machine, is used by scientists working in solid state physics, chemistry, biology, etc., for experiments needing intense illumination. It is sometimes concentrated in a certain frequency range by 'wiggling' the beam as it leaves the machine with the help of a few magnets so as to follow a shape like the outline of a camel's back. Such a device is called a *wiggler*. Many wiggles in succession, say 50 or more, serve to concentrate the radiation spatially in a narrow cone, and spectrally in a narrow frequency interval. The beam is made wavy and waves are produced, and for this reason a multi-period wiggler is called an *undulator* (this term was introduced by Motz in 1951, and means wave-maker, from '*unda*', Latin for 'wave').

In the simplest case the undulator structure consists of a succession of magnet pairs of alternating polarity, as illustrated in Fig. 1.1. With suitable pole pieces the magnetic field near the axis is very nearly sinusoidal; the spatial period of the sine wave will be designated as l_u. The exact determination of the orbits of high-energy electrons of velocity v passing between the

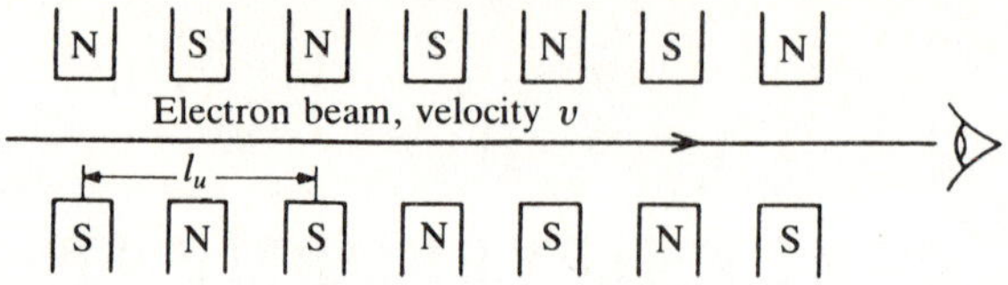

FIG. 1.1. The undulator structure: the electron beam traverses an arrangement of magnets, the fields of which alternate in direction with spatial period l_u.

pole pieces leads to near-sinusoidal orbits in a plane perpendicular to that of the figure. The elliptic functions representing these orbits (Combe and Feix 1953*a*) can be approximated by

$$x = x_0 + A \sin 2\pi z/l_u, \qquad y = y_0, \qquad z = z_0 + v_z t. \tag{1.1a–c}$$

Here v_z is the longitudinal velocity of the electrons inside the undulator and A is the amplitude of their transverse motion in the x–z plane. By means of a simple argument based on special relativity (Motz 1951) it is easy to determine the fundamental frequency of the radiation emitted by a single electron during its passage.

Imagine an observer attached to the electron's frame of reference, moving with speed v_z (the E-frame). He looks at the magnet structure and sees it contracted by a Lorentz factor $\gamma_z = (1 - \beta_z^2)^{-1/2}$, where β_z, as usual, represents v_z/c. For such an observer the wavelength of the electron's oscillation is shortened from l_u to $l'_u = l_u/\gamma_z$, and the frequency of its radiation is increased from $f = v_z/l_u$ to $f' = \gamma_z v_z/l_u$. An observer in the laboratory frame of reference (L-frame), placed as the eye in Fig. 1.1, sees the frequency radiated by the electron which is moving towards him Doppler-shifted by a further factor of γ_z so that, altogether, the observed frequency is proportional to γ_z^2.

To appreciate the obtainable effect, consider the relativistic relation[1]

$$\mathscr{E}_k/m_e c^2 = \gamma - 1, \tag{1.2}$$

where $\mathscr{E}_k$ is the kinetic energy of the electron and m_e its rest mass. The value of $m_e c^2$ is approximately 0.5 MeV and therefore, for 2 MeV electrons, γ^2 is 25, which shows that the frequency increase can be large.

Actually, the frequency increase factor turns out to be twice this, as can be seen from the more detailed argument below which applies to the L-frame and requires no appeal to special relativity.

Figure 1.2 represents what happens when an electron passes through a magnetic-field structure of spatial period l_u, moving from A to B with axial velocity v_z. Consider the radiation emitted at an angle ϑ to the line AB. Just as in any other optics problem, we consider the condition for the formation of a phase front BD, normal to the direction of propagation. The phase difference, expressed here as a time delay between the radiation from B

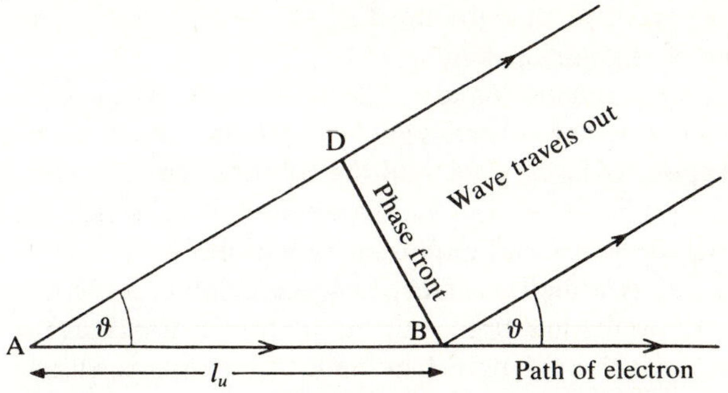

FIG. 1.2. Derivation of the condition for synchronism. The electron moves from A to B while the wave moves from A to D.

and from A, must equal the wave period, τ, or any multiple thereof. Thus we must have

$$\frac{l_u}{v_z} - \frac{l_u \cos \vartheta}{c} = \frac{l_u}{v_z}\left(1 - \frac{v_z}{c}\cos \vartheta\right) = p\tau, \tag{1.3}$$

where p is an integer.

The variation of the emitted frequency $f = 1/\tau$ with ϑ is given by

$$f = \frac{pv_z}{l_u[1 - (v_z/c)\cos \vartheta]}. \tag{1.4}$$

According to the definition of γ_z, $\gamma_z^{-2} = 1 - (v_z/c)^2 = (1 - v_z/c)(1 + v_z/c)$. For relativistic electrons the speed v_z is close to the speed of light in vacuum, c, so that $1 + v_z/c \simeq 2$ and $2(1 - v_z/c) \simeq \gamma_z^{-2}$. Thus, for $\vartheta = 0$,

$$f = 2p\gamma_z^2 v_z/l_u. \tag{1.5}$$

It is seen that moderately relativistic electrons can radiate a frequency in the millimetre-wave range (100 GHz) along the axis when passing through a structure with a spatial period of, say, 2 cm. For generation of waves at one micrometre, $\gamma = 100$, i.e. a 50 MeV beam, is needed. There exist several other devices for generating millimetre waves using oscillating electron beams, e.g. the ubitron of Phillips (1960), based on a patent by Gorn, the cyclotron maser of Hirshfield *et al.* (1964, 1965), and others. However, it is impossible to scale the wavelength of radiation generated by these devices down into the infrared, because the magnetic fields required would be of the order of $10^3 \div 10^4$ T.

Equation (1.3) shows that the radiated frequency is a maximum along the axis and shades off continuously towards lower values with increasing ϑ. It is easy to see that the frequency of any harmonic is reduced by, say, a factor of two when $\vartheta = \gamma_z^{-1}$, so that the most of the radiation is concentrated in a narrow cone of angular opening $\vartheta_r = 1/\gamma_z$.

It is interesting to consider eqn (1.3) for the case when $p = 0$. In vacuum there is no solution for ϑ because $v_z < c$, but in a medium with refractive index n the speed of light is c/n, and the solution, $\cos \vartheta = c/nv_z$, determines the angle under which Cherenkov radiation is emitted, which can therefore be regarded as a special case of undulator radiation.

The total energy radiated is most easily calculated in the E-frame, as that radiated by an oscillating dipole during the time it traverses the length L' of the undulator, which has length L in the L-frame; that is, all quantities being now distinguished by primes,

$$W' = \frac{1}{6\pi\varepsilon_0}\left(\frac{e\dot{v}'}{c^2}\right)^2 \frac{L'}{\beta_z}. \tag{1.6}$$

This result is then transformed into the L-frame by means of the formulae

$$W = (W' - c\beta_z p'_{z'})/(1 - \beta_z^2)^{1/2}, \qquad p_z = [p'_{z'} - \beta_z W'/c]/(1 - \beta_z^2)^{1/2}, \tag{1.7a,b}$$

where the momentum $p'_{z'}$ and energy W' are in the E-frame, and therefore $p'_{z'} = 0$.

The magnetic field does not produce any force on the electron in the E-frame, but the Lorentz transformation

$$E'_{x'} = (E_x + \beta_z B_y)/(1 - \beta_z^2)^{1/2}$$

yields an electric force in the x-direction which, in this frame, does produce an acceleration,

$$\dot{v}' = -e\beta_z B_y/m_e(1 - \beta_z^2)^{1/2}, \tag{1.8}$$

so that in the L-frame the energy becomes

$$W = \frac{1}{6\pi\varepsilon_0}\left(\frac{e^2}{m_e c^2}\right)^2 \beta_z N l_u \gamma_z^2 B^2 \tag{1.9}$$

for an undulator of length Nl_u.

1.2 The spectral distribution of energy

It is important to know the energy radiated into different directions as a function of frequency (Motz 1951; Combe and Feix 1953*b*). According to Motz (1951), a suitable starting point for the analysis is the expression for the time-averaged energy $P_{k\omega}(\boldsymbol{r})\,\mathrm{d}\omega$ emitted into a frequency range between ω and $\omega + \mathrm{d}\omega$ by a current with density distribution $\boldsymbol{J}(\boldsymbol{r}')$. Let the observer be situated at a point with radius vector $\boldsymbol{r}$ drawn from the origin of the co-ordinate system. Let the distance $\boldsymbol{r}' - \boldsymbol{r}$ be large compared to the radiating current system, and let $\boldsymbol{k}$ be a vector of magnitude $k = \omega/c$, which points from the current system to the observer at $\boldsymbol{r}$. Then the average radiated energy is given by[2]

$$P_{k\omega}(\boldsymbol{r})\,\mathrm{d}\omega = \frac{k^2}{2\pi r^2 c}\left|\int \boldsymbol{J}_{\perp k\omega}(\boldsymbol{r}')e^{-i\boldsymbol{k}\cdot\boldsymbol{r}'}\,\mathrm{d}\tau'\right|^2 \mathrm{d}\omega, \tag{1.10}$$

where $\boldsymbol{J}_{\perp k\omega}(\boldsymbol{r}')$ is the Fourier amplitude of the current density vector projected on to the plane perpendicular to $\boldsymbol{k}$. For a derivation of (1.10) see Schiff (1949), p. 264. On defining

$$\Omega_\vartheta = 2\pi v_z/l_u(1 - \beta_z \cos\vartheta), \tag{1.11}$$

$$Z = \left(\frac{\omega}{\Omega_\vartheta}\right)\frac{eB(1-\beta_z^2)^{1/2}l_u \sin\vartheta\cos\varphi}{2\pi m_e c^2(1-\beta_z\cos\vartheta)}, \tag{1.12}$$

$$S_p = \int_{-N\pi}^{N\pi} \exp[i(\omega/\Omega_\vartheta - p)u]\mathrm{d}u = \frac{2\sin[(\omega/\Omega_\vartheta - p)N\pi]}{\omega/\Omega_\vartheta - p}, \tag{1.13}$$

where ϑ and φ are spherical angular co-ordinates and Nl_u is the length of the undulator, the result of the calculation carried out by Motz (1951), the energy radiated per unit area into the pth harmonic of the fundamental frequency Ω_ϑ, is given by:

$$\frac{1}{(1-\beta_z\cos\vartheta)^2}\left(\frac{\omega}{\Omega_\vartheta}\right)^2\frac{e^2\beta_z^2 S_p^2}{4\pi^2 r^2 c}$$

$$\times\left\{\sin^2\vartheta[\mathrm{J}_p(Z)]^2 + \tfrac{1}{4}\sin^2\chi[\mathrm{J}_{p+1}(Z) + \mathrm{J}_{p-1}(Z)]^2\,\frac{e^2B^2l_u^2(1-\beta_z^2)}{4\pi^2(m_e c^2)^2}\right\}, \tag{1.14}$$

where $\mathrm{J}_p(Z)$ is a Bessel function of order p, and

$$\sin^2\chi = 1 - \cos^2\varphi\,\sin^2\vartheta. \tag{1.15}$$

S_p, of course, has a large maximum in the narrow frequency range

$$\omega/\Omega_\vartheta = p \pm \pi/N \tag{1.16}$$

and subsidiary maxima. It is seen that ω is restricted to small frequency bands centred at the harmonics of the fundamental frequency Ω_ϑ.

Approximating the Bessel function J_p by the first term of its Taylor expansion, $Z^p/p!2^p$, we see that the energy in the higher harmonics is small compared to that in the fundamental when $e\beta_z Bl_u$ is small compared to $m_e c^2$, and that the total energy radiated into the fundamental ($p = 1$) under angles ranging from 0 to ϑ becomes:

$$\frac{1}{8\pi}\left(\frac{e^2}{m_e c^2}\right)^2 \beta_z^3 B^2 N l_u (1 - \beta_z^2)$$

$$\times \int_0^{2\pi} \mathrm{d}\varphi \int_0^{\vartheta} \mathrm{d}\vartheta \left[\frac{\sin^5\vartheta \cos^2\varphi}{(1 - \beta_z \cos\vartheta)^5} + \frac{\sin\vartheta - \sin^3\vartheta \cos^2\varphi}{(1 - \beta_z \cos\vartheta)^3}\right]\left(\frac{\omega}{\Omega_\vartheta}\right)^2. \qquad (1.17)$$

The energy distribution has a minimum for $\varphi = 90°$, i.e. in the direction perpendicular to the plane of the transverse oscillation. The frequency spread depends on the angle under which the radiation is received. Suppose that we aim at a frequency spread of 5 per cent; then

$$\Delta\Omega_\vartheta/\Omega_\vartheta \simeq -\beta_z \Delta \cos\vartheta/(1 - \beta_z \cos\vartheta) \simeq \beta_z \vartheta^2/2(1 - \beta_z) = \tfrac{1}{20}, \qquad (1.18)$$

$$\vartheta^2 = \tfrac{1}{20}\,\frac{2(1 - \beta_z)}{\beta_z} \simeq \frac{1}{20\gamma_z^2}.$$

The result of the integration (1.17) shows that the energy received into a cone of opening ϑ is approximately 1/20 of the total radiated energy given by (1.9), which is of the order of the magnetic-field energy swept out by an electron with the classical radius $e^2/m_e c^2$.

1.3 The effect of bunching the beam

So far we have only looked at the energy radiated by a single electron. The radiation due to N electrons is proportional to the square of the total radiated field emitted by all the electrons of the beam. Let electrons numbered $i = 1, 2, 3, \ldots, N$ emit fields E_i. The power will be proportional to

$$(E_1 + E_2 + \cdots + E_N)^2 = E_1^2 + E_2^2 + \cdots + E_N^2 + 2E_1E_2 + \cdots + 2E_1E_3 + \cdots$$

If the phases of the fields are random all the double products will vanish on averaging, and the power will be proportional to the number of electrons, N. But when the phases are *not* random, the $N(N - 1)$ double products yield, for large N, a power proportional to N^2, which is much greater than N.

It is easy to show that the expectation value is

$$\langle E^2 \rangle = N \int E^2(r) p(r)\, \mathrm{d}r + N(N - 1)\left[\int E(r) p(r)\, \mathrm{d}r\right]^2 \qquad (1.19)$$

if $p(r_i)$ is the probability for the i-th electron to be at a distance r_i from a point where it creates the field $E(r_i)$. Suppose that this probability is given by

$$p(r) = \begin{cases} 1/(2r_0) & \text{for } -r_0 \leqslant r \leqslant r_0, \\ 0 & \text{for } r > r_0 \text{ and } r < r_0, \end{cases}$$

and that the field varies as $A \cos kr$. We then find that $[\int E(r)p(r)dr]^2$ becomes $[(\sin kr_0)/kr_0]^2$, which vanishes for short wavelengths (large values of k) but furnishes a term proportional to $N(N-1)$ in eqn (1.19) for a wavelength λ comparable to r_0; for example, if $k = 2\pi/\lambda$ and $\lambda = 4r_0$.

It follows (Motz 1951) that very large energies could be radiated if it proved possible to bunch the electrons so as to concentrate them into short packets with linear size comparable to the wavelength. We shall mention below experiments with a linear accelerator which had this objective. In the millimetre range this is possible, but fabrication of a device with magnet spacings comparable to, say, a wavelength in the infrared is, to say the least, very difficult. There is no transverse space left for the beam, even though such longitudinal spacings occur naturally in crystals or superlattices. The alternating fields may be electrical ones, and indeed passage through certain channels in solids yields undulator radiation but, of course, the attenuation in solids is high. Passage of a beam near a finely ruled grating is also possible, and Smith and Purcell (1953) have analysed such a device. This, of course, is of theoretical interest but, again, beam currents are bound to be small and high-density beams would tend to burn up the grating.

A different way to provide bunching which does not suffer from wavelength limitations is by means of electromagnetic radiation rather than static fields. However, although it was realized a few years later (Motz and Nakamura 1959*b*) that external radiation could be used for this purpose, thus effectively transforming the undulator into a radiation amplifier, it was not recognized until much later (the laser had not yet been invented) that bunching at however small a wavelength could be generated by the output radiation beam itself if suitable feedback had been provided. This is the principle which eventually gave rise to the free-electron laser.

1.4 Radiation in a rectangular waveguide

Since experiments in the millimetre and submillimetre region are carried out using waveguides, we briefly recall that the phase velocity v_p in a guide is given by

$$v_p = f\lambda_g, \tag{1.20}$$

where λ_g is the wavelength in the guide, defined by

$$\frac{2\pi}{\lambda_g} = k_g = \left(\frac{\omega^2}{c^2} - \alpha^2\right)^{1/2} \tag{1.21}$$

($\omega = 2\pi f$ is the angular frequency). In a rectangular guide with sides of length a and b

$$\alpha^2 = \left(\frac{m\pi}{a}\right)^2 + \left(\frac{n\pi}{b}\right)^2, \tag{1.22}$$

where m and n are integers.

We now want to determine, following the standard electron-tube approach of Motz and Walsh (1962), the frequencies ω radiated into the guide, and for this purpose we rewrite (1.3) for $\vartheta = 0$, replacing c by the phase velocity v_p of eqn (1.20):

$$\frac{l_u}{v_z} - \frac{l_u}{v_p} = \frac{2\pi p}{\omega} = \frac{2\pi}{\omega_u} - \frac{2\pi v_z k_g}{\omega \omega_u}, \tag{1.23}$$

where ω_u was defined as

$$\omega_u = 2\pi v_z / l_u. \tag{1.24}$$

Therefore

$$\omega - v_z k_g = p\omega_u \tag{1.25}$$

is equivalent to (1.3). The propagation constant of the guide, k_g, is given by (1.21), so that

$$k_g^2 = (\omega - p\omega_u)^2 / v_z^2 = \omega^2/c^2 - \alpha^2. \tag{1.26}$$

This is an equation for ω, the solution of which is given by:

$$\omega = p\omega_u \gamma_z^2 \pm [(p\omega_u)^2 - (\alpha c/\gamma_z)^2]^{1/2} \beta_z \gamma_z^2. \tag{1.27}$$

Figure 1.3 is helpful for finding the frequencies graphically. It is a plot of the propagation constant k_g as a function of the angular frequency ω. The line

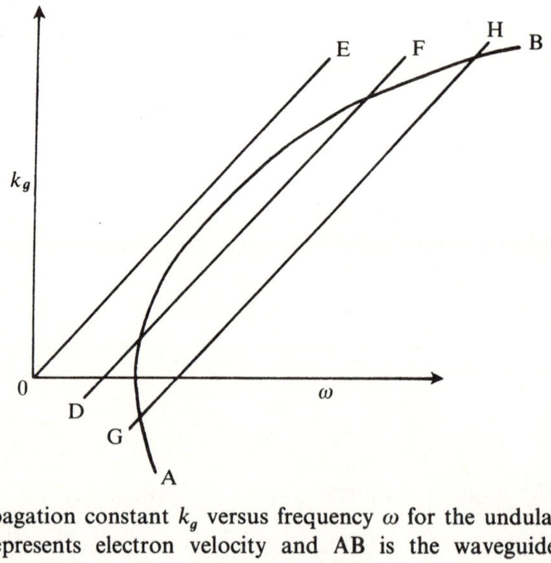

FIG. 1.3. Propagation constant k_g versus frequency ω for the undulator in a waveguide. OE represents electron velocity and AB is the waveguide transmission characteristic. The parallel lines OE, DF, and GH are separated by ω_u along the ω axis.

AB is the waveguide propagation curve for a particular mode, i.e. for a given α, as given by (1.21) and (1.22). The distance OA corresponds to the cut-off frequency αc, where the wavelength of the guide becomes infinite.

The line OE represents the velocity of the electron, i.e. $\omega = v_z k_g$. AB is asymptotic to the line $\omega = ck_g$ for values of ω/c and k_g which are much larger than α. Equation (25) in the form

$$\omega = v_z k_g + p\omega_u$$

is represented by the lines DF and GH for $p = 1$ and $p = 2$. The intersection points of these lines with AB give the roots of eqn (26). There will be two real roots provided that $p\omega_u > \alpha c/\gamma_z$. If $p\omega_u > \alpha c$ the lower-frequency root will represent a wave travelling in the direction opposite to that of the beam.

In the case of

$$p\omega_u = \alpha c/\gamma_z$$

the two roots coincide, so that

$$\omega = p\omega_u \gamma_z^2.$$

At this point the group velocity $\mathrm{d}\omega/\mathrm{d}k_g$ is equal to the electron velocity.

The power flow from the electron to the undulator radiation was calculated by Motz and Nakamura (1959*a*) with some care. It is given by

$$P_{mnp} = \frac{e^2 c |F_{mnp}|^2 p\omega_u [1 + \gamma_z^2 (p\omega_u a/\pi mc)^2]}{\varepsilon_0 [(p\omega_u)^2 - c^2 \alpha_{mn}^2/\gamma_z^2]^{1/2}}, \tag{1.28}$$

with F_{mnp} given by

$$F_{mnp} = \frac{-\mathrm{i}}{(1 + \delta_{m,0})ab} (\mathrm{e}^{\mathrm{i}m\pi x_0/a} - (-1)^p \mathrm{e}^{-\mathrm{i}m\pi x_0/a}) \mathrm{J}_p\left(\frac{m\pi A}{a}\right) \sin\left(\frac{n\pi y_0}{b}\right), \tag{1.29}$$

where δ_{mn} is a Kronecker symbol and J_p is a Bessel function of order p. Equation (28) gives the sum of the forward- and backward-travelling powers which are in the ratio of the respective frequencies (Motz and Nakamura 1959*a*). Since the backward frequency is small compared to the forward one, it is a good approximation to the forward power.

For $m = 0$, the limit for $m \to 0$ of the product

$$(p\omega_u a/\pi mc)^2 [\mathrm{J}_p(m\pi A/a)]^2$$

must be taken.

The analysis of Motz and Nakamura (1959*a*) proceeds in terms of the Hertz potential $\boldsymbol{\Pi}$, from which the electric and magnetic fields can be obtained by the differential operations

$$\mu_0 \boldsymbol{h} = \frac{1}{c^2} \nabla \wedge \frac{\partial \boldsymbol{\Pi}}{\partial t}, \tag{1.30}$$

$$\boldsymbol{e} = \nabla(\nabla \cdot \boldsymbol{\Pi}) - \frac{1}{c^2}\frac{\partial^2 \boldsymbol{\Pi}}{\partial t^2}. \tag{1.31}$$

Maxwell's equations are satisfied if

$$\nabla^2 \boldsymbol{\Pi} - \frac{1}{c^2}\frac{\partial \boldsymbol{\Pi}}{\partial t^2} = -\frac{\boldsymbol{p}}{\varepsilon_0}, \tag{1.32}$$

where the polarization vector $\boldsymbol{p}$ is defined by

$$\boldsymbol{j} = \partial \boldsymbol{p}/\partial t, \qquad \rho = -\mathrm{div}\,\boldsymbol{p}, \tag{1.33}$$

$\boldsymbol{j}$ and ρ being current and charge density.

The boundary condition at the waveguide walls is:

$$\boldsymbol{\Pi}_{\parallel} = 0, \qquad \nabla \cdot \boldsymbol{\Pi} = 0, \tag{1.34}$$

where $\boldsymbol{\Pi}_{\parallel}$ is the tangential component of $\boldsymbol{\Pi}$.

The operator on the left of (1.32) determines a set of orthogonal and normalized eigenfunctions:

$$\varphi_{mnx} = \frac{2}{[(1+\delta_{m0})ab]^{1/2}} \cos\left(\frac{m\pi x}{a}\right)\sin\left(\frac{n\pi y}{b}\right), \tag{1.35a}$$

$$\varphi_{mny} = \frac{2}{[(1+\delta_{n0})ab]^{1/2}} \sin\left(\frac{m\pi x}{a}\right)\sin\left(\frac{n\pi y}{b}\right), \tag{1.35b}$$

$$\varphi_{mnz} = \frac{2}{(ab)^{1/2}} \sin\left(\frac{m\pi x}{a}\right)\sin\left(\frac{n\pi y}{b}\right), \tag{1.35c}$$

with α_{mn} given by eqn (22), are eigenfunctions for the three components of $\boldsymbol{\Pi}$ satisfying the boundary conditions. The solution of (1.32) is found by the Green function technique for current and charge densities:

$$\boldsymbol{j} = (e\omega_u A \cos \omega_u t, 0, ev_z)\Delta, \qquad \rho = e\Delta, \tag{1.36a,b}$$

with

$$\Delta = \delta(x - A \sin \omega_u t - x_0)\delta(y - y_0)\delta(z - v_z t - z_0). \tag{1.36c}$$

The Green function is denoted as G_ν ($\nu = x, y, z$) and is given by

$$G_\nu(\boldsymbol{r}, t, \boldsymbol{r}', t') = \frac{1}{4\pi^2}\sum_{m,n}\int \mathrm{d}\omega \int \mathrm{d}\beta \, \frac{\varphi_{mn\nu}(x, y)\varphi_{mn\nu}(x', y')\mathrm{e}^{\mathrm{i}[\omega(t-t') - \beta(z-z')]}}{\alpha_{mn}^2 + \beta^2 - \omega^2/c^2}. \tag{1.37}$$

The Hertz potential is given by

$$\varepsilon_0 \Pi_\nu = \int G_\nu(\boldsymbol{r}, t, \boldsymbol{r}', t') p_\nu(\boldsymbol{r}', t')\, \mathrm{d}^3 \boldsymbol{r}'\, \mathrm{d}t'.$$

The integral of (1.37) has to be worked out by contour integration and the electric field is then given by (1.31).

The power flow from the electron to the guide field is calculated from the work done,

$$P(t) = -\int \boldsymbol{E}(\boldsymbol{r}, t) \cdot \boldsymbol{j}(\boldsymbol{r}, t) \, \mathrm{d}^3\boldsymbol{r}, \tag{1.38}$$

by the electric field on the current density per unit of time. The result has already been given in eqn (1.28).

1.5 Early experiments

1.5.1 *Millimetre- and submillimetre-wave generation*

Early experiments at Stanford (Motz *et al.* 1953) were aimed at millimetre- and submillimetre-wave generation. The source of high-energy electrons was a special linear accelerator powered by 10 cm wavelength microwaves, designed for the injection system of the Mark III Stanford Linac. In a linac, acceleration occurs by continuous interaction of a longitudinal electric field with the electrons of the beam, which keep in step with the wave. The electrons are, as it were, riding on the wave at a phase which ensures phase-stability, i.e. lagging electrons are accelerated and leading ones are retarded. The microwave in the standard accelerator sections is slightly retarded by inserting discs with holes for the beam which are non-uniformly spaced, so that the wave travels faster as the electrons accelerate. A special problem arises at the first section, the bunching section, where the beam is initially very slow. This problem is overcome by a special design which, at the same time, aims at producing electron bunches that are as short as possible compared to 10 cm, in order to produce, finally, an optimally monochromatic high-energy beam.

The arrangement used for millimetre-wave generation by means of the undulator is shown in Fig. 1.4. The undulator consisted of an assembly of

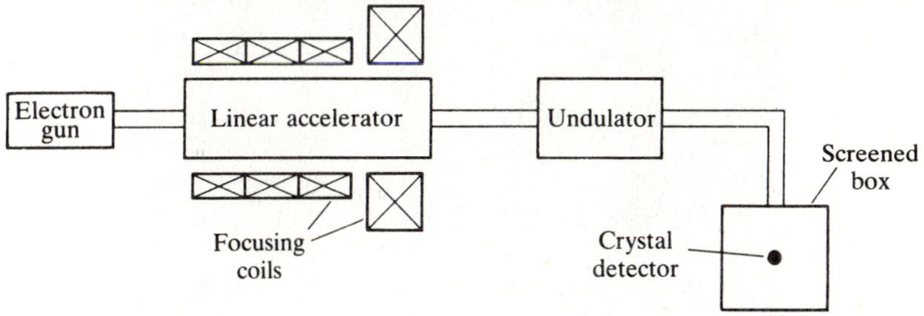

FIG. 1.4. Block diagram of the millimetre-wave generation experiment (from Motz *et al.* 1953).

magnetron magnets, with pole pieces inserted to produce the spatially alternating field. A waveguide was placed between the pole pieces. The magnet assembly is shown in Fig. 1.5, and the construction details of the magnets can be seen in Fig. 1.6.

The magnetic-field distribution in the uniform section near the axis may be approximated as

$$B_y = B\cos(2\pi z/l_u), \qquad B_x = B_z = 0. \tag{1.39}$$

Since

$$\ddot{x} = (e/\gamma m_e)v_z B_y, \tag{1.40}$$

$$\dot{x} = (e/\gamma m_e)\int v_z B_y\,\mathrm{d}t = (e/\gamma m_e)\int B_y\,\mathrm{d}z. \tag{1.41}$$

Taking a point where $x = 0$ as the origin of z, the orbit is given by

$$x = (eBl_u^2/4\pi^2 m_e v_z \gamma)[1 - \cos(2\pi z/l_u)]. \tag{1.42}$$

Early experiments were carried out by Motz *et al.* (1953), Motz and Mallory (1955), Combe (1956), Bernashevsky *et al.* (1960), and Motz and Walsh (1962).

Experimenters also noted the presence of millimetre-wave radiations which were not due to the undulator, but originated in the 10 cm linac. They occur even when the undulator magnetic field is switched off. The millimetre-wave

FIG. 1.5. The magnetic undulator partially assembled (from Motz *et al.* 1953).

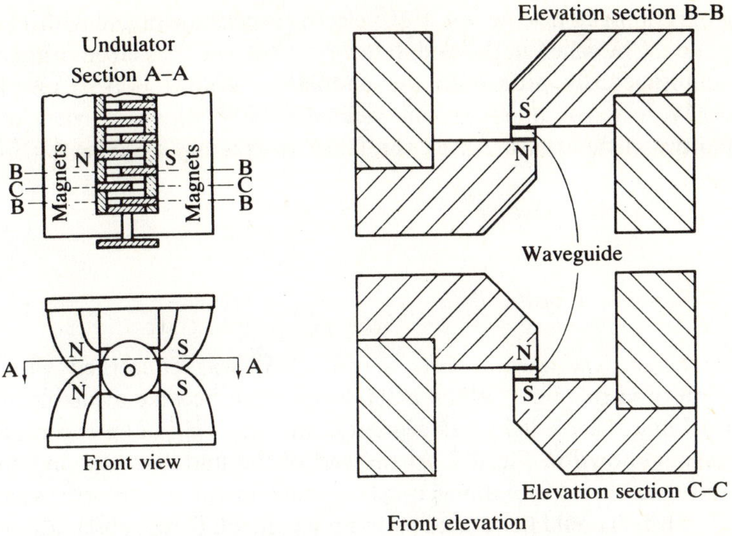

FIG. 1.6. Details of undulator construction (from Motz *et al.* 1953).

power radiated by the undulator observed in the experiment of Motz *et al.* (1953), carried out with a 3–5 MeV linac buncher, was of the order of 1 W. The undulator had 50 pole gaps with $l_u = 40$ mm.

It has been explained before that, in a region of wavelengths of the same order as the bunch length, the power, P, radiated is proportional to the square of the number of electrons per bunch, i.e. to the current of the beam. Theoretically, for perfect bunching:

$$P = AN + BN^2. \tag{1.43}$$

Experimentally it was found (Motz and Walsh 1962) that

$$P \sim N^{1.68} \tag{1.44}$$

in the wavelength region between 6 mm and 8 mm, indicating that the bunch length was approximately 10 mm, i.e. 10 per cent of the accelerator wavelength.

From (1.28) one can find that the power radiated in the H_{01} mode for $p = 1$ is given by

$$P_{011} = 1.9 \times 10^{-15}(Bl_u)^2 \frac{v_z}{abc}\left[\frac{1}{[(v_z/c)^2 - (l_u/2b\gamma_z)^2]^{1/2}} + 1\right], \tag{1.45}$$

which for one electron gives:

$$P = 4.3 \times 10^{-16}\ \text{W}.$$

With a current of 20 mA, i.e. 4.4×10^7 electrons per bunch, and with 11 field periods, i.e. 11 bunches, in the undulator, $N = 4.8 \times 10^8$ electrons altogether were simultaneously present in the undulator. With $P \sim N^{1.7}$ one finds $P = 0.12$ W, whereas measurement yielded 0.55 W which, in view of the uncertainties of the measurement, is a fairly satisfactory agreement.

1.5.2 *Generation of visible light*

Another early experiment (Motz *et al.* 1953) was concerned with the generation of visible light. When a 100 MeV beam became available at the Stanford Linac, visible light was generated with the help of an experimental arrangement shown in Fig. 1.7. At the end of the undulator the evacuated waveguide was closed by a thin glass plate and a mirror was placed, as shown in Fig. 1.7, which could be rotated by remote control. Direct observation was not possible because of the high neutron and gamma-ray level. Therefore a photomultiplier was used to verify that light was indeed reflected from the mirror when it was inclined at 45°. If a colour film had been put there, a picture such as that of Fig. 1.8 would have been obtained. Instead, the colour was checked with the help of an interference filter with a 100 Å wide pass band centred on 4400 Å. Varying the beam voltage from 80 to 120 MeV, maximum transmission was obtained between 90 and 95 MeV. Polarization was investigated by means of two polaroid filters which could be rotated

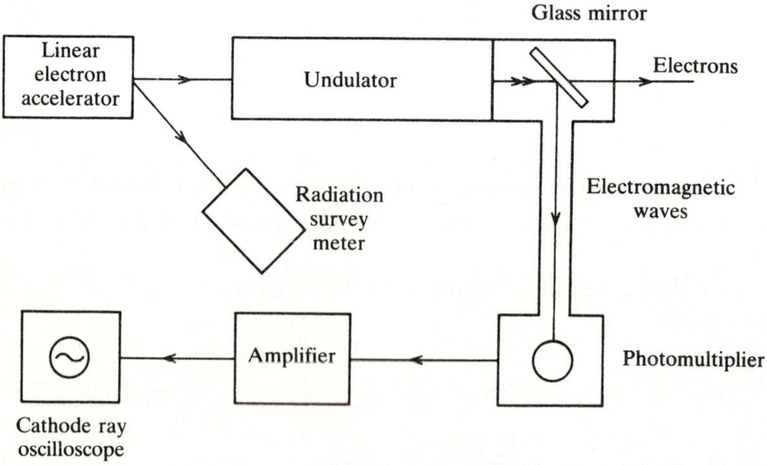

FIG. 1.7. Detector arrangement for the light-generation experiment (from Motz *et al.* 1953).

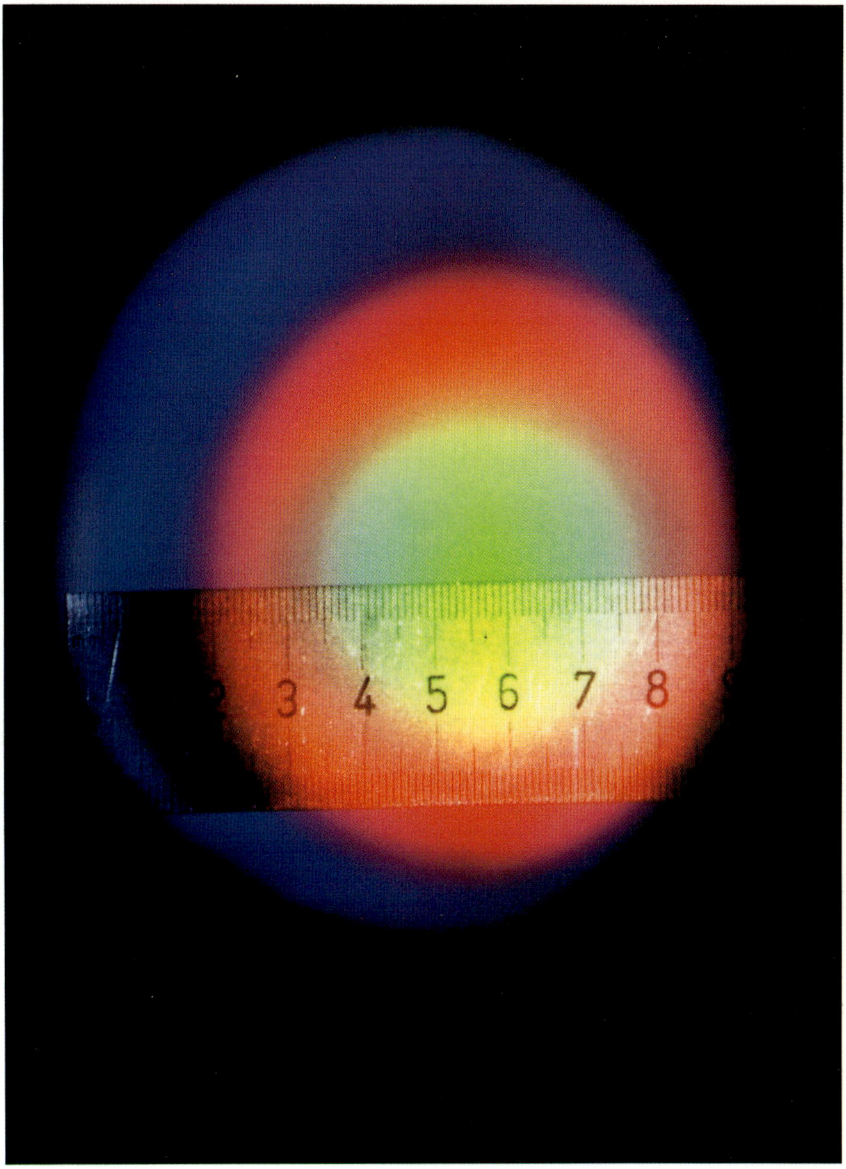

FIG. 1.8. Spontaneous radiation from the LELA undulator on the ADONE storage ring (courtesy of the LELA group of Istituto Nazionale di Fisica Nucleare, INFN, Napoli, Italy).

alternately into the light-beam line. It was found to be mainly perpendicular to the gap field, as theoretically predicted.

1.6 The twister structure

The undulator structure described so far, the linear one, produces linear polarization. It is also possible to generate circularly polarized waves with the help of a structure consisting of a rectangular electromagnet coil twisted in the shape of a double helix, as illustrated in Fig. 1.9. This produces a field $\boldsymbol{B}$ on its axis such that:

$$B_x = B_0 \cos k_u z, \qquad B_y = B_0 \sin k_u z, \qquad B_z = 0, \qquad (1.46\text{a–c})$$

where $k_u = 2\pi/l_u$ and l_u is again the spatial period. The equations of motion of an electron in the fields (1.46) are:

$$\frac{\mathrm{d}}{\mathrm{d}t}(\gamma m_e v_x) = -ev_z B_y = -ev_z B_0 \sin k_u z, \qquad (1.47\text{a})$$

$$\frac{\mathrm{d}}{\mathrm{d}t}(\gamma m_e v_y) = ev_z B_x = ev_z B_0 \cos k_u z, \qquad (1.47\text{b})$$

$$\frac{\mathrm{d}}{\mathrm{d}t}(\gamma m_e v_z) = e(v_x B_y - v_y B_x), \qquad (1.47\text{c})$$

which can be integrated to give:

$$v_x = \frac{eB_0}{\gamma m_e k_u} \cos k_u z, \qquad v_y = \frac{eB_0}{\gamma m_e k_u} \sin k_u z, \qquad v_z = \text{constant}. \qquad (1.48\text{a–c})$$

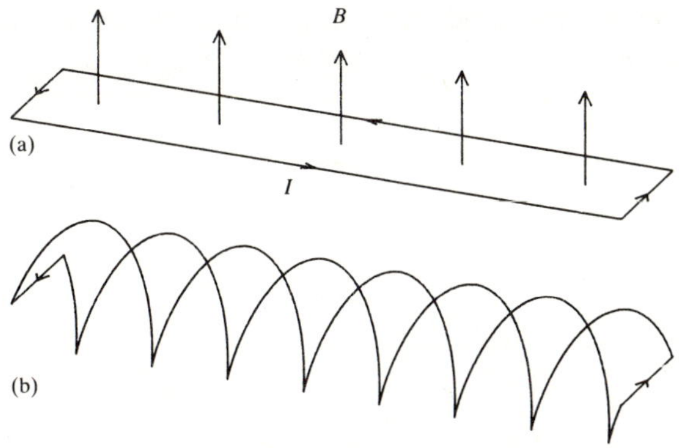

FIG. 1.9. Twister structure. A rectangular electromagnet coil (a) producing a constant field B from a current I is twisted into a double helix (b). The resulting $\boldsymbol{B}$-field has a constant amplitude and a helicoidally changing direction.

We must, of course, have:

$$v^2 = v_z^2 + v_x^2 + v_y^2 = v_z^2 + \frac{e^2 B_0^2 l_u^2}{\gamma^2 m_e^2 4\pi^2}. \tag{1.49}$$

The velocity v may be expressed in terms of γ as

$$v = c(1 - 1/\gamma^2)^{1/2}, \tag{1.50}$$

and γ in turn by

$$\gamma = \left(1 - \frac{v^2}{c^2}\right)^{-1/2} = \frac{eV}{m_e c^2} + 1 = \frac{V}{V_e} + 1, \tag{1.51}$$

where $eV_e = m_e c^2$, and eV is the electron energy.

Electrons with velocities given by (1.48) describe helical orbits with constant v_z. In the twister the orbiting electrons can be made to interact with a nearly synchronous circularly polarized wave, e.g. a laser field so polarized.

It is useful to re-derive the condition for synchronous movement by considering Fig. 1.10, which shows a twister, or undulator, section and a wave of wavelength λ entering and leaving the structure. After the passage let the leading edge of the wave at A end up at a distance λ from B′ at A′, while the trailing originally at B ends up at B′. Thus, in a time l_u/v_z, the wave has moved through a distance $l_u + \lambda$ with speed c, so that

$$l_u/v_z = (l_u + \lambda)/c,$$

and since $f = c/\lambda$ this may be written

$$f = (v_z/l_u)(1 - v_z/c)^{-1}, \tag{1.52}$$

which is in accordance with (1.4) for $p = 1$ and $\vartheta = 0$.

During the passage of the wave, the electric vector turns through a full circle. Thus the electric field maintains approximately a constant angle relative to the vector tangent to the electron orbit.

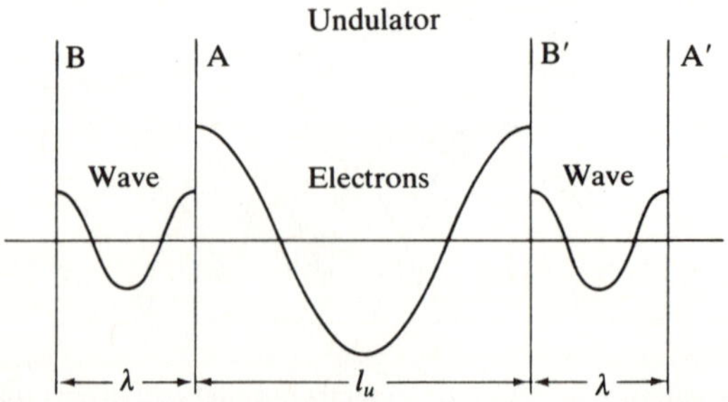

FIG. 1.10. Passage of a wave with wavelength λ across the undulator with spatial period l_u.

The circularly polarized wave may either be a signal to be amplified or a laser field for particle acceleration.

1.7 The undulator as an electron accelerator

In an undulator, electrons may be accelerated by means of a laser beam (M. Chodorow, R. Kompfner, and H. Motz, A laser electron accelerator, unpublished). Consider a twister structure and a circularly polarized laser beam of frequency f given by (1.52), so that it moves synchronously with electrons of axial velocity v_z. There will be electrons with a phase relative to the wave such that the electric field, $\boldsymbol{E}_w$, of the wave is parallel to their azimuthal velocity $\boldsymbol{u} = (v_x, v_y)$, having modulus $u = (v_x^2 + v_y^2)^{1/2} = eB_0/\gamma m_e k_u$.

The field will do work on the electron at a rate:

$$\frac{\mathrm{d}W}{\mathrm{d}t} = eE_w u = v_z \frac{\mathrm{d}W}{\mathrm{d}z} = ev_z \frac{\mathrm{d}V}{\mathrm{d}z}.$$

Thus the electron energy eV increases steadily along the structure. Electrons of an electron beam at different phases will be less accelerated or even decelerated, and we shall consider the behaviour of the entire beam in subsequent chapters. In a Gaussian laser beam in a confocal optical cavity of length L developing power P,

$$E_w = (\pi P/L\lambda)^{1/2}(\mu_0/\varepsilon_0)^{1/4} \tag{1.53}$$

and the energy change of an optimally phased particle is given by

$$\frac{\mathrm{d}V}{\mathrm{d}z} = \frac{cl_u B_0}{2\pi V}\left(\frac{\pi P}{L\lambda}\right)^{1/2}\left(\frac{\mu_0}{\varepsilon_0}\right)^{1/4}. \tag{1.54}$$

Even if some electrons lose energy, it can already be seen that some will steadily gain energy, provided that the wave still keeps in step as they accelerate. This may be achieved by means of a device in which either the magnetic field or the twister wavelength or both are gradually changed to maintain synchronism. If l_u alone is varied with maximum magnetic field, it may be shown (Chodorow *et al.* loc. cit.; Motz 1979) that l_u must vary according to

$$l_u(z) \sim (2\lambda)^{1/3}\left[\frac{2\pi V(z)}{cB_0}\right]^{2/3}. \tag{1.55}$$

Alternatively, if it is intended to maintain synchronism by tapering the magnetic field, it may be shown (Chodorow *et al.* loc. cit.; Motz 1979) that this may be achieved with a $B(z)$ varying along the machine according to

$$B(z) = \frac{2\pi}{cl_u}\left[\left(\frac{2\lambda}{l_u} E_w z\right)^2 + \frac{4\lambda}{l_u}\frac{cB(0)}{k_u} E_w z + c^2 \frac{B^2(0)}{k_u^2}\right]^{1/2}. \tag{1.56}$$

With today's available laser power an acceleration of a GeV/metre can, in principle, be achieved.

1.8 The undulator as an amplifier: analogy with the travelling-wave tube

So far we have only discussed what may be called spontaneous radiation by an undulating beam of electrons. We now want to show that under the influence of an incoming radiation this beam can be made to radiate in such a way as to amplify the incoming electromagnetic wave. The incoming wave stimulates the undulating electrons to amplify it. The words 'stimulated' and 'spontaneous' can be used in the well known technical sense, although we mostly deal with the classical limit of quantum theory. Classical electronic devices may often usefully be looked at in this way. The amplification mechanism has many facets. Some are properties of single undulating electrons; others rely on properties of the entire beam. If the radiation acts in such a way as to bunch the beam more closely, it follows from the foregoing discussion that the radiation will be intensified.

The undulator as an amplifier bears a close analogy to the travelling-wave tube invented by Kompfner (1947). The underlying principle may be explained with reference to Fig. 1.11, which shows a wire bent into the shape of a helix and an electron beam passing along the axis of the helix. The object of the arrangement is to make the electron beam move in synchronism with an electromagnetic wave. In free space such a wave moves with the velocity of light, c, and synchronism with the beam is impossible. In the linac the wave is slowed down by irises inserted in a waveguide; in the case of the undulator the beam is slowed down by the undulation so that it catches up with a later phase of the wave. Kompfner realized that an electromagnetic wave can be guided by the spiralling wire so that the axial velocity is $c \cos \alpha = v$, where α is the pitch of the helix. The wave on the helix continuously interacting with the beam will cause an a.c. perturbation I of the d.c. beam current. This in turn introduces a perturbation of the wave guided by the helix, and for near synchronism between wave and beam the two perturbations will mutually

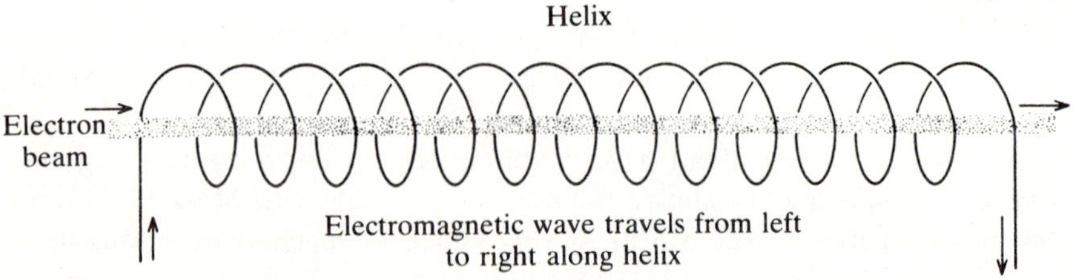

FIG. 1.11. Interaction between the electron beam and the radio frequency fields of the wave travelling along the helix in a travelling-wave tube amplifier.

reinforce each other. The result is that the wave is amplified at the expense of the d.c. energy of the beam.

We shall now sketch the theory of the device for small signals varying in space and time as $\exp(i\omega t - \Gamma z)$. We shall find that the propagation constant Γ is complex, and if it has a positive real part, the wave is growing spatially. In his book Pierce (1950) gives a cogent detailed explanation in terms of a current interacting with a transmission line which has propagation characteristics representing those of the helix. The equation of motion for the electrons leads to the relation between the a.c. beam current I and the a.c. line voltage V in terms of the d.c. beam current I_0, the d.c. beam velocity v, and the d.c. beam voltage V_0 given by $eV_0 = mv^2/2$:

$$I = (\mathrm{i}I_0 k_e \Gamma V)/2V_0(\mathrm{i}k_e - \Gamma)^2, \tag{1.57}$$

where $k_e = \omega/v$, and $\mathrm{i} = (-1)^{1/2}$.

The beam current I, on the other hand, induces a voltage V on the line:

$$V = -\Gamma\Gamma_1 K_c I/(\Gamma^2 - \Gamma_1^2), \tag{1.58}$$

where K_c is the characteristic impedance of the line. Equations (1.56) and (1.57) have to be simultaneously satisfied, and we must have

$$\frac{\mathrm{i}K_c I_0 k_e}{2V_0(\Gamma_1^2 - \Gamma^2)(\mathrm{i}k_e - \Gamma)^2} = 1, \tag{1.59}$$

where Γ_1 is the propagation constant of the circuit without the beam.

The phase velocity of the wave on the circuit is $\mathrm{i}\omega/\Gamma_1$, and for synchronism with the beam $\mathrm{i}\omega/\Gamma_1 = v$ or $\Gamma_1 = \mathrm{i}k_e$.

For small departure from synchronism we can put

$$-\Gamma = -\mathrm{i}k_e + \xi = -\Gamma_1 + \xi. \tag{1.60}$$

For small ξ, i.e. for $|\xi| \ll k_e$, one obtains from (1.59)

$$\xi^3 = -\mathrm{i}k_e^3 I_0 K_c/4V_0. \tag{1.61}$$

Pierce sets

$$K_c I_0/4V_0 = C^3, \qquad \xi = k_e C\delta, \tag{1.62}$$

so that

$$\delta = (-\mathrm{i})^{1/3} = \exp(-1/6 - 2n/3)\pi\mathrm{i}. \tag{1.63}$$

Thus three waves can exist, the values of δ being

$$\delta_1 = \mathrm{e}^{-\mathrm{i}\pi/6} = \sqrt{3}/2 - \mathrm{i}/2, \quad \delta_2 = \mathrm{e}^{-\mathrm{i}5\pi/6} = -\sqrt{3}/2 - \mathrm{i}/2, \quad \delta_3 = \mathrm{e}^{\mathrm{i}\pi/2} = \mathrm{i}.$$

The variation in the z-direction of the three waves is:

$$\exp(-\Gamma z) = \exp[(-\mathrm{i} + C\delta)(\omega/v)z]. \tag{1.64}$$

The first wave is one with increasing amplitude which travels more slowly than the electrons. The second one also travels more slowly and is decreasing, while the third one is unattenuated and travels faster than the electrons. The rate of increase of the growing wave is given in decibels per metre by:

$$20 \times 2.303 \times C(\sqrt{3}/2)(2\pi/\lambda). \tag{1.65}$$

The gain of an undulator was treated in a similar fashion by Motz and Nakamura (1959*b*). These authors adopt the same beam equation as Pierce (1950) and Kompfner (1947) but obtain the undulator–waveguide circuit equation from their previous analysis of undulator radiation in a waveguide (Motz and Nakamura 1959*a*), which was described in Section 1.4. Using the notation of Section 1.4, the ratio of the electric field to the current E/I is deduced to be:

$$\frac{E}{I} = \frac{|F_{mnp}|^2}{\mathrm{i}\omega\varepsilon_0} \frac{\Gamma'^2 + \omega^2/c^2 - p^2k_u^2(\omega^2a^2/c^2\pi^2m^2 - 1)}{\alpha^2 - \Gamma'^2 - \omega^2/c^2}, \tag{1.66}$$

where $\Gamma' = \mathrm{i}pk_u + \Gamma$. Combining this with (1.57) and putting $V_0/I_0 = R_0$ leads to

$$\frac{I}{E} = \frac{-\mathrm{i}k_e/2R_0}{(k_e + pk_u - \mathrm{i}\Gamma')^2} = \frac{\Gamma'^2 + \omega^2/c^2 - \alpha^2}{\mathrm{i}K_uk_g^3}, \tag{1.67}$$

where K_u is an abbreviation for

$$K_u = \frac{|F_{mnp}|^2}{\omega\varepsilon_0k_g^2}\left[\Gamma'^2 + \frac{\omega^2}{c^2} - (pk_u)^2\left(\frac{\omega^2a^2}{c^2\pi^2m^2} - 1\right)\right] \tag{1.68}$$

and, as before,

$$k_g^2 = \frac{\omega^2}{c^2} - \alpha^2.$$

We can thus write:

$$\frac{k_g^2 - (\mathrm{i}\Gamma')^2}{K_uk_g^3} = \frac{k_e}{2R_0(k_e' - \mathrm{i}\Gamma')^2}, \tag{1.69}$$

with

$$k_e' = k_e + pk_u. \tag{1.70}$$

This has the form of Pierce's equation. Following him we put

$$\mathrm{i}\Gamma' = k_e' + \xi \simeq k_g = \left(\frac{\omega^2}{c^2} - \alpha^2\right)^{1/2} \tag{1.71}$$

and obtain, for his gain parameter C,

$$C^3 = K_u/4R_0 = |F_{mnp}|^2\left\{\alpha_{mn}^2 - (pk_u)^2\left(\frac{\omega^2 a^2}{m^2\pi^2 c^2} - 1\right)\right\}\Big/(4\omega\varepsilon_0 k_g^3 V_0/I_0).$$

Amplification will take place for approximate synchronism when

$$(\omega + p\omega_u)/v \simeq (\omega^2/c^2 - \alpha_{mn}^2)^{1/2} \tag{1.72}$$

which is the same as (1.26), at frequencies given approximately by (1.27). The velocity of the amplified wave is smaller than that given by (1.20).

Solutions for the three waves must fulfil further conditions. The sum of the voltages must be the line voltage, and initial conditions are prescribed for a.c. beam currents and velocities.

Motz and Nakamura (1959*b*) give as an estimate,

$$Ck_g \simeq (A^2\alpha^2 I_0/a^3 b\omega\varepsilon_0 V_0)^{1/3}, \tag{1.73}$$

where A, the amplitude of the oscillation, is given by

$$A = eB_0 l_u^2/4\pi^2 m_e v. \tag{1.74}$$

Motz and Nakamura (1959*b*) treated the special case of a non-relativistic undulating electron beam amplifying radiation in a guide. A similar treatment applying to a relativistic beam in free space was given by Kroll (1978) and Kroll *et al.* (1978).

1.9. The ubitron

A device that used an undulator to achieve amplification of electromagnetic radiation in the microwave spectral region was built by Phillips in several versions during the years 1958–65; he named it UBITRON (an acronym for Undulating Beam Interaction elecTRON-tube). According to Phillips (1988), he came across the idea of electromagnetic amplification by an undulating electron beam in 1957, while he was trying to understand the causes of a spurious oscillation occurring in the operation of a newly developed travelling-wave tube, which used an undulating magnetic field for focusing purposes. The oscillation, in fact, turned out to be unrelated to the undulating field, but in the meantime Phillips had realized that an interaction mechanism previously unexploited in travelling-wave tubes was available, and that it held the potential for large power output.

He started an experimental programme to build an amplifier tube based on this interaction, and by the end of 1958 he had the first experimental ubitron, which did not work properly because of a beam instability, but was sufficient to elicit the interest of the US army and air force in the new device. In the following years, Phillips identified and eliminated the instability (Phillips

1960), by modifying the waveguide used in the ubitron so as to filter out unwanted propagation modes, and built several working prototypes under a military-funded programme aimed at developing high-power microwave weapons. Although the ubitron achieved microwave power generation at the megawatt level, which was quite a success at the time, the military eventually lost interest in microwave weapons and, by cutting funds, in 1965 put an end to Phillips' work in this field.

Phillips' ubitron evidenced a facet of undulator interaction quite different from the one that had appeared in Motz's previous experiments. In fact, the ubitron was even physically a quite different device, a microwave tube a few centimetres in diameter and about half a metre long, and used electrons of an energy around 100 keV obtained through a corresponding potential difference applied across a cathode and an anode, without the size of an electron accelerator. Whereas Motz had discovered the unique property of the undulator interaction mechanism of being scalable to the sub-microwave and visible spectral range by changing its resonance wavelength through its dependence on the γ of relativistic electrons, Phillips experimentally proved its capability of providing, at microwave wavelengths, a higher power output and efficiency than competing travelling-wave tubes.

Both uses of the undulator still survive today, and while some researchers seek to extend the operation of modern descendants of the undulator called free-electron lasers to lower and lower wavelengths, others build modern descendants of the ubitron (which are also, perhaps confusingly, called free-electron lasers) that produce higher and higher microwave power outputs from mildly relativistic (say, around 1 MeV) electron beams.

1.10 The free-electron laser

In 1971 Madey initiated an important step by proposing that an undulator structure working as an amplifier at optical wavelengths could replace the active medium between the mirrors of a laser optical cavity. He called such a device a *free-electron laser.*

Madey gave an interesting calculation of gain based on an approximate treatment of relativistic quantum theory (Madey 1971*a*,*b*; Madey *et al.* 1973). He argued that the sinusoidal magnetic field of the undulator, independent of time in the L-frame, would be seen by electrons in the E-frame as a photon flux from which they could be scattered, and related the transition rates of this process to the more easily calculable rates of Compton scattering. Transforming back into the L-frame, the Compton cross-section was used to calculate the transition rate from all initial electron and photon states to a given final state, a desired cavity mode. He thus obtained a theory very similar to that of the stimulated Compton effect, which had been dealt with already in 1933 by Kapitza and Dirac, and was being studied at the same time by other authors (Pantell *et al.* 1968; Sukhatme and Wolff 1973), as a possible means of amplifying coherent optical radiation.

In 1979 Madey gave a much greater generality to his quantum-mechanical gain formula by being able to show that gain of an undulator used as an amplifier is proportional to the derivative with respect to electron energy (or, within an acceptable approximation, with respect to radiation frequency) of its spontaneous emission, a result which has since become known as Madey's theorem. In particular, Madey proved that, to second order in perturbation theory,

$$\langle \mathscr{E}_f - \mathscr{E}_i \rangle = \frac{1}{2}\frac{\partial}{\partial \mathscr{E}_i}\langle (\mathscr{E}_f - \mathscr{E}_i)^2 \rangle, \tag{1.75}$$

where $\mathscr{E}_i$ and $\mathscr{E}_f$ are the initial and final energies of an electron, before the entrance and after the exit of the undulator, and angular brackets denote an average over all the electrons in the input beam. Since it can also be shown that $-\langle \mathscr{E}_f - \mathscr{E}_i \rangle$ is proportional to the stimulated and $\langle (\mathscr{E}_f - \mathscr{E}_i)^2 \rangle$ to the spontaneous undulator emission, eqn (75) affords a very quick way to calculate gain. For the undulator considered in Section 1.2, since the most rapidly varying factor in the expression (1.14) of spontaneous radiated power is S_p^2, where $S_p \sim \sin x/x$, with $x = N\pi(\omega/\Omega_3 - p)$ being a decreasing function of energy through Ω_3, gain turns out to be approximately proportional to

$$\frac{\mathrm{d}}{\mathrm{d}x}\left(\frac{\sin x}{x}\right)^2. \tag{1.76}$$

Madey derived his theorem, just as his previous results, from quantum mechanics. The question of whether the quantum treatment is necessary is an interesting one. We shall see in the next chapter that it may be answered by a simple argument modelled after that devised by Schiff (1952) for synchrotron radiation. However, at the time when they began to study the free-electron laser, Madey and his collaborators were convinced that at optical wavelengths the functioning of the undulator could only be described by quantum theory and, although Madey *et al.* (1973) remark that Planck's constant, h, cancels out in their final gain formula, they could not reconnect their results with those of Motz and Nakamura (1959*b*). It was later shown by Hopf *et al.* (1976) and Kroll (1978) that a completely classical interpretation of both the stimulated Compton effect and undulator amplification is possible down to wavelengths well beyond the optical range and that, with a proper distinction of operating regimes, the results of Madey *et al.* (1973) were consistent with those of Motz and Nakamura (1959*b*). A proof of Madey's theorem based on classical mechanics was independently given by Krinsky *et al.* (1982) and by Kroll (1982).

The limitations of the perturbation methods used in Madey's quantum approach are severe: they do not allow the calculation of the modifications of particle orbits due to the radiation process, as the many-photon problem involved would need higher-order perturbations which, to say the least, could only be attacked by a difficult diagram technique.

1.11 Colson's theory of the single-electron gain in a short twister

A classical theory of the operation of the free-electron laser which was free from those limitations was independently given by Colson (1977*a,b*; 1978), Bambini and Renieri (1978) and Hopf *et al.* (1978). According to Colson, the equations of motion of electrons passing through a twister written down above (1.47) are augmented by including a circularly polarized electromagnetic wave,

$$\boldsymbol{E}_w = E_1(\cos\alpha, -\sin\alpha, 0), \qquad \boldsymbol{B}_w = B_1(\sin\alpha, \cos\alpha, 0) \quad (1.77\text{a,b})$$

with $\alpha = k_w z - \omega_w t + \varphi$, where φ is a phase constant, and we study the combined action of this field and the field of the magnets (1.46) on the electrons.

The equation of motion in the sum of the wave field (1.77) and the magnet field (1.46), which we denote simply by $\boldsymbol{E}$, $\boldsymbol{B}$, is:[3]

$$\frac{\mathrm{d}}{\mathrm{d}t}(\gamma\boldsymbol{\beta}) = \frac{e}{mc}[\boldsymbol{E} + \boldsymbol{\beta} \wedge \boldsymbol{B}], \tag{1.78}$$

where, as before, $\boldsymbol{\beta} = \boldsymbol{v}/c$. We already have the solution (1.48) in the absence of the wave field, which has constant velocity, v_0, and constant γ, γ_0, and put

$$\boldsymbol{\beta} = \boldsymbol{\beta}_0 + \boldsymbol{\beta}_1. \tag{1.79}$$

We are particularly interested in the change of energy $mc^2\gamma$ of the electron because its decrease is equal to the increase of field energy, i.e. to the amplification.

It is given by the work done by $\boldsymbol{E}$ as:

$$\frac{\mathrm{d}\gamma}{\mathrm{d}t} = \frac{e}{mc}\boldsymbol{\beta}\cdot\boldsymbol{E}. \tag{1.80}$$

We choose $v = \beta c$ to be the synchronous velocity given by

$$\beta_{0z} = \omega_w/(\omega_w + k_u c) \tag{1.81}$$

(which is (1.52) in another guise) and remember that the orientation of $\boldsymbol{E}$ with respect to the orbit tangent remains constant, so that the work done according to $\dot{\gamma} = (e/mc)\boldsymbol{\beta}\cdot\boldsymbol{E}_E$ persists in the same sign over many magnet periods. The relative orientation and the work done is different for different entrance positions z_0 within a radiation wavelength λ_w. The initial condition may be chosen so that the beam predominantly loses energy to the radiation field in one pass through a finite-length magnet. This is the gain mechanism.

To solve (1.78) we separate $\boldsymbol{\beta}_1$ into parallel and transverse parts, as $\boldsymbol{\beta}_1 = \boldsymbol{\beta}_{\parallel} + \boldsymbol{\beta}_{\perp}$. The transverse components of (1.78) can be integrated trivially by noting that the right side is a perfect time derivative. Then using this $\boldsymbol{\beta}_{\perp}$ and $\gamma(t)$ from $\gamma^{-2} = 1 - \boldsymbol{\beta}\cdot\boldsymbol{\beta}$, reference to transverse motion may be eliminated. The two equations given by (1.80) and the parallel component of (1.78) are identical non-linear second-order ordinary differential equations in $z(t)$. All this is exact.

In this way one finds the equation:

$$\ddot{z} = -(e/mc)^2(E_1B_0/k_u\gamma^2)\sin[(k_u + k_w)z - \omega_w t + \varphi - \pi/2]. \quad (1.82)$$

We look for a solution z in the form

$$z = z_0 + \beta_{0z}ct + \delta(z), \quad (1.83)$$

with small $\delta(z)$. Colson (1977*a*,*b*; 1978) defines a quantity,

$$\Delta\omega = k_u\beta_{0z}c - \omega_w(1 - \beta_{0z}) \quad (1.84)$$

which is zero, according to (1.81), when the electron moves synchronously with the wave, and is therefore analogous to the parameter ξ which in the travelling-wave tube type theory of the previous chapter measures the departure from synchronism.

The phase of the sine-function of (1.82) may now be written

$$\xi = \Delta\omega t + (k_u + k_w)[z_0 + \delta(z)] + \varphi - \pi/2$$

and (1.82) becomes

$$\ddot{\xi} = -\Omega^2 \sin \xi(t), \quad (1.85)$$

where $\Omega^2 = 2e^2B_0E_1/\gamma_z^2m_e^2c^2$ with $\gamma_z^2 = \gamma_0^2/(1 + e^2B_0^2l_u^2/4\pi^2m_e^2c^4)$, which is the equation of a pendulum in the large-amplitude regime.

It may be integrated to yield

$$\dot{\xi}^2 = 2\Omega^2(\cos \xi - \cos \xi_0) + \Delta\omega^2. \quad (1.86)$$

This equation could describe the motion of a ball rolling on a sinusoidal 'potential hill' surface. One can see it oscillating in a valley or, if the energy is high enough, getting over the next hill and carrying on the motion over hills and dales. All this is seen from the phase plot of Fig. 1.12, which shows the 'velocity' parameter $\dot{\xi}/\Omega$ plotted against the linear excursion ξ. The exact solutions are in terms of elliptical integrals, but a great deal of information can be obtained by perturbation analysis.

The behaviour is further illustrated by the phase plots of Fig. 1.13(a–f). In (a) the curves with numbers 1–4 show locations in phase space of particles with uniform initial distribution at later times in units of $1/\Omega$. In Figs. 1.13(b–f) the horizontal lines marked 0 indicate the initial velocity parameter

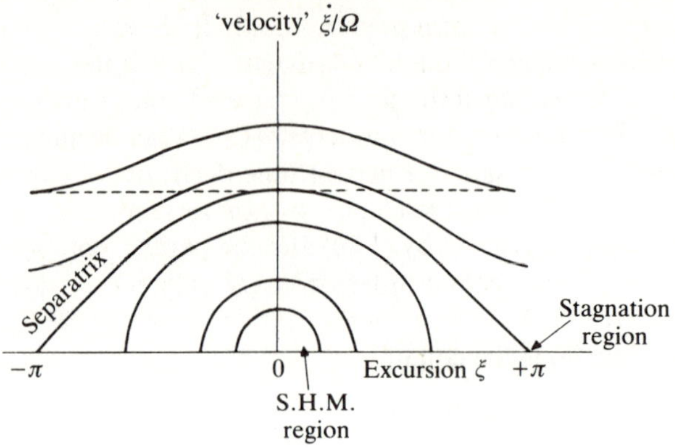

FIG. 1.12. Phase plot for the oscillation of a pendulum. The broken horizontal line represents $\dot{\xi}/\Omega = 2$ (from Lawson 1980; © 1980 Gordon and Breach Science Publishers SA).

in steps of 0.5 up to 2.5. Clearly, the behaviour is so complicated that numerical calculations would not easily elicit Colson's analytical formula derived below.

He expands in terms of a small parameter $\Omega/\Delta\omega$ and defines the gain $G(t)$ as the fractional change in radiation energy. He expands (1.86) to fourth order and obtains:

$$G(t) = \frac{4e^4 B_0^2 \rho_e l_u}{(\Delta\omega\gamma_0 mc)^3}\left(1 - \cos\Delta\omega t - \frac{\Delta\omega t}{2}\sin\Delta\omega t\right), \tag{1.87}$$

where ρ_e is the electron density. If electrons have the synchronous velocity ($\Delta\omega = 0$) there is no gain at any time. For off-resonance electrons the gain oscillates with frequency and has a small, increasing, amplitude proportional to t. In an undulator of finite length Nl_u the gain evolves for a fixed time $T = Nl_u/\beta_z c$. Maximum gain is achieved by choosing

$$\Delta\omega \simeq 2.6056\ \mathrm{T}^{-1}.$$

Equation (1.87) may be rewritten to show that the gain is proportional to

$$f(x) = \frac{\mathrm{d}}{\mathrm{d}x}\left(\frac{\sin x}{x}\right)^2 \tag{1.88}$$

with $x = -\Delta\omega T/2$, which is the same as the result given by Madey's theorem.

The assumption that $\Omega \ll \Delta\omega$ means that the analysis is restricted to weak fields or large γ_0.

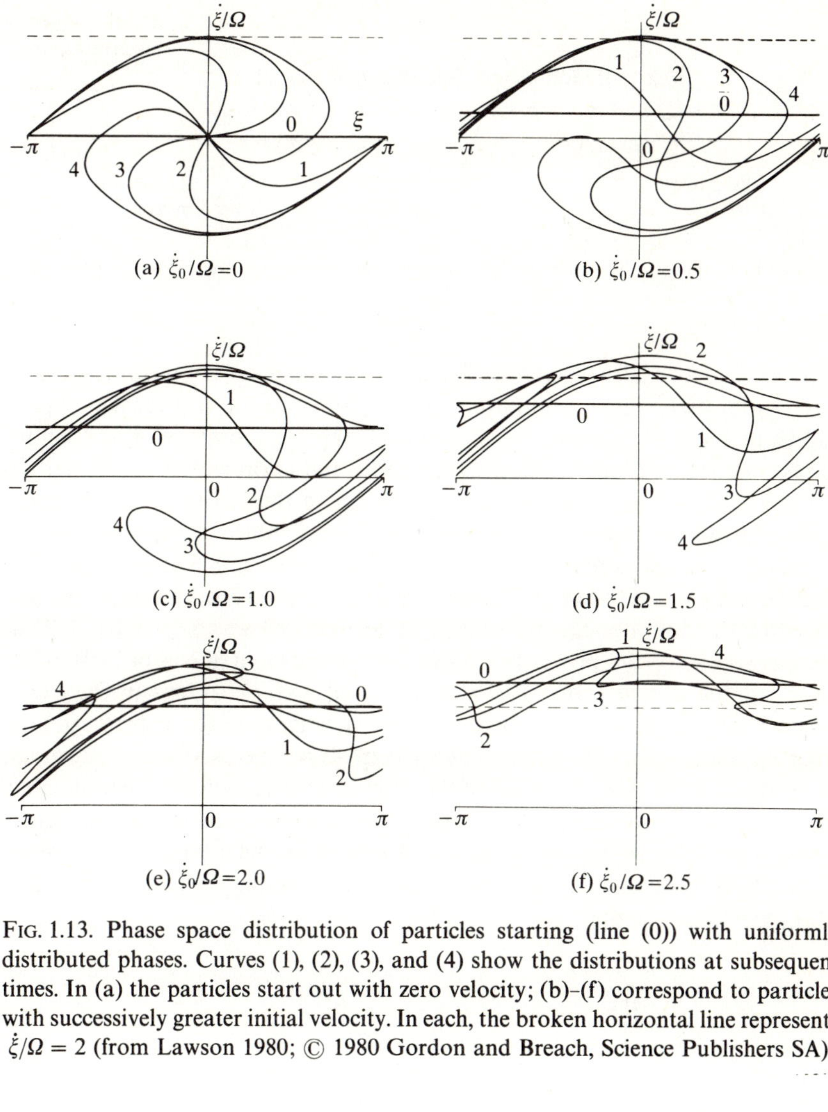

FIG. 1.13. Phase space distribution of particles starting (line (0)) with uniformly distributed phases. Curves (1), (2), (3), and (4) show the distributions at subsequent times. In (a) the particles start out with zero velocity; (b)–(f) correspond to particles with successively greater initial velocity. In each, the broken horizontal line represents $\dot{\xi}/\Omega = 2$ (from Lawson 1980; © 1980 Gordon and Breach, Science Publishers SA).

1.12 The first free-electron laser experiment

In 1976 Elias *et al.* attempted an experiment to operate an FEL at the CO_2-laser wavelength of 10.6 μm. They succeeded in obtaining amplification of the beam of an external CO_2 laser but failed to achieve self-oscillation of the FEL because of beam misalignment and interference with the accelerator structures due to an over-large radiation-beam cross-section.

In 1977 Deacon, Elias, Madey, Schwettman and Smith tried again at the smaller wavelength of 3.4 μm, and obtained the first self-oscillation of an FEL

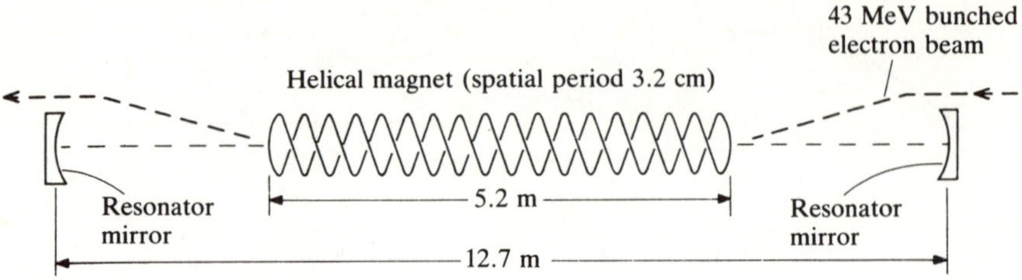

FIG. 1.14. Schematic diagram of the free-electron laser oscillator (from Deacon *et al.* 1977).

consisting of a twister structure placed between mirrors in an optical cavity. The twister had a large number of periods (160) and the arrangement that is shown in Fig. 1.14. A bunch of electrons of length 1.4 mm from the Stanford Superconducting Linac emitted spontaneous radiation with a wavelength in the infrared (3.4 μm). The radiation built up in the cavity as the bunches moved in approximate synchronism with the forward component of the standing wave and amplified it. The bunches arrived in 8.6 ns intervals. The beam energy was 43 MeV. In terms of time the bunch length was 4.7 ps and therefore the duty cycle (ratio of time on to time off) was 5.05×10^{-5}. Thus the measured average power of 0.34 W corresponds to a power of 7 kW while on. Since the mirror transmission was 1.5 per cent, the power in the cavity was 500 kW. This is an impressive figure, particularly for a first attempt. Figure 1.15(a) shows the emission spectrum above the oscillation threshold; the width is 8 nm. The spectrum of the spontaneous emission observed below the threshold is shown in Fig. 1.15(b). The superconducting helix structure had a spatially periodic field with a 0.24 T peak. In the cryogenic laboratory at Stanford it was natural to choose a superconducting coil, although this is by no means necessary.

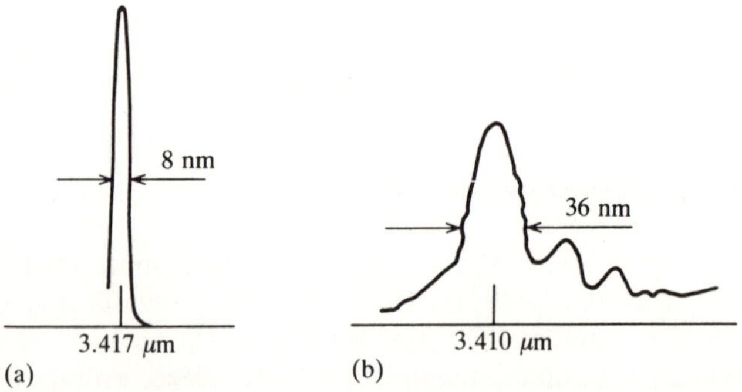

FIG. 1.15. Emission spectrum of the Stanford 1977 free-electron laser: (a) above the oscillation threshold; (b) spontaneous radiation below the threshold (from Deacon *et al.* 1977).

From the figures given, it is seen that the bunch length was large compared to the wavelength. Since the mean deceleration of a synchronous beam is zero when the arrival time is random with respect to the laser phase, the gain mechanism explained above must be involved. Note that the peak of the spontaneous radiation is shifted with respect to the laser line in the direction expected.

Notes

1. γ in eqn (1.2) is actually $(1 - |\boldsymbol{\beta}|^2)^{-1/2}$ and not $(1 - \beta_z^2)^{1/2}$. Although the difference between γ and γ_z may be noticeable, as will be seen later, it is not important for the present order-of-magnitude considerations.
2. For consistency with Motz (1951), cgs units are used in this section.
3. For consistency with Colson (1977*a*,*b*; 1978), cgs units are used in this section.

References

Bambini, A. and Renieri, A. (1989). The free-electron laser: a single-particle classical model. *Lett. Nuovo Cimento* **21**, 399–404.

Bernashevsky, G. A., Vystavkin, A. N. and Lomize, L. G. (1960). In *Millimetre Waves*. Polytechnic Press, New York.

Colson, W. B. (1977*a*). One-body electron dynamics in a free-electron laser. *Phys. Lett.* **64A**, 190–92.

Colson, W. B. (1977*b*). Free electron laser theory. Ph.D Dissertation, Stanford University.

Colson, W. B. (1978). One-body analysis of free-electron lasers. *FEL77*, 157–96.

Combe, R. (1956). Thése de l'Université de Paris.

Combe, R. and Feix, M. (1953*a*). Mouvement d'un électron dans un ondulateur magnétique. *C. R. Acad. Sci. Paris* **237**, 1318–20.

Combe, R. and Feix, M. (1953*b*). Frequences et puissance des ondes rayonnées dans un ondulateur magnetique. *C. R. Acad. Sci. Paris* **237**, 1660–62.

Deacon, D. A. G., Elias, L. R., Madey, J. M. J., Ramian, G. J., Schwettman, H. A. and Smith, T. I. (1977). First operation of a free-electron laser. *Phys. Rev. Lett.* **38**, 892–4.

Elias, L. R., Fairbank, W. M., Madey, J. M. J., Schwettman, H. A., and Smith, T. I. (1976). Observation of stimulated emission of radiation by relativistic electrons in a spatially periodic transverse magnetic field. *Phys. Rev. Lett.* **36**, 717–20.

Hirshfield, J. L. and Wachtel, J. M. (1964). Electron cyclotron maser. *Phys. Rev. Lett.* **12**, 533–6.

Hirshfield, J. L., Bernstein, I. B. and Wachtel, J. M. (1965). Cyclotron resonance interaction of microwaves with energetic electrons. *IEEE J. Quantum Electron.* **QE-1**, 237–45.

Hopf, F. A., Meystre, P., Moore, G. T. and Scully, M. O. (1978). Nonlinear theory of free-electron devices. *FEL77*, 41–114.

Hopf, F. A., Meystre, P., Scully, M. O. and Louisell, W. H. (1976). Classical theory of a free-electron laser. *Opt. Commun.* **18**, 413–16.

Kapitza, P. L. and Dirac, P. A. M. (1933). *Proc. Camb. Phil. Soc.* **29**, 297.

Kompfner, R. (1947). *Proc. IRE* **35**, 124.

Krinsky, S., Wang, J. M. and Luchini, P. (1982). Madey's gain-spread theorem for the free-electron laser and the theory of stochastic processes. *J. Appl. Phys.* **53**, 5453–8.

Kroll, N. M. (1978). The free-electron laser as a travelling wave tube amplifier. *FEL*77, 115–56.

Kroll, N. M. (1982). A note on the Madey gain-spread theorem. *FEL*81, 315–23.

Kroll, N. M. and McMullin, W. A. (1978). Stimulated emission from relativistic electrons passing through a spatially periodic transverse magnetic field. *Phys. Rev. A* **17**, 300–8.

Lawson, J. D. (1980). Coherent interaction between waves and particle streams. *Particle Accelerators* **10**, 73–81.

Madey, J. M. J. (1971*a*). Stimulated emission of bremsstrahlung in a periodic magnetic field. *J. Appl. Phys.* **42**, 1906–13.

Madey, J. M. J. (1971*b*), Ph.D. Dissertation, Stanford University.

Madey, J. M. J. (1979). Relationship between mean radiated energy, mean squared radiated energy, and spontaneous power spectrum in a power series expansion of the equations of motion in a free electron laser. *Nuovo Cimento* **50B**, 64–88.

Madey, J. M. J., Schwettman, H. A. and Fairbank, W. M. (1973). A free electron laser. *IEEE Trans. Nucl. Sci.* **20**, 980–2.

Motz, H. (1951). Applications of the radiation from fast electron beams. *J. Appl. Phys.* **22**, 527–35.

Motz, H. (1979). Undulators and 'free-electron lasers'. *Contemp. Phys.* **20**, 547–68.

Motz, H. and Mallory, K. B. (1955). Generation of submillimeter waves. *J. Appl. Phys.* **26**, 1384.

Motz, H. and Nakamura, M. (1959*a*). Radiation of an electron in an infinitely long waveguide. *Ann. Phys.* **7**, 84–131.

Motz, H. and Nakamura, M. (1959*b*). The generation of submillimeter waves and fast wave amplification. *Proc. Symp. Millimeter Waves, Polytechnic Institute of Brooklyn, March 31 – April 2*, pp. 155–63.

Motz, H. and Walsh, D. (1962). Dependance of power radiated on beam current in a magnetic undulator. *J. Appl. Phys.* **33**, 978–81.

Motz, H., Thon, W. and Whitehurst, R. N. (1953). Experiments on radiation by fast electron beams. *J. Appl. Phys.* **24**, 826–33.

Pantell, R. H., Soncini, G. and Puthoff, H. E. (1968). Stimulated photon–electron scattering. *IEEE J. Quantum Electron.* **QE-4**, 905–7.

Phillips, R. M. (1960). The Ubitron, a high-power travelling-wave tube based on a periodic beam interaction in unloaded waveguide. *IRE Trans. Electron. Dev.* **7**, 231.

Phillips, R. M. (1988). History of the ubitron. *FEL*87, 1–9.

Pierce, J. R. (1950). *Travelling-wave tubes.* Van Nostrand, New York.

Schiff, L. I. (1949). *Quantum mechanics.* McGraw-Hill, New York.

Schiff, L. I. (1952). Quantum effects in the radiation from accelerated relativistic electrons. *Am. J. Phys.* **20**, 474–8.

Smith, S. J. and Purcell, E. M. (1953). Visible light from localized surface charges moving across a grating. *Phys. Rev.* **92**, 1069.

Sukhatme, V. P. and Wolff, P. W. (1973). Stimulated Compton scattering as a radiation source—theoretical limitations, *J. Appl. Phys.* **44**, 2331–4.

2

OVERVIEW OF THE GENERAL OPERATIONAL PRINCIPLES

In the following chapters a detailed theory of the workings of undulators and free-electron lasers based on Hamiltonian mechanics will be presented. It will be useful to precede it here with an exposition of the various points of view from which the problem may be surveyed.

2.1 Cherenkov, synchrotron, wiggler and undulator radiation: a geometrical overview

2.1.1 *Cherenkov radiation*

When an electron passes through a dielectric medium with constant velocity a polarization is produced momentarily and the medium re-radiates transferred energy. One can see from the simple Huygens construction of Fig. 2.1 that a wavefront will form in the direction BC, making an angle $90° - \vartheta$ with the direction of electron travel, such that

$$c/n = v \cos \vartheta, \tag{2.1}$$

where n is the refractive index of the medium.

Equation (2.1) defines a thin conical shell containing the radiation. The phenomenon has some analogy with the shock waves emitted by a projectile or the (two-dimensional) bow waves of a ship. Energy is radiated into the entire frequency band for which n is larger than unity. It stops at X-rays, where $n < 1$. The electron velocity is, of course, not quite constant, as the energy must come from the particle, but the energy loss and speed change are small. Since the particle does not slow down appreciably, inertia effects are negligible, and therefore the radiated power does not depend on the mass and is the same for all particles of equal charge and speed (protons, mesons, etc.). The amount of power radiated per unit time by the particle is given by

$$W = \frac{e^2}{4\pi\varepsilon_0 c^2} \int [1 - c^2/n^2(\omega)v^2]\omega \, d\omega, \tag{2.2}$$

where the integral is extended over all frequencies for which $nv/c > 1$.

Radiation effects produced by acceleration or deceleration of charges differ markedly from Cherenkov radiation; the radiation intensity depends on e/m.

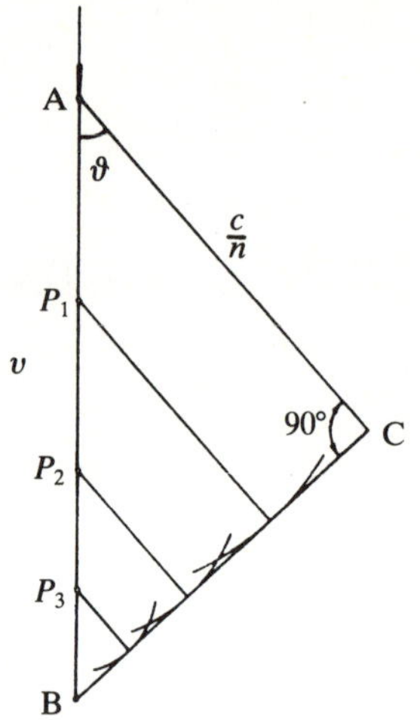

FIG. 2.1. Huygens construction for Cherenkov cone. Spherical waves starting at successive positions A, P_1, P_2,... along the electron path form a wavefront B–C.

2.1.2 *Radiation from fast accelerated electrons*

In contrast with Cherenkov radiation, we shall see that radiation from accelerated charges is contained within a narrow cone centred on the instantaneous direction of movement of the charge, with an angular opening

$$\vartheta_r = 1/\gamma, \tag{2.3}$$

where $\gamma = (1 - v^2/c^2)^{-1/2}$. The wavelength radiated by charges in orbits which are circles of radius R ranges from high values down to a wavelength,

$$\lambda_0 \simeq R/\gamma^3. \tag{2.4}$$

The angular characteristics of the radiation can be inferred from Fig. 2.2, which shows an extended electron e moving along the line AB with speed v. The potential at a distant observation point P is measured by the accumulation of contributions collected by the spherical surface CD as it traverses e while collapsing radially with speed c. The time that is required for this is inversely proportional to the difference between the collapse speed and the component of the electron velocity along the radial line eP, i.e. $c - v\cos\vartheta$. This dependence on the angle ϑ will hold even for the limit of a point charge.

The relativistic relation between energy $\mathscr{E}$ and speed is

$$\mathscr{E}/mc^2 = \gamma = (1 - v^2/c^2)^{-1/2}. \tag{2.5}$$

Since v/c is very close to unity and $1 - v/c \simeq 1/(2\gamma^2)$ we may, for small angles ϑ, put

$$c - v\cos\vartheta \simeq c(1 - v/c + v\vartheta^2/2c) \simeq c(\vartheta_r^2 + \vartheta^2)/2, \tag{2.6}$$

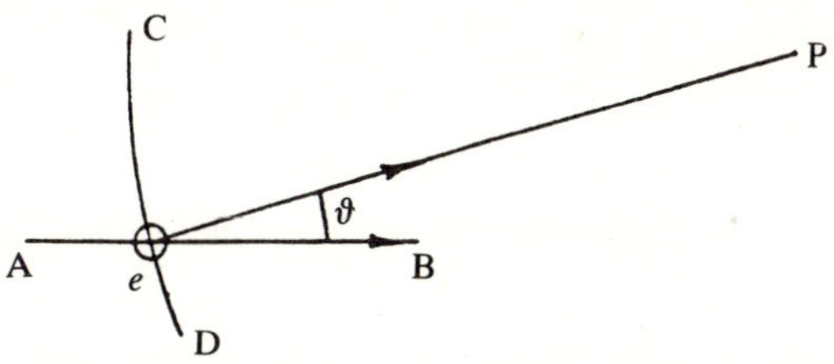

FIG. 2.2. An extended electron e moves along the line AB with speed v. The potential at a distant observation point P is measured by the accumulation of contributions collected by the spherical surface CD as it traverses e while collapsing radially with speed c.

where $\vartheta_r = 1/\gamma$. Thus the potential arriving at P is measured by the value of $1/(\vartheta_r^2 + \vartheta^2)$ at the earlier time when the spherical surface traversed the charge. This means that the radiation is most intense inside a cone with angular opening of the order of ϑ_r that is centred on the instantaneous direction of the motion of the electron. For a 300 MeV electron $\vartheta_r \simeq 0.1°$ and, in general, ϑ_r is inversely proportional to the energy E.

2.1.3 *Frequency characteristics*

The radiation that is emitted by the electron during a time interval t passes by a distant observer in the forward direction much faster because of the relativistic Doppler effect. Figure 2.3 illustrates emission by an electron moving through the distance AB equal to vt in time t with speed v. Thus the radiation emitted by the electron as it moves from A to B lies in the shell between the surfaces CD and EF at a later time. For a distant observer

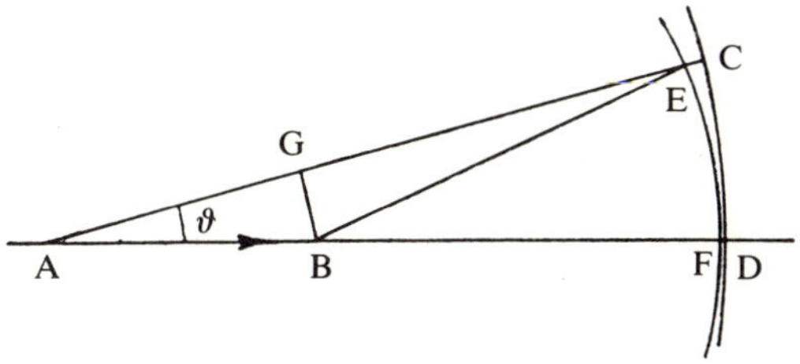

FIG. 2.3. An electron moves from A to B with speed v in time t. At the time $t + T$ after the electron is at A, or time T after it is at B, the radiation it emits between A and B lies between the spherical surfaces CD and EF. AC is equal to $c(t + T)$, BE is equal to cT, and BG is perpendicular to AC.

$(T \gg t)$ we can put

$$\begin{aligned} \mathrm{CE} &= \mathrm{AC} - \mathrm{AG} - \mathrm{GE} \simeq \mathrm{AC} - \mathrm{AG} - \mathrm{BE} \\ &= c(t+T) - vt\cos\vartheta - cT \\ &= ct[1-(v/c)\cos\vartheta] \simeq ct(\vartheta_r^2 + \vartheta^2). \end{aligned}$$

Thus, within the forward cone $\vartheta < \vartheta_r$ the thickness of the pulse that goes past an observer is less than the distance of travel of the electron during emission by a factor of $1/\gamma^2$.

So far the electron orbit has not been specified. If we assume it to be a circular arc of radius R centred at O (Fig. 2.4), subtending an angle $2\vartheta_r$ at the centre, the central ray will follow the direction AR_2 tangent to the arc at A, and the radiation will mainly be located between the rays AR_1 and AR_3, each making an angle ϑ_r with the central ray. Similarly, at B emission will concentrate intensity between BS_1 and BS_3 which form an angle ϑ_r with BS_2, tangent to the arc at B. It is seen that AR_3 and BS_1 are parallel, i.e. they coincide, and we have drawn them separately only for clarity. At a long distance from A and B the observer sees radiation only from the segment AB while he is within the forward cone of opening ϑ_r. The thickness of the pulse at a large distance is approximately equal to ct/γ^2, as we have seen, where t is the transit time of the electron between A and B which, in turn, is found to be $R\vartheta_r/c = R/c\gamma$ by inspection of the figure, so that the pulse has a thickness of order $\lambda_0 \simeq R/\gamma^3$ when it passes the observer. The duration of the pulse will thus be of order $R/c\gamma^3$ and a Fourier analysis will give frequencies extending from low ones up to about $c\gamma^3/R$. Such a radiation was first observed from synchrotrons in 1948 (although the theory had already been given by Schott in 1912) and is now generally known as synchrotron radiation.

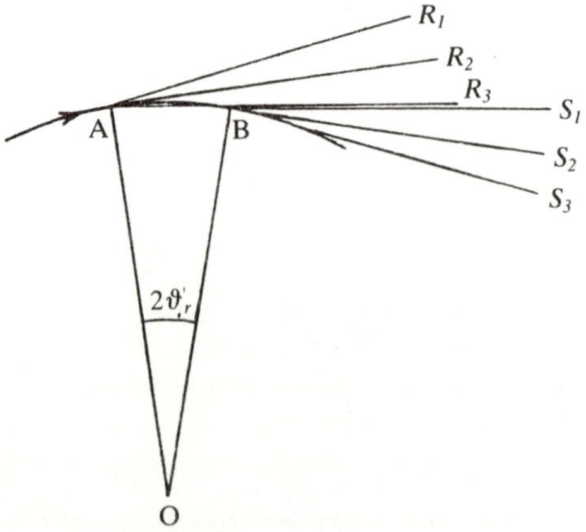

FIG. 2.4. A distant observer stationed along the line AR_3 or BS_1 receives radiation only when the electron is between A and B, since otherwise he lies outside the forward cone of angular opening ϑ_r.

The power radiated is high, and constitutes an important energy loss which has to be made good by the radio-frequency generators that supply the machine's power. The power radiated is approximately the field energy $B^2/2\mu_0$ swept out by an electron with the classical radius $r_e = e^2/4\pi\varepsilon_0 m_e c^2$ along its path, multiplied by γ^2, i.e. as though the electron's charge was squashed down laterally to γ times its radius. The precise formula is

$$P = \tfrac{2}{3}(r_e c\mathscr{E}^2 F^2)/(m_e c^2)^3, \tag{2.7}$$

where F is the force acting on the electron, i.e. ecB, and $\mathscr{E}$ is the electron energy.

2.1.4 *Synchrotron radiation*

For a synchrotron in which the particles are accelerated in a circular orbit to an energy $\mathscr{E}$, $B \sim \mathscr{E}$ and, according to eqn (2.7), the energy loss per turn is proportional to the fourth power of $\mathscr{E}$. For electrons it is given by

$$\Delta\mathscr{E} = C\mathscr{E}^4/R, \tag{2.8}$$

where

$$C = 4\pi r_e/3(m_e c^2)^3 = 88.5 \text{ keVm/GeV}^4. \tag{2.9}$$

Thus, for electrons of 2 GeV in a circular orbit with radius 8 m, the machine would suffer a power loss of 177 keV per turn, equivalent to a power radiated of 177 kW when the current is one Ampére! This limits either synchrotron energy or current for economic reasons.

On the other hand, when the synchrotron is to be used as a powerful source of light only a fraction $\vartheta_r/2\pi$, i.e. $1/(2\pi\gamma)$, of the power reaches out in any given direction (although in a practical beam transport system suitable optics can make a fraction as high as 10 or even 100 ϑ_r available at a single experimental outlet).

2.1.5 *Undulator radiation*

From this point of view it becomes clear that one would do better by letting the beam describe a succession of arcs, up and down, or rather the sinusoidal orbit of the undulator. When there are only a few arcs or 'wiggles' the device is called a wiggler. The electron describes a shallow sinusoidal path and with average speed not very different from c it will cover the length L of the undulator in a time $T = L/c$.

Returning to the previous estimate of the thickness of the shell of radiation given by ct/γ^2, where t is the time employed to go past one period l_u of the undulator, it is seen that the wavelength of the radiation from an undulator ranges down from long wavelengths to l_u/γ^2. A wiggler has a broad spectrum, and an undulator with many periods a much sharper one.

The simplest undulator structure, consisting of a succession of magnets of alternating polarity, has been considered in the first chapter and was illustrated in Fig. 1.1. The frequencies radiated from such a structure were derived from a geometrical construction, shown in Fig. 1.2, according to which a phase front is formed when the phase difference between rays is a multiple of π. The special case of $p = 0$ can only occur in a dielectric medium with refractive index $n > 1$, so that the speed of light is c/n and eqn (1.3) leads to

$$\cos \vartheta = c/nv.$$

This is just the Cherenkov angle; thus Cherenkov radiation may be regarded as a special case of undulator radiation.

2.2 Limits of the validity of the classical description

2.2.1 *Synchrotron radiation*

Classical theory assumes that the electron is a point charge and can only be valid when the quantum mechanical calculation can be cast into such a form that the electron behaves as a localized charge. Localization to a point is, however, not only impossible but also unnecessary. It is sufficient that it be possible to construct a quantum-mechanical wave packet such that each of its dimensions is small in comparison with the reciprocal of the corresponding component of the propagation vector of the emitted radiation. We have learned that the emission is mainly contained within an angle ϑ_r which the forward direction makes with the propagation vector of magnitude $2\pi/\lambda_0$.

Thus the longitudinal dimension of the electron wave packet must be small compared to λ_0 and the transverse dimension small in comparison to λ_0/ϑ_r.

The longitudinal dimension cannot be made smaller than the De Broglie wavelength $h/\gamma m_e c$ of the electron, since otherwise shorter wavelength components would be needed and the momentum and energy of the electrons would be increased. The conditions

$$h/m_e c\gamma \ll \lambda_0, \qquad h/\lambda_0 \ll m_e c\gamma \tag{2.10a,b}$$

are equivalent and the second one gives the alternative criterion that the momentum of the emitted photon must be small compared to the electron momentum. For synchrotron radiation, with $\lambda_0 = R/\gamma^3$, we obtain the condition

$$\gamma^2 \ll R/\lambda_C, \tag{2.11}$$

where $\lambda_C = h/m_e c \simeq 2.43 \times 10^{-12}$ m is the Compton wavelength of the electron. Thus, for a synchrotron radius of 1 m, γ must be small compared to 10^6 for classical theory to be valid.

This discussion did not take the spreading of the wave packet into account. The packet must maintain its dimensions during the time of the radiation process which, as we have seen, is roughly equal to $R/c\gamma$. Thus, in the direction of motion, which we shall call the z-direction, the packet must remain small compared to $\lambda_0 = R/\gamma^3$ during the time $R/c\gamma$. In the two dimensions perpendicular to z the packet must remain small compared to $\lambda_0/\vartheta_r = R/\gamma^2$ during that time.

A longitudinal spread δz, by Heisenberg's principle, requires a momentum spread $\delta p_z = \hbar/\delta z$. Differentiating the relativistic relation $p_z = m_e v_z/(1 - v_z^2/c^2)^{1/2}$ gives

$$\delta p_z = m_e \delta v_z/(1 - v_z^2/c^2)^{3/2}$$

and makes the longitudinal velocity spread equal to

$$\delta v_z \simeq \delta p_z/m_e\gamma^3 \simeq \hbar/(m_e\gamma^3\delta z).$$

This velocity spread causes an additional co-ordinate spread during the emission time $R/\gamma c$, equal to

$$\delta v_z R/\gamma c \simeq R\hbar/(m_e c\gamma^4\delta z).$$

In deriving the previous criterion (2.10) we have assumed that $\delta z = h/\gamma m_e c$. However, we now see that with this value the packet would have spread by an amount R/γ^3 during the emission time, thus becoming as large as the radiated wavelength and making classical theory invalid.

Schiff (1952) solves this difficulty by pointing out that the wave packet has been chosen with too small an initial spread. Instead, he minimizes the final spread of the packet,

$$\Delta z = \delta z + \hbar R/m_e c\gamma^4\delta z, \tag{2.12}$$

by choosing an optimal value for δz. It is easily seen that both this optimal δz and Δz are roughly equal to $(\hbar R/m_e c\gamma^4)^{1/2}$. We now expect classical theory to be valid when

$$\Delta z \ll \lambda_0, \qquad \text{i.e.} \qquad (\hbar R/m_e c\gamma^4)^{1/2} \ll R/\gamma^3.$$

This requires

$$\gamma \ll (m_e cR/\hbar)^{1/2} = (R/\bar{\lambda}_C)^{1/2}, \tag{2.13}$$

which is equivalent to eqn (2.10). A similar argument applied to the perpendicular direction again leads to the same criterion.

2.2.2 *Undulator radiation*

In order to apply the results derived for synchrotron radiation to the undulator we can use the expression (1.5) of the radiated frequency, which gives the wavelength $\lambda_0 = l_u/2\gamma^2$. According to the first criterion used above in eqn (2.10), this gives, for the limit of γ beyond which quantum effects are expected, $\gamma \ll l_u/\lambda_C$. However, this first criterion does not take into account that the packet now spreads over the whole length L of the undulator. In order to apply the second criterion we must now replace the emission time $R/\gamma c$ in eqn (2.12) by L/c, and in this way obtain

$$\Delta z = \delta z + \hbar L/m_e c\gamma^3 \delta z. \tag{2.14}$$

According to eqn (2.14) the minimum spread Δz, corresponding to an optimal choice of δz, is $(\hbar L/m_e c\gamma^3)^{1/2}$ and therefore the limit on γ resulting from the condition that $\Delta z \ll \lambda_0$ is given by

$$\gamma \ll l_u^2/L\bar{\lambda}_C = l_u/N\bar{\lambda}_C, \tag{2.15}$$

where $N = L/l_u$ is the number of periods of the undulator. Notice that this limit on γ is, in the case of the undulator, $1/N$ times smaller than the one obtained without considering wave packet spreading.

Equation (2.15) may also be restated as a lower limit on output wavelength λ_0, by expressing γ as $(l_u/2\lambda_0)^{1/2}$:

$$\lambda_0 \gg N^2\bar{\lambda}_C^2/l_u. \tag{2.16}$$

For the typical period of a magnetic undulator, of the order of centimetres, and a number of periods N of the order of 100, the resulting limits on γ and λ_0 are about 10^8 and 10^{-16} cm, respectively. The functioning of such an undulator is therefore well described by classical mechanics under all practically conceivable operating conditions.

2.3 Analogy with Compton (Thomson) Scattering

It has already been seen, at the very beginning of Chapter 1, that the phenomenon of undulator radiation may be looked at from the vantage point of an observer attached to the moving electron. In the frame of reference of this observer, the E-frame, the undulator is seen to approach the observer with a speed which, in the case of a highly relativistic electron, is near the speed of light. In terms of quantities in the E-frame which moves with speed $v_z \simeq c$, denoted by primes, the axial co-ordinate z in the L-frame may be expressed as

$$z = \frac{z' + v_z t'}{(1 - v_z^2/c^2)^{1/2}} \simeq \gamma(z' + ct'). \tag{2.17}$$

Therefore the static undulator field which is a sinusoidal function of z/l_u will look, in the E-frame, like a radiation field of wavelength l_u/γ and frequency $\gamma c/l_u$, running across the electron in the backward (decreasing z) direction, and one can then see the interaction process as a Compton collision.

As is well known (see, e.g., Heitler 1954) the quantum-mechanical, Compton, electron–photon scattering process does not deviate appreciably from classical, Thomson, scattering until the energy of the photon becomes comparable with the electron rest mass, i.e. down to wavelengths of the order of the Compton wavelength, $\lambda_C \simeq 10^{-12}$ m. This does not happen, for an undulator period l_u of the order of centimetres, up to $\gamma \simeq 10^{10}$, an even higher limit than the one imposed by eqn (2.15). Therefore the backward-propagating undulator wave will be scattered in the E-frame into a forward wave of the same frequency, $\gamma c/l_u$, and this will appear in the L-frame with a Doppler-shifted frequency $f_0 = 2\gamma^2 c/l_u$. In an infinitely long undulator, appearing in the E-frame as a perfect plane wave, scattered radiation would have exactly this frequency. In a finite undulator N-period long, appearing in the E-frame as a wave train of N periods, scattered radiation has the same spectral power distribution as this wave train, i.e. one proportional to $[(\sin x)/x]^2$, with $x = N\pi(f - f_0)/f_0$. This is probably the simplest possible explanation of the $[(\sin x)/x]^2$ spectral distribution of undulator radiation.

Useful as it is for a qualitative understanding, the interpretation of undulator operation as a kind of Compton scattering has its practical applicability limited by the perturbation methods that underlie the usual description of scattering processes. Perturbation methods are unable to describe the distortion of electron trajectories that takes place at high magnetic fields, leading to the generation of harmonics of the fundamental emission frequency, nor, when the undulator is used as an amplifier, its operation in the large-gain or large-signal regimes. On the other hand, since the wavelength of the wave seen by the electron in the E-frame in the place of the undulator is far greater than the Compton wavelength, classical radiation formulae may be used freely, and in Chapter 4 we shall review them and use them to calculate the intensity of undulator radiation directly in the laboratory frame, without appealing to multiple frames of reference or to perturbation methods.

Probably the greatest merit of the analogy between undulator emission and Compton scattering today is that it may be worked backwards, transferring the knowledge that has been accumulated on undulators on to the process of Compton scattering. In this respect it is interesting to observe that a backward-propagating electromagnetic wave of wavelength l_u/γ and frequency $\gamma c/l_u$, which we have seen to be approximately equivalent to the undulator field in the E-frame, when transformed back into the L-frame appears as an electromagnetic wave of wavelength $2l_u$ and frequency $c/2l_u$. Therefore it can be stated that, within the above approximation, the effects upon a relativistic electron of an oscillating static field of period l_u and of a backward-propagating electromagnetic wave of wavelength $2l_u$ are identical. This result was already known before the invention of undulators and free-electron lasers, as the Weizsäcker–Williams approximation (Heitler 1954).

2.4 Analogy with Raman scattering

When its density is high, the electron beam passing through the undulator behaves as an electron plasma (a cold electron plasma in the simplest case) and radiation is no longer scattered by single electrons but rather by *plasma waves*. Since in the interaction with plasma waves the frequency of the incident radiation changes, just as in atomic Raman scattering, this phenomenon too is given the name of Raman scattering.

In the E-frame, where the electron plasma is at rest, Langmuir plasma waves can exist at the plasma frequency $\omega'_p = (e^2\rho'_e/\varepsilon_0 m_e)^{1/2}$, ρ'_e being the electron number density, and the undulator equivalent wave of angular frequency $\omega'_0 = \gamma c k_u$ (where $k_u = 2\pi/l_u$) is scattered into two waves of frequency $\omega'_0 \pm \omega'_p$. These will then appear in the laboratory as two waves with angular frequency

$$2\gamma\omega'_0 \pm 2\gamma\omega'_p = 2\gamma^2 c k_u \pm 2\gamma(ej/\gamma c\varepsilon_0 m_e)^{1/2}, \tag{2.18}$$

where we have set $\rho'_e = \rho_e/\gamma = j/\gamma e c$, j being the e-beam current density in the laboratory, where the electrons have a velocity close to c.

In order that the difference between Compton and Raman scattering may be observed in undulator emission, the Raman frequency shift which, relative to centre frequency, is $(ej/\gamma^3 c^3 k_u^2 \varepsilon_0 m_e)^{1/2}$, must exceed the linewidth due to finite undulator length, i.e. $1/N = l_u/L$. This condition may be written as a higher limit on γ:

$$\gamma^3 \lesssim jL^2 e/c^3\varepsilon_0 m_e. \tag{2.19}$$

Since $c^3\varepsilon_0 m_e/e \simeq 1.4\,\text{kA}$ it can be seen that the Raman effect influences undulator operation only for high current densities and moderate values of γ. Its influence will be discussed in Chapter 6.

2.5 Stimulated emission

Quantum mechanics makes it particularly clear that every spontaneous-radiation process has absorption and stimulated-emission counterparts, by giving explicit relationships between the transition probabilities associated with these processes; so one speaks, for instance, of the stimulated Compton and Raman effects. However, the above is not actually a quantum-mechanical property of nature, requiring only the principle of detailed balance, which descends from microscopic invariance with respect to reflection of time, and some additional statistical postulates which are independent of quantum mechanics. Therefore the concept of stimulated emission may be formulated equally well in classical mechanics, even if it is not usually given this name.

When an electron radiates, the radiated energy may be calculated as the work done by the electric field on the electron while retarding it. Similarly, the energy imparted in accelerating the electron is the work done by the electron on the field, thereby weakening it. By appeal to the reciprocity theorem of electrodynamics, which is the classical analogue of the principle of detailed balance as applied to photons, it may be shown that the energy radiated into a given radiation mode, e.g. an outgoing spherical wave, may be computed as the work done by such a wave, assumed as a 'virtual' wave, on the electron. This already shows that there is a relationship between the radiation of an electromagnetic wave by the electron and the modifications that a similar wave driven onto the electron itself produces upon its motion. Combining this relation with Madey's formula (1.75) relating the mean and mean squared electron energy variations, which can be obtained quite generally from Hamiltonian mechanics, will be the way in which we calculate, in Chapter 5, the 'stimulated' amplification of an external wave by the electron beam passing through the undulator. The result will be valid for any amplitude of the undulator magnetic field, but will be limited by the fact that one electron at a time is considered and the amplitude of the electron energy variation, and therefore of the impinging wave, is assumed to be small. We express these two assumptions by calling the range of validity of the above description the 'small-gain, small-signal regime'. The interaction between electrons manifests itself when the gain of the free-electron laser is not small, as then the electromagnetic field 'seen' by each electron can no longer be identified with the input wave but must be calculated as the self-consistent field generated, according to Maxwell's equations, by the electrons themselves. In this regime of operation, which may be called large-gain, it becomes preferable to adopt the approach of continuum mechanics and describe the electrons through their distribution function in phase space, the evolution of which is determined by Boltzmann's equation. In this description, which will be dealt with in Chapter 6, coupled electron-density and electromagnetic waves are obtained as the solution of the coupled Maxwell and Boltzmann equations, and the growth rate of these waves accounts for FEL amplification.

2.6 Picture of oscillating electron in moving frame

When, on the other hand, the effects upon the electron motion of the input wave field become comparable with those of the undulator, it becomes no longer possible to treat the former as a small perturbation. Output radiation then ceases to depend linearly on input and we speak of a large-signal, or saturation, regime. As long as interactions between electrons may be neglected and the electromagnetic field acting upon a single electron may be identified with the sum of the input wave and the undulator field, i.e. in the small-gain case, a Lorentz reference frame, called the 'synchronous' frame, may be found in which the input wave and the undulator equivalent radiation have the same frequency ω' and are seen to approach the electron from opposite sides. When the two waves are described by vector potentials A_u, A_w, the electron moves in the combined vector potential field

$$\begin{aligned} \boldsymbol{a} &= \mathrm{Re}[A_u \mathrm{e}^{\mathrm{i}k_u z} + A_w \mathrm{e}^{\mathrm{i}(\omega t - kz)}] \\ &= \mathrm{Re}[A_u \mathrm{e}^{\mathrm{i}(\omega' t' + k_u' z')} + A_w \mathrm{e}^{\mathrm{i}(\omega' t' - k_w' z')}]. \end{aligned} \tag{2.20}$$

It will be shown that the forces acting on the electron in this field may be approximately derived from a potential, variously called ponderomotive or quasi-potential, given by the average square of $\boldsymbol{a}$. It may be seen from eqn (2.20) that the square of $\boldsymbol{a}$ contains a term independent of t', and sinusoidally varying with z'. When the speed of the electron in the laboratory is such as to bring it to approximate rest in the synchronous frame (which is the usual undulator resonance condition in yet another guise), the electron may become trapped in one of the wells of the sinusoidal quasi-potential and start to oscillate back and forth in it. A study of these oscillations will enable us to describe, in Chapter 7, the saturation regime of the FEL.

2.7 Tapering the undulator

When both input power and gain of an FEL are high, the field acting upon each electron is again the self-consistent field generated by all the other electrons, i.e. a wave the speed of which is different from the speed of light in vacuum and possibly varies along the undulator; at the same time, because of the high power extraction, an electron's energy may diminish noticeably during its motion across the undulator. Similarly, when the undulator is used to couple energy from the input wave into the electron beam to accelerate the latter, both the wave and the electron properties may vary significantly between the entrance and exit of the undulator.

We have learned that it is possible to maintain synchronism over a fairly long range of energy loss or gain by tapering the undulator structure, either by varying the period or the field amplitude. It is then advantageous to consider, as we shall do in Chapter 8, a non-inertial synchronous frame in which a constant position z' corresponds to a point which maintains a constant difference between the phase of the undulator field and that of the laser wave forming a potential well.

This is now a frame with continuously changing speed, and the electron can be trapped in the potential well as in a bucket, and be carried along with it towards a different energy. Depending on the design we can have accelerating buckets, thus making an electron accelerator, or decelerating buckets, thus making an FEL. This is a new energy-transfer mechanism, completely different from those acting in the other regimes considered, and is intrinsically non-linear, so that the amplitude of the input wave field is an integral part of the design. This means that a tapered undulator is generally better suited as a power amplifier than as an oscillator FEL, because it has good efficiency under design conditions but works worse than an untapered one at input amplitudes other (particularly, lower) than the design value.

2.8 Is the free-electron laser a laser?

As we have seen that quantum effects play hardly any role in the functioning of a free-electron laser, one might ask whether this device should be called a laser at all. If any device producing coherent optical radiation by means of an amplifier placed in an optical cavity is called a laser, then the answer is definitely yes. But if one has to stand by the original meaning of the acronym LASER, Light Amplification by Stimulated Emission of Radiation, the answer is doubtful. In a conventional laser there is an energy pump, which may be optical, electrical, chemical, etc., producing an inverted population of quantum energy levels in the active medium. The stimulated transition from the higher to the lower level is then the mechanism that radiation is amplified by. In a free-electron laser there is no pump and, although the words 'stimulated emission' may be given a classical meaning and, on the other hand, in a certain interpretation (see, e.g., Motz 1979) it may be maintained that the device relies on a pre-existing inverted electron population near the maximum of a Maxwell distribution of electron velocities, it is undeniable that the undulator functioning as an amplifier is much more similar to microwave electronic tubes, travelling-wave tubes or even klystrons and magnetrons, than to lasers and masers.

Although one should be well aware that a free-electron laser is no more alike to a conventional laser than a travelling-wave tube is to a maser, use of the word 'laser' may be justified by the fact that, over time, it has outgrown its original meaning and may today be interpreted to denote any powerful source of coherent optical radiation rather than one based on a specific mechanism, and therefore we have no better name to propose for the free-electron laser. However, there is a tendency today to use the term 'free-electron laser' somewhat liberally. Whereas Madey initially coined it for a device which would generate light from an ultra-relativistic (high-γ) electron beam passing through an undulator with mirrors at its ends, the term has subsequently been adopted to designate more or less any new kind of electromagnetic oscillator based on an optical resonator and an electron beam, whether ultrarelativistic or not and whether assisted by an undulator or another coupling device. It is, nevertheless, clear that too loose a definition of the word 'laser' is as useless as a too pedantic one, and certainly not every emitter of electromagnetic waves powered on free electrons can be called a free-electron laser, for such a definition would include even the triode. Since it would be impossible to treat all such devices in detail in a single book, we shall adhere to the strictest definition and limit the scope of this book to the interaction of an ultrarelativistic electron beam with an undulator, with only incidental excursions on related devices. The ultrarelativistic electron beam gives the free-electron laser its unique capability of producing visible light (so that, at least as far as initials are concerned, it is correct to call it Laser rather than Maser), even if it may also be used at larger or smaller wavelengths. This feature also gives a characteristic look to the theory because of the approximations that are most appropriate to it, which will be discussed in the following chapters.

References

Heitler, W. (1954). *The quantum theory of radiation.* Oxford University Press.

Motz, H. (1979). Is the free-electron laser a laser? *Phys. Lett.* **71A**, 41–3.

Schiff, L. I. (1952). Quantum effects in the radiation from accelerated relativistic electrons. *Am. J. Phys.* **20**, 474–8.

3

MOTION OF AN ELECTRON IN AN UNDULATING MAGNETIC FIELD

The first step in studying the interaction of an electron beam with an undulator is, of course, the calculation of the electron trajectories in its field. In this chapter we apply to the undulator problem a Hamiltonian formulation which will allow us to gain an advantage from the ultrarelativistic character of the electron motion we are interested in. The required basic elements of Hamiltonian mechanics are summarized, for the reader's ease of reference, in Appendix A.

3.1 Generalized Hamilton equations

The relativistic relation between energy $\mathscr{E}$ and momentum $\boldsymbol{P}$ of an electron in the presence of an electromagnetic field of vector potential $\boldsymbol{a}$ and scalar potential φ (such that the electric field is $\boldsymbol{e} = -\partial\varphi/\partial\boldsymbol{r} - \partial\boldsymbol{a}/\partial t$ and the magnetic induction field is $\boldsymbol{b} = \text{curl}\,\boldsymbol{a}$) is given by (A.12; see Appendix A)

$$(\mathscr{E} - e\varphi)^2 - c^2(\boldsymbol{P} - e\boldsymbol{a})^2 = (m_e c^2)^2. \tag{3.1}$$

The above relation is all that is needed to obtain the relativistic equations of motion in Hamiltonian form; in fact, considering the electromagnetic potentials as known functions of co-ordinates and time, and naming Hamiltonian $\mathscr{H}$ the energy expressed through the above equation as a function of momentum, co-ordinates, and time, one can obtain the electron motion from the Hamilton equations (A.13):

$$\mathrm{d}\boldsymbol{r}/\mathrm{d}t = \partial\mathscr{H}/\partial\boldsymbol{P}, \qquad \mathrm{d}\boldsymbol{P}/\mathrm{d}t = -\partial\mathscr{H}/\partial\boldsymbol{r}. \tag{3.2}$$

The compact notation afforded by Hamiltonian formalism allows many useful transformations to be effected easily upon the equations of motion. In particular, a needed transformation will be the adoption of an independent variable other than time. Let $\Lambda(\boldsymbol{r}, \boldsymbol{P}, t, \mathscr{E}) = 0$ be an equation implicitly expressing the relation between energy and momentum; that is, any one of the infinitely many equations equivalent to eqn (3.1); we are going to show that the equations

$$\mathrm{d}\boldsymbol{r}/\mathrm{d}\lambda = \partial\Lambda/\partial\boldsymbol{P}, \qquad \mathrm{d}t/\mathrm{d}\lambda = -\partial\Lambda/\partial\mathscr{E}; \tag{3.3a}$$

$$\mathrm{d}\boldsymbol{P}/\mathrm{d}\lambda = -\partial\Lambda/\partial\boldsymbol{r}, \qquad \mathrm{d}\mathscr{E}/\mathrm{d}\lambda = \partial\Lambda/\partial t; \tag{3.3b}$$

$$\Lambda(\boldsymbol{r}, \boldsymbol{P}, t, \mathscr{E}) = 0, \tag{3.3c}$$

where λ is a new independent variable, are equivalent to eqns (3.2). Equations (3.3) will be referred to as the generalized Hamilton equations. In fact, eliminating λ from eqns (3.3) one has, for instance, $d\boldsymbol{r}/dt = (d\boldsymbol{r}/d\lambda)/(dt/d\lambda) = -(\partial\Lambda/\partial\boldsymbol{P})/(\partial\Lambda/\partial\mathscr{E})$ and, according to the theory of implicit differentiation, this is just $\partial\mathscr{E}/\partial\boldsymbol{P}$ when $\mathscr{E}$ is implicitly given as a function of $\boldsymbol{P}$ by $\Lambda(\boldsymbol{r}, \boldsymbol{P}, t, \mathscr{E}) = 0$; that is, $\partial\mathscr{H}/\partial\boldsymbol{P}$. The other Hamilton equations can be recovered in a similar way.

Equations (3.3) can have as many different forms as different equivalent equations $\Lambda(\boldsymbol{r}, \boldsymbol{P}, t, \mathscr{E}) = 0$ exist, and each form corresponds to a different independent variable λ. If, for instance, Λ is taken to be the difference between the right-hand and left-hand sides of eqn (3.1), λ turns out to be proportional to the relativistic proper time; if Λ is taken to be $\mathscr{H}(\boldsymbol{r}, \boldsymbol{P}, t) - \mathscr{E}$, λ coincides with world time t, as the equation $dt/d\lambda = -\partial\Lambda/\partial\mathscr{E} = 1$ shows, and the other equations reduce to the standard Hamilton equations (3.2). If Λ is chosen to be $P_z - \mathscr{K}(\boldsymbol{x}, \boldsymbol{P}_\perp, t, \mathscr{E}, z)$, where $\boldsymbol{x}$ and $\boldsymbol{P}_\perp$ are the projections of $\boldsymbol{r}$ and $\boldsymbol{P}$ on to a plane normal to the z-axis, and $\mathscr{K}(\boldsymbol{x}, \boldsymbol{P}_\perp, t, \mathscr{E}, z)$ is the expression of P_z obtained upon extracting it from eqn (3.1), λ turns out to be the co-ordinate z, as the equation $dz/d\lambda = \partial\Lambda/\partial P_z = 1$ shows, and eqns (3.3) specialize to:

$$d\boldsymbol{x}/dz = -\partial\mathscr{K}/\partial\boldsymbol{P}_\perp, \qquad dt/dz = \partial\mathscr{K}/\partial\mathscr{E}; \tag{3.4a}$$

$$d\boldsymbol{P}_\perp/dz = \partial\mathscr{K}/\partial\boldsymbol{x}, \qquad d\mathscr{E}/dz = -\partial\mathscr{K}/\partial t. \tag{3.4b}$$

Equations (3.4) are frequently used in the theory of particle accelerators (see, e.g., Courant and Snyder 1958). Since they will be used throughout most of this book, it is worthwhile to give the explicit expression of $\mathscr{K}$:

$$\mathscr{K} = ea_z \pm [(\mathscr{E} - e\varphi)^2/c^2 - (\boldsymbol{P}_\perp - e\boldsymbol{a}_\perp)^2 - m_e^2c^2]^{1/2}. \tag{3.5}$$

For ease of notation we shall adopt a partially dimensionless system of units such that $e = m_e = c = 1$, substituting t for ct (with the dimensions of a length), γ for $\mathscr{E}/m_ec^2$ (dimensionless), $\boldsymbol{P}$ for $\boldsymbol{P}/m_ec$ (dimensionless), φ for $e\varphi/m_ec^2$ (dimensionless), and $\boldsymbol{a}$ for $e\boldsymbol{a}/m_ec$ (dimensionless).[1] In this system of units eqn (3.5) becomes

$$\mathscr{K} = a_z \pm [(\gamma - \varphi)^2 - (\boldsymbol{P}_\perp - \boldsymbol{a}_\perp)^2 - 1]^{1/2}. \tag{3.6}$$

In eqns (3.5) and (3.6) the sign of the square root is undetermined. In fact, one should be cautioned that using z as the independent variable is only possible if $z(t)$ is an invertible function, that is if the axial velocity dz/dt does not change sign. When this is the case one sign can be chosen for the square root, according to the initial conditions, and maintained throughout the motion.

3.2 The Hamilton–Jacobi equation and the variational formulation

The Hamilton–Jacobi equation and the least action variational principle corresponding to the forms (3.3) and (3.4) of the Hamilton equations of motion may also be useful.

The standard Hamilton–Jacobi equation is (A.25)

$$\partial S/\partial t = -\mathscr{H}(\boldsymbol{r}, \partial S/\partial \boldsymbol{r}, t), \tag{3.7}$$

where $\mathscr{H}(\boldsymbol{r}, \boldsymbol{P}, t)$ is the Hamiltonian of the system and $S(\boldsymbol{r}, t)$ the, unknown, action function. When an independent variable other than time is chosen to represent the motion, the Hamilton–Jacobi equation undergoes a barely formal modification: in fact, given any equation $\Lambda(\boldsymbol{r}, \boldsymbol{P}, t, \mathscr{E}) = 0$ implicitly representing the relation $\mathscr{E} = \mathscr{H}(\boldsymbol{r}, \boldsymbol{P}, t)$, the equation

$$\Lambda(\boldsymbol{r}, \partial S/\partial \boldsymbol{r}, t, -\partial S/\partial t) = 0 \tag{3.8}$$

is obviously equivalent to eqn (3.7). Indeed, this is perhaps the simplest way to show that time and co-ordinates are on the same footing in the equations of motion and therefore their roles are interchangeable (provided that due care is taken of a different sign arising in the definition of energy and momentum).

Let us now introduce the Lagrangian of the system $\mathscr{L}_t(\boldsymbol{r}, \mathrm{d}\boldsymbol{r}/\mathrm{d}t, t)$ (which we have qualified by the subscript t to denote that it refers to the use of time as the independent variable), through the equation (A.14)

$$\mathscr{L}_t = \boldsymbol{P}\cdot \mathrm{d}\boldsymbol{r}/\mathrm{d}t - \mathscr{H}, \tag{3.9}$$

where, we recall, $\boldsymbol{P}$ must be eliminated by means of the equations $\mathrm{d}\boldsymbol{r}/\mathrm{d}t = \partial\mathscr{H}/\partial\boldsymbol{P}$ in order to express the Lagrangian as a function of $\boldsymbol{r}$, $\mathrm{d}\boldsymbol{r}/\mathrm{d}t$, and t. Hamilton's least action principle states that the motion of a mechanical system corresponds to an extremum of the action integral (A.2)

$$S = \int_{O_i}^{O_f} \mathscr{L}_t(\boldsymbol{r}, \mathrm{d}\boldsymbol{r}/\mathrm{d}t, t)\, \mathrm{d}t, \tag{3.10}$$

among all continuous and differentiable curves $\boldsymbol{r}(t)$ that run from a given point $O_i = (\boldsymbol{r}_i, t_i)$ to a given point $O_f = (\boldsymbol{r}_f, t_f)$ of space–time.

The Lagrangian must be given a different form when an independent variable λ other than time is adopted. To obtain this form, just note that the action integral itself must be independent of how the motion is represented. Therefore, eqn (3.10) may be rewritten as

$$S = \int_{O_i}^{O_f} \mathscr{L}_t \frac{\mathrm{d}t}{\mathrm{d}\lambda}\, \mathrm{d}\lambda = \int_{O_i}^{O_f} \mathscr{L}_\lambda\, \mathrm{d}\lambda, \tag{3.11}$$

thus defining the Lagrangian $\mathscr{L}_\lambda$ corresponding to the independent variable λ as

$$\mathscr{L}_\lambda = \mathscr{L}_t\, \mathrm{d}t/\mathrm{d}\lambda. \tag{3.12}$$

In order to obtain $\mathscr{L}_\lambda$ as a function of $\boldsymbol{r}$, $\mathrm{d}\boldsymbol{r}/\mathrm{d}\lambda$, t and $\mathrm{d}t/\mathrm{d}\lambda$, eqn (3.12) could be combined with eqn (3.9) and the relation between the Hamiltonian $\mathscr{H}$ and the function $\Lambda(\boldsymbol{r}, \boldsymbol{P}, t, E)$, but a more direct way is to use the differential relation (A.16) $dS = \mathscr{L}_t\, dt = \boldsymbol{P}\cdot d\boldsymbol{r} - \mathscr{E}\, dt$. Since dS is also equal to $\mathscr{L}_\lambda\, d\lambda$, because of eqn (3.12), one obtains

$$\mathscr{L}_\lambda = \boldsymbol{P}\cdot d\boldsymbol{r}/d\lambda - \mathscr{E}\, dt/d\lambda \tag{3.13}$$

which is the equation, similar to eqn (3.9), that connects $\mathscr{L}_\lambda$ with the function $\Lambda(\boldsymbol{r}, \boldsymbol{P}, t, \mathscr{E})$ of eqn (3.3c). In using eqn (3.13), as in the analogous case of eqn (3.9), $\boldsymbol{P}$ and $\mathscr{E}$ must be eliminated by means of the equations (3.3a) $d\boldsymbol{r}/d\lambda = \partial\Lambda/\partial\boldsymbol{P}$, $dt/d\lambda = -\partial\Lambda/\partial\mathscr{E}$.

In particular, when $\Lambda = P_z - \mathscr{H}(\boldsymbol{x}, \boldsymbol{P}_\perp, t, \mathscr{E}, z)$, as was assumed in deriving eqns (3.4), eqn (3.13) becomes

$$\mathscr{L}_z = \mathscr{H} + \boldsymbol{P}_\perp\cdot d\boldsymbol{x}/dz - \mathscr{E}\, dt/dz. \tag{3.14}$$

3.3 Ultrarelativistic approximation

The expression (3.6) of $\mathscr{H}$ which, we recall, represents the axial component of momentum expressed as a function of energy, transverse momentum, co-ordinates, and time, when γ is much greater than 1, can be simplified by expansion in a series of reciprocal powers of γ. Assuming, with no loss of generality, the positive sign for the square root appearing in eqn (3.6), the result of such an expansion up to terms of order γ^{-1} is

$$\mathscr{H} \simeq \gamma + a_z - \varphi - \frac{1 + (\boldsymbol{P}_\perp - \boldsymbol{a}_\perp)^2}{2\gamma}. \tag{3.15}$$

The right-hand side of eqn (3.15) will be referred to as the ultrarelativistic Hamiltonian, even if it represents axial momentum rather than energy. Its use is justified whenever γ^2 is much larger than $1 + (\boldsymbol{P}_\perp - \boldsymbol{a}_\perp)^2$; that is, when the axial velocity of the electron is close to the speed of light.

The ultrarelativistic Hamiltonian can also be obtained through another approximation which, though less straightforward, lets it appear from a different angle. Instead of studying the electron motion in the laboratory frame, write the equations of motion in a frame $\boldsymbol{x}$, z', t' moving in the increasing-z direction with a constant speed U close to the electron's speed (for instance, the E-frame or the synchronous frame). These are, of course, still eqns (3.2), (3.3), or (3.4), provided that the energy and momentum, as well as the e.m. potentials, are duly transformed. The laboratory axial co-ordinate and time will then be expressed as functions of the new ones by

$$z = (z' + Ut')/(1 - U^2)^{1/2}, \qquad t = (t' + Uz')/(1 - U^2)^{1/2}. \tag{3.16}$$

Now, supposing that the electromagnetic field acting upon the electron is the superposition of a static field in the laboratory frame, a function of $\boldsymbol{x}$ and z, and a paraxial wave travelling in the increasing-z direction, a function of $\boldsymbol{x}$, z

and $t - z$, write U as $1 - (1 - U)$ and neglect in the expressions of z and $t - z$ higher powers of $1 - U$:

$$z \simeq [2(1 - U)]^{-1/2}(t' + z'), \qquad t - z \simeq [(1 - U)/2]^{1/2}(t' - z'). \quad (3.17)$$

We shall show by means of the Hamilton–Jacobi equation that the exact equations of motion in the z', t'-frame, joined with the approximate co-ordinate transformation (3.17), are equivalent to the ultrarelativistic equations of motion written in the laboratory z, t-frame, provided that the electromagnetic potentials φ and a_z are also transformed according to suitable formulas ($\boldsymbol{a}_\perp$ is unchanged by the transformation, as are transverse co-ordinates and momentum).

In fact, the derivatives of the action function after the change of variables (3.17) become

$$\frac{\partial S}{\partial t'} = [2(1 - U)]^{-1/2}\frac{\partial S}{\partial z} + \{[2(1 - U)]^{-1/2} + [(1 - U)/2]^{1/2}\}\frac{\partial S}{\partial t}, \quad (3.18a)$$

$$\frac{\partial S}{\partial z'} = [2(1 - U)]^{-1/2}\frac{\partial S}{\partial z} + \{[2(1 - U)]^{-1/2} - [(1 - U)/2]^{1/2}\}\frac{\partial S}{\partial t}. \quad (3.18b)$$

Inserting these expressions into the exact Hamilton–Jacobi equation $(\partial S/\partial t' + \varphi')^2 - (\partial S/\partial z' - a'_{z'})^2 = 1 + (\partial S/\partial \boldsymbol{x}' - \boldsymbol{a}'_\perp)^2$ (where $\boldsymbol{x}' = \boldsymbol{x}$ and $\boldsymbol{a}'_\perp = \boldsymbol{a}_\perp$) gives

$$2\left(\frac{\partial S}{\partial z} + \varphi - a_z\right)\frac{\partial S}{\partial t} + 2\left(\frac{\partial S}{\partial t}\right)^2 = 1 + \left(\frac{\partial S}{\partial \boldsymbol{x}} - \boldsymbol{a}_\perp\right)^2, \quad (3.19)$$

provided that the transformed potentials φ' and $a'_{z'}$ are defined as $\varphi' = -a'_{z'} = [2(1 - U)]^{-1/2}(\varphi - a_z)$. Equation (3.19) is the Hamilton–Jacobi equation corresponding to the Hamiltonian (3.15), as can be verified after the substitutions $\gamma = -\partial S/\partial t$, $\boldsymbol{P}_\perp = \partial S/\partial \boldsymbol{x}$, and $\mathcal{H} = \partial S/\partial z$.

Notice that for $\mathrm{d}z/\mathrm{d}t = 0$ eqns (3.17) give $\mathrm{d}z'/\mathrm{d}t' = -1$, while for $\mathrm{d}z/\mathrm{d}t = 1$ they give $\mathrm{d}z'/\mathrm{d}t' = 1$. The interpretation of eqns (3.17) is then that the ultrarelativistic approximation makes an object, standing still in the laboratory, appear in the moving frame as running across the electron with the speed of light (equal to 1 in our system of units), rather than with the speed U, albeit preserving the relativistic invariance of the speed of light for waves propagating in the same direction as the electron.

3.4 Ideal sinusoidal undulator

The prototype magnetic field configuration of an undulator, which we shall refer to as 'ideal' undulator, is given by

$$\boldsymbol{b}_\perp = \mathrm{Re}(\boldsymbol{B}_u \mathrm{e}^{\mathrm{i}k_u z}) \quad (0 \leqslant z \leqslant L), \qquad b_z = 0 \quad (3.20)$$

and is generated by the potentials

$$a_\perp = \mathrm{Re}(A_u e^{ik_u z}) \quad (0 \leqslant z \leqslant L), \qquad a_z = \varphi = 0, \tag{3.21}$$

where $A_u = -\hat{z} \wedge B_u/ik_u$ is a constant complex vector in the transverse plane, with Cartesian components A_{u1} and A_{u2} along the axes x_1 and x_2, and $k_u = 2\pi/l_u$ is a real constant, the undulator wavenumber; $a_\perp$ is assumed to be zero for $z < 0$ and $z > L$. Equations (3.20) and (3.21) may represent fields of any polarization. In particular, if $A_{u2} = \pm iA_{u1}$ we say that the undulator field is circularly polarized, because then the vector $a_\perp$ rotates, clockwise with increasing z if $A_{u2} = iA_{u1}$, and anticlockwise with increasing z if $A_{u2} = -iA_{u1}$. Circularly polarized fields have a property which will be useful later; namely, A_u^2, that is the scalar product $A_u \cdot A_u$, is zero. This can be easily proved by observing that $A_u^2 = A_{u1}^2 + A_{u2}^2$ and, if $A_{u2} = \pm iA_{u1}$, then $A_u^2 = A_{u1}^2 - A_{u1}^2 = 0$. Actually, the scalar product of any two circularly polarized complex vectors with the same orientation is zero, because this scalar product can be written as the product of a scalar times the square of one of the vectors, and this square is zero. (Notice that when A_u is a complex vector A_u^2 and $|A_u|^2$ are different. $|A_u|^2$, which equals $|A_{u1}|^2 + |A_{u2}|^2$, and will be simply denoted in the following as A_u^2, is only zero when both A_{u1} and A_{u2} are.)

Although the potential of eqn (3.21) is not consistent with the equations governing a magnetic field in a vacuum (it could, in principle, be obtained in a material medium with a suitably tapered magnetic permeability), it is a convenient starting point for studying undulators and free-electron lasers, and will almost always be assumed to represent the undulator field in the next chapters; at the end of this chapter we shall consider the limits of this assumption.

One can easily obtain the motion of an electron in the field (3.20) from the generalized Hamilton equations (3.4), using either the exact Hamiltonian $\mathscr{H}$ of eqn (3.6) or, if γ is high enough, the approximate one of eqn (3.15). Since the e.m. vector potential of eqn (3.21) is a function of z only, and z is the independent variable, the Hamiltonian $\mathscr{H}$ turns out to depend on no one of the Lagrangian co-ordinates t and x, and therefore γ and $P_\perp$ are constant, as may be seen from eqns (3.4b). On integrating eqns (3.4a) one then finds:

$$x = x_i - \int_0^z \frac{\partial \mathscr{H}}{\partial P_\perp} dz, \qquad t = t_i + \int_0^z \frac{\partial \mathscr{H}}{\partial \gamma} dz, \tag{3.22}$$

where the initial conditions x_i, t_i are the arrival location and time of the electron at the undulator entrance section, $z = 0$. Putting the expression (3.6) of $\mathscr{H}$ into eqns (3.22) gives

$$x = x_i + \int_0^z \frac{P_\perp - a_\perp}{[(\gamma - \varphi)^2 - (P_\perp - a_\perp)^2 - 1]^{1/2}} dz, \tag{3.23a}$$

$$t = t_i + \int_0^z \frac{\gamma - \varphi}{[(\gamma - \varphi)^2 - (P_\perp - a_\perp)^2 - 1]^{1/2}} dz, \tag{3.23b}$$

where,[2] according to eqn (3.21), $\varphi = a_z = 0$ and $(\boldsymbol{P}_\perp - \boldsymbol{a}_\perp)^2 = \boldsymbol{P}_\perp^2 + A_u^2/2 + \mathrm{Re}(-2\boldsymbol{P}_\perp \cdot \boldsymbol{A}_u \mathrm{e}^{\mathrm{i}k_u z} + A_u^2 \mathrm{e}^{2\mathrm{i}k_u z}/2)$. As the denominator of the fraction appearing under the sign of integral in eqns (3.23) is the square root of a quadratic polynomial in trigonometric functions, the integral itself is an elliptic integral, and can be either put in standard form and referred to tables (e.g., Abramowitz and Stegun 1968, ch. 17) or calculated numerically.

It so happens, however, that in the cases of interest for the applications of undulators and free-electron lasers, eqn (3.23) contains a large parameter, γ, and can be approximated by a series expansion in $1/\gamma$. Such an expansion is nothing other than the ultrarelativistic approximation introduced in the previous section; in fact, when the ultrarelativistic Hamiltonian of eqn (3.15) is put in eqns (3.22) instead of the exact one all the integrals become expressible through trigonometric functions, as

$$\boldsymbol{x} = \boldsymbol{x}_i + \frac{\boldsymbol{P}_\perp}{\gamma} z - \mathrm{Re}\left[\frac{\boldsymbol{A}_u}{\mathrm{i}\gamma k_u}(\mathrm{e}^{\mathrm{i}k_u z} - 1)\right], \tag{3.24a}$$

and

$$t = t_i + \left(1 + \frac{1 + \boldsymbol{P}_\perp^2 + A_u^2/2}{2\gamma^2}\right) z - \mathrm{Re}\left[\frac{\boldsymbol{P}_\perp \cdot \boldsymbol{A}_u}{\mathrm{i}\gamma^2 k_u}(\mathrm{e}^{\mathrm{i}k_u z} - 1)\right]$$
$$+ \mathrm{Re}\left[\frac{A_u^2}{8\mathrm{i}\gamma^2 k_u}(\mathrm{e}^{2\mathrm{i}k_u z} - 1)\right]. \tag{3.24b}$$

3.5 The electron frame of reference

In eqns (3.24) $\boldsymbol{x}(z)$ and $t(z)$ are both expressed by the sum of a linearly increasing and a bounded oscillating term. Therefore, the electron motion can be considered to be the superposition of a uniform translation and an oscillation. Let us discuss the properties of the translation motion first, represented by the equations

$$\boldsymbol{x} = \boldsymbol{x}_i + (\boldsymbol{P}_\perp/\gamma) z, \tag{3.25a}$$

$$t = t_i + [1 + (1 + \boldsymbol{P}_\perp^2 + A_u^2/2)/2\gamma^2] z. \tag{3.25b}$$

One can notice that the angle of the trajectory, $\mathrm{d}\boldsymbol{x}/\mathrm{d}z$, is unaltered by the presence of the undulator field (i.e. independent of $\boldsymbol{A}_u$), while the coefficient of z in the expression of $t(z)$, the reciprocal of the axial translation speed $\mathrm{d}z/\mathrm{d}t$, increases with $\boldsymbol{A}_u$. The change in speed is the same as would have been obtained, in the absence of any magnetic field, by replacing the electron mass, m_e, in eqn (3.5) by $m_e M$, where

$$M = (1 + A_u^2/2)^{1/2}. \tag{3.26}$$

In fact the dimensionless Hamiltonian of eqn (3.6) would then become $\mathcal{H} = \pm(\gamma^2 - \boldsymbol{P}_\perp^2 - M^2)^{1/2}$, and the ultrarelativistic one of eqn (3.15) $\mathcal{H} = \gamma - (\boldsymbol{P}_\perp^2 + M^2)/2\gamma$. The reciprocal velocity would then be $\mathrm{d}t/\mathrm{d}z = \partial\mathcal{H}/\partial\gamma = 1 + (\boldsymbol{P}_\perp^2 + M^2)/2\gamma^2$, which is the same as the coefficient of z in the second of eqns (3.25) when M is chosen according to eqn (3.26).

Indeed, as shown in Appendix B, it is a general property of the motion of charged particles in strong, rapidly oscillating electromagnetic fields that the effect of the field is equivalent to a mass increase by the factor M (3.26) (Kibble 1966; Luchini and Solimeno 1982). We shall have occasion to use this property in Section 3.8. Therefore $m_e M$, or just M in dimensionless form, may be called the effective mass of the electron.

Let us note, to assist the reader when consulting the literature, that the dimensionless potential A_u, or sometimes $A_u/\sqrt{2}$, is also met under the name of deflection parameter K. (In fact, when the electron motion is nonrelativistic the value of A_u gives the order of magnitude of the oscillating deflection angle of the electron trajectory. This angle, however, becomes A_u/γ under relativistic conditions.) Correspondingly, the expression of the effective mass becomes either $(1 + K^2/2)^{1/2}$ or $(1 + K^2)^{1/2}$.

For the effective mass increase to be substantial, the dimensionless vector potential A_u must have a magnitude comparable to or greater than unity, which means that the oscillating magnetic induction field $\boldsymbol{b}$ of eqn (3.20) must be comparable to or greater than $k_u m_e c/e$; that is, for instance, about 0.4 T for an undulator period of $l_u = 2.5$ cm. A field intensity of 0.4 T is easily achievable with either permanent magnets or electromagnets, so that the electron effective-mass increase in an undulator is usually quite considerable.

In order to study the oscillatory electron motion, let us now introduce a new reference frame, the E-frame, in uniform translatory motion relative to the laboratory frame with the speed corresponding to eqns (3.25). This will produce the effect of eliminating the translational motion, and in the new frame the electron will appear to perform a bounded orbit.

The exact form of the Lorentz transformation leading to such a frame is complicated and not particularly interesting. A considerable simplification can be obtained, in the ultrarelativistic case, by adopting the approximate transformation formulae (3.17). Let us restrict our discussion to the case $\boldsymbol{P}_\perp = 0$ (when $\boldsymbol{P}_\perp \neq 0$, it can be made zero by rotating the laboratory reference frame first). On choosing $U = 1/(1 + M^2/2\gamma^2) \simeq 1 - M^2/2\gamma^2$, eqns (3.17) give

$$z = \frac{\gamma}{M}(t' + z'), \qquad t - z = \frac{M}{2\gamma}(t' - z');$$

i.e.

$$z' = \left(\frac{\gamma}{M} + \frac{M}{2\gamma}\right)z - \frac{\gamma}{M}t, \qquad t' = \frac{\gamma}{M}t - \left(\frac{\gamma}{M} - \frac{M}{2\gamma}\right)z. \tag{3.27}$$

The transverse co-ordinates $\boldsymbol{x}$ are unaffected by the transformation, so we can set $\boldsymbol{x}' = \boldsymbol{x}$. In the $\boldsymbol{x}'$, z', t'-frame eqns (3.24) become

$$\boldsymbol{x}' = \boldsymbol{x}_i - \mathrm{Re}\left[\frac{\boldsymbol{A}_u}{\mathrm{i}\gamma k_u}(\mathrm{e}^{\mathrm{i}k_u z} - 1)\right], \tag{3.28a}$$

$$z' = -\frac{\gamma}{M}t_i - \mathrm{Re}\left[\frac{\boldsymbol{A}_u^2}{8\mathrm{i}M\gamma k_u}(\mathrm{e}^{2\mathrm{i}k_u z} - 1)\right], \tag{3.28b}$$

$$t' = \frac{\gamma}{M}t_i + \frac{M}{\gamma}z + \mathrm{Re}\left[\frac{\boldsymbol{A}_u^2}{8\mathrm{i}M\gamma k_u}(\mathrm{e}^{2\mathrm{i}k_u z} - 1)\right], \tag{3.28c}$$

where z acts as a parameter. The transverse co-ordinates $\boldsymbol{x}' = \boldsymbol{x}$ are periodic functions with spatial frequency k_u of the parameter z, describing an oscillation with amplitude

$$\Delta x_u = A_u/\gamma k_u, \tag{3.29}$$

whereas the axial co-ordinate z' is a periodic function of z with twice the frequency. Therefore, the projection of the electron trajectory on the x_1-x_2 plane is an ellipse, reducing to a straight-line segment in the case of linear and to a circle in the case of circular polarization, while its side view, in either the x_1-z or the x_2-z plane, has a characteristic figure-of-eight shape, looking like a Lissajous figure on an oscilloscope. The typical appearance of these electron orbits is shown in Fig. 3.1. Notice that the ratio of axial to transverse extension of the oscillation is $A_u^2/8MA_u = A_u^2/8A_u(1 + A_u^2/2)^{1/2}$, and is contained between zero (when either the undulator is circularly polarized or $A_u \to 0$) and $1/4\sqrt{2}$ (when the undulator is linearly polarized and $A_u \to \infty$).

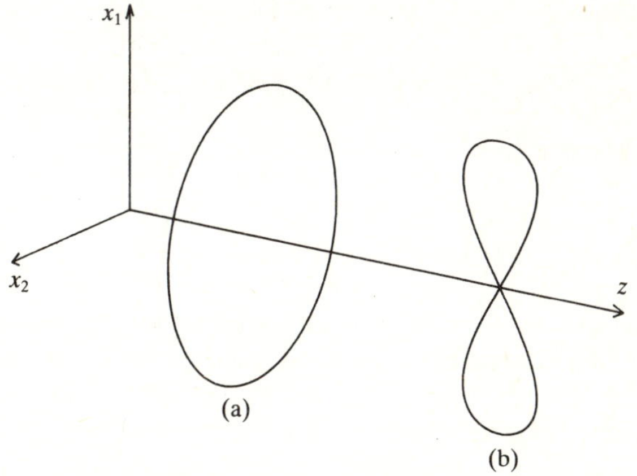

FIG. 3.1. Electron orbits in the E-frame, in a circularly polarized (a) and in a linearly polarized (b) undulator.

A shape similar to a figure-of-eight is also obtained from eqns (3.23) when γ is not large, i.e. outside the range of validity of the ultrarelativistic approximation, but the figure-of-eight then becomes slimmer and slimmer with decreasing γ, tending to approach the shape of a vertical segment at non-relativistic electron speeds.

3.6 Electromagnetic undulator

It has been explained in the previous chapter that the motion of an ultrarelativistic electron in a static undulating field is very similar to the motion of the same electron in a counterstreaming electromagnetic wave. Therefore the same effects which can be obtained with a magnetostatic undulator can also be obtained with such a wave, and one then speaks of an electromagnetic undulator. While this would not usually be the first choice, because it is practically impossible to obtain the same magnetic field intensity from an electromagnetic wave as from static magnets, an electromagnetic undulator has been proposed as a means of obtaining very small undulator periods, particularly in connection with the two-stage FEL which we shall describe later.

Regarding the electron motion, the electromagnetic undulator is in a certain sense simpler to study than the static one, because it gives rise to exactly sinusoidal trajectories (in a suitable parametric representation of the motion) and does not need any approximation. This can be easily seen from the Hamilton–Jacobi equation (see, e.g., exercise 2 of section 47 of Landau and Lifschitz (1975); cf. also Appendix B).

Consider the Hamilton–Jacobi equation (3.8) for undulator dimensionless potentials given by

$$\boldsymbol{a}_\perp = \mathrm{Re}[\boldsymbol{A}_u \mathrm{e}^{\mathrm{i}k_u(t+z)}] = \mathrm{Re}(\boldsymbol{A}_u \mathrm{e}^{\mathrm{i}k_u\tau}), \qquad a_z = \varphi = 0; \tag{3.30}$$

that is, the equation

$$\left(\frac{\partial S}{\partial t}\right)^2 - \left(\frac{\partial S}{\partial z}\right)^2 - \left(\frac{\partial S}{\partial \boldsymbol{x}} - \boldsymbol{a}_\perp\right)^2 = 1. \tag{3.31}$$

Since the transverse co-ordinates $\boldsymbol{x}$ do not appear explicitly, transverse momentum $\boldsymbol{P}_\perp$ is a constant of motion. In terms of the action function, a solution exists of the form $S(t, z, \boldsymbol{x}) = S'(t, z) + \boldsymbol{P}_\perp \cdot \boldsymbol{x}$. As may easily be verified, the equation for S' is

$$\left(\frac{\partial S'}{\partial t}\right)^2 - \left(\frac{\partial S'}{\partial z}\right)^2 = 1 + (\boldsymbol{P}_\perp - \boldsymbol{a}_\perp)^2. \tag{3.32}$$

As the vector potential depends only on $\tau = t + z$, it may be useful to operate a change of variables such that τ is one of the new variables. Assuming t and τ as new variables, one has

$$\left(\frac{\partial S'}{\partial t}\right)_z = \left(\frac{\partial S'}{\partial t}\right)_\tau + \left(\frac{\partial S'}{\partial \tau}\right)_t, \qquad \left(\frac{\partial S'}{\partial z}\right)_t = \left(\frac{\partial S'}{\partial \tau}\right)_t, \tag{3.33}$$

and eqn (3.32) becomes

$$\left(\frac{\partial S'}{\partial t}\right)^2 + 2\frac{\partial S'}{\partial t}\frac{\partial S'}{\partial \tau} = 1 + (\boldsymbol{P}_\perp - \boldsymbol{a}_\perp)^2. \tag{3.34}$$

Since the vector potential is a function of τ only, a solution exists of the form $S'(t, \tau) = S''(\tau) - \gamma' t$, γ' being a new constant of motion (a kind of energy or momentum, not to be confused, however, with the true energy, which is $-(\partial S/\partial t)_z$ rather than $-(\partial S/\partial t)_\tau$ and is not constant). The equation for S'' is

$$\frac{\partial S''}{\partial \tau} = \frac{\gamma'}{2} - \frac{1 + [\boldsymbol{P}_\perp - \boldsymbol{a}_\perp(\tau)]^2}{2\gamma'}. \tag{3.35}$$

Equation (3.35) can be easily integrated and, thereafter, the motion may be expressed, according to the general rules of the Hamilton–Jacobi method (see Appendix A), by the equations $\partial S/\partial \boldsymbol{P}_\perp =$ constant and $\partial S/\partial \gamma' =$ constant.

As may have already been recognized, if it is remembered that in the Hamilton–Jacobi formalism $\mathscr{H} = P_z = \partial S/\partial z$, eqn (3.35) is very similar to the expression (3.15) of the ultrarelativistic Hamiltonian, with τ in the place of z. It is therefore not surprising that the solution, too, is very similar to eqn (3.24):

$$\boldsymbol{x} = \boldsymbol{x}_i + \frac{\boldsymbol{P}_\perp}{\gamma'}\tau - \mathrm{Re}\left[\frac{\boldsymbol{A}_u}{\mathrm{i}\gamma' k_u}(\mathrm{e}^{\mathrm{i}k_u\tau} - 1)\right], \tag{3.36a}$$

$$t = t_i + \left(\frac{1}{2} + \frac{1 + P_\perp^2 + A_u^2/2}{2\gamma'^2}\right)\tau - \mathrm{Re}\left[\frac{\boldsymbol{P}_\perp \cdot \boldsymbol{A}_u}{\mathrm{i}\gamma'^2 k_u}(\mathrm{e}^{\mathrm{i}k_u\tau} - 1)\right]$$
$$+ \mathrm{Re}\left[\frac{A_u^2}{8\mathrm{i}\gamma'^2 k_u}(\mathrm{e}^{2\mathrm{i}k_u\tau} - 1)\right]. \tag{3.36b}$$

Let us emphasize that eqns (3.36) are exact and valid for any values of the parameters, whereas eqns (3.24) are only an approximation for $\gamma \gg 1$. The similarity of the results shows once again that the undulator is seen by an ultrarelativistic electron as a counterstreaming electromagnetic wave.

3.7 Limitations of the ideal undulator model

The ideal magnetic field of eqn (3.20) is not a solution of Maxwell's equations in vacuum and therefore cannot be realized exactly. In this section we want to investigate how close one can theoretically get to eqns (3.20) and (3.21), and what modifications of the electron motion must be expected in a real field. (The practical means of generating a suitable magnetic field will be considered in Chapter 10.)

Let us state the problem in the following way: consider a magnetic field, the vector potential of which equals that given by eqn (3.21) on some straight line that we assume as z-axis. How close can $\boldsymbol{a}$ remain to this form in a neighbourhood of the given line? In order to answer this question, let us expand the static vector potential $\boldsymbol{a}(\boldsymbol{x}, z)$ in a Taylor series in $\boldsymbol{x}$;

$$\boldsymbol{a} = \boldsymbol{a}(0, z) + \frac{\partial \boldsymbol{a}}{\partial x_\alpha} x_\alpha + \frac{1}{2} \frac{\partial^2 \boldsymbol{a}}{\partial x_\alpha \, \partial x_\beta} x_\alpha x_\beta + \cdots \tag{3.37}$$

where α, $\beta = 1$, 2, and summation over repeated indices is implied. By assumption, $\boldsymbol{a}(0, z) = \mathrm{Re}(A_u \mathrm{e}^{\mathrm{i}k_u z})$. By choosing, in addition, $\boldsymbol{a}_\perp$ to be an even function of $\boldsymbol{x}$ (a_z must then be an odd function), the first derivatives of $\boldsymbol{a}_\perp$ can be made to vanish. One would like the higher derivatives to vanish as well, but this is forbidden by the field equations, $\Delta_2 \boldsymbol{a} = 0$ and div $\boldsymbol{a} = 0$. In particular, for the second-order coefficients the equation $\Delta_2 \boldsymbol{a} = 0$ requires that

$$\frac{\partial^2 \boldsymbol{a}_\perp}{\partial x_1^2} + \frac{\partial^2 \boldsymbol{a}_\perp}{\partial x_2^2} = -\frac{\partial^2 \boldsymbol{a}_\perp}{\partial z^2} = \mathrm{Re}(k_u^2 A_u \mathrm{e}^{\mathrm{i}k_u z}). \tag{3.38}$$

Therefore, the second-order coefficients must at least be of the order of $k_u^2 A_u$, which means that the field must depart sensibly from the form (3.20) at a distance from the axis of the order of l_u, as could also be inferred from a dimensional argument.

From the above considerations we immediately obtain a first constraint on the validity of the ideal undulator model: it is necessary that the electrons move in the transverse direction only over a range of distances small compared to l_u. The amplitude of the transverse oscillations of the electron trajectory is $\Delta x_u = A_u/\gamma k_u$, according to eqn (3.29), so that this constraint is satisfied if

$$A_u/\gamma \ll 1. \tag{3.39}$$

Condition (3.39) is certainly well verified in laser devices working with highly ultrarelativistic electrons; it must be checked case-by-case in microwave devices working with electrons of a few MeV energy, but even then the lower γ is often counterbalanced by an A_u smaller than unity.

In addition to requiring that the oscillations of a single electron should not drive it outside the region in which the magnetic field is approximately that of the ideal sinusoidal undulator, it must be ensured that the cross-section of the whole electron beam does not exceed the size of this region. For an electron beam with an initial cross-section diameter σ_e spreading in free space with an angular divergence ϑ_e, this second constraint requires that

$$\sigma_e \ll l_u, \qquad \vartheta_e \ll l_u/L = 1/N. \tag{3.40, 3.41}$$

The limit of eqn (3.41), however, may be relaxed if the electron beam is subjected to some kind of focusing, which may be due either to external quadrupole magnets or, as will be seen in the next section, to the undulator itself. We shall also see later that the conditions for optimal operation of a free-electron laser put an upper bound on ϑ_e which is usually more stringent than eqn (3.41), so that the one-dimensional ideal undulator is, under most conditions, an acceptable model.

3.8 Betatron oscillations

Even if conditions (3.39)–(3.41) are fulfilled, as they will be supposed to be in most of the following, the unavoidable small lateral variation of the vector potential of a real field can, after several periods, have a non-negligible accumulated effect on the electron trajectory. This effect can be studied using the concept of effective mass.

Consider, quite generally, a vector potential $\boldsymbol{a} = \boldsymbol{a}_0(z) + \boldsymbol{a}_1(\boldsymbol{x}, z)$, where $\boldsymbol{a}_0$ is given by eqn (3.21) and $\boldsymbol{a}_1$ is a small perturbation, for instance the quadratic term appearing in eqn (3.37). The electron motion in the field represented by this vector potential can be assumed to be given, approximately, by the expressions

$$\boldsymbol{x} = \boldsymbol{x}_g(z) - \mathrm{Re}\left(\frac{\boldsymbol{A}_u}{\mathrm{i}\gamma k_u}\,\mathrm{e}^{\mathrm{i}k_u z}\right), \tag{3.42a}$$

$$t = t_g(z) - \mathrm{Re}\left(\frac{\boldsymbol{P}_\perp(z)\cdot\boldsymbol{A}_u}{\mathrm{i}\gamma^2 k_u}\,\mathrm{e}^{\mathrm{i}k_u z}\right) + \mathrm{Re}\left(\frac{A_u^2}{8\mathrm{i}\gamma^2 k_u}\,\mathrm{e}^{2\mathrm{i}k_u z}\right), \tag{3.42b}$$

where $\mathrm{d}\boldsymbol{x}_g/\mathrm{d}z$, $\mathrm{d}t_g/\mathrm{d}z$ and $\boldsymbol{P}_\perp$ are slowly varying functions of z (i.e. their derivatives are small quantities). In fact, as may be seen from eqns (3.24), when $\boldsymbol{a}_1$ vanishes these functions become constant, and therefore their derivatives may be assumed to be of the order of $\boldsymbol{a}_1$.

The point whose motion is represented by $\boldsymbol{x}_g(z)$ and $t_g(z)$ takes on the name of 'guiding centre', because by construction the electron performs a bound orbit about this point. It is shown in Appendix B that the motion of the guiding centre is approximately ruled by the Hamiltonian $\mathscr{H} = (\boldsymbol{P}^2 + M^2)^{1/2}$, similar to that of a free particle of mass M, where $M(\boldsymbol{x}, z) = (1 + \langle \boldsymbol{a}^2 \rangle)^{1/2}$, the angular brackets denoting an average over the 'fast' (with respect to the motion of the guiding centre) oscillation with period l_u. M is the already introduced 'effective mass' of the electron.

Notice that M is not an intrinsic property of the particle, like the true rest mass, but rather a scalar quantity characterizing the long-period interaction between the particle and the magnetic field. In this sense it can be considered as the relativistic extension of the concept of potential energy. Indeed, in a non-relativistic theory the same interaction would be described by an 'effective potential'.

Let us add, for the sake of generality, that only the assumption of slow variation of the velocity and momentum of the guiding centre on the scale established by l_u is necessary for the motion of the guiding centre to be capable of being described in terms of an effective mass. The ultrarelativistic assumption is not needed.

Considering, in particular, the expression (3.37) of the vector potential, obtained from the Taylor expansion with respect to the transverse coordinates of a general 'real' undulator field, we find, up to second order in $\boldsymbol{x}$,

$$M^2 = 1 + A_u^2/2 + A_{\alpha\beta} x_\alpha x_\beta, \tag{3.43}$$

where

$$A_{\alpha\beta} = \left\langle \boldsymbol{a}_\perp(0, z) \cdot \frac{\partial^2 \boldsymbol{a}_\perp}{\partial x_\alpha \, \partial x_\beta} + \frac{\partial a_z}{\partial x_\alpha} \frac{\partial a_z}{\partial x_\beta} \right\rangle$$

and is a symmetric tensor. In general, $A_{\alpha\beta}$ is a function of z; if, however, the magnetic field is assumed to be generated by a periodic magnet structure, and is therefore itself periodic in z with period l_u (though not coincident in all space with that represented by eqn (3.22)), $A_{\alpha\beta}$ becomes a constant. We shall now determine the motion of the guiding centre in the case that $A_{\alpha\beta}$ is constant.

For this purpose the Hamilton–Jacobi method may be used. (In the rest of this section the subscript g is omitted: $\boldsymbol{x}$ and t always refer to the guiding centre.) The Hamilton–Jacobi partial differential equation (3.7) corresponding to the guiding-centre Hamiltonian $(\boldsymbol{P}^2 + M^2)^{1/2}$ is

$$\left(\frac{\partial S}{\partial t}\right)^2 = \left(\frac{\partial S}{\partial z}\right)^2 + \frac{\partial S}{\partial x_\alpha} \frac{\partial S}{\partial x_\alpha} + 1 + \frac{A_u^2}{2} + A_{\alpha\beta} x_\alpha x_\beta. \tag{3.44}$$

Since the Hamiltonian is independent of t and z, $\partial S/\partial t = -\gamma$ and $\partial S/\partial z = P_z$ are two constants of motion. On assuming the principal directions of the symmetric tensor $A_{\alpha\beta}$ as axes x_1 and x_2, so that $A_{12} = 0$, it becomes possible to separate variables in the Hamilton–Jacobi equation, thus seeking a solution in the form $S = -\gamma t + P_z z + S_1(x_1) + S_2(x_2)$. The resulting ordinary differential equations for S_1 and S_2 are

$$(\partial S_\alpha/\partial x_\alpha)^2 + A_{\alpha\alpha} x_\alpha^2 = R_\alpha \qquad \text{(summation on } \alpha \text{ is } \textit{not} \text{ implied here)}, \tag{3.45}$$

where $R_{1,2}$ are two additional constants of motion tied to γ and P_z by the equation

$$\gamma^2 = R_1 + R_2 + P_z^2 + 1 + A_u^2/2. \tag{3.46}$$

The solution of eqn (3.45) is straightforward; the expression of the motion is then obtained, according to the general Hamilton–Jacobi method, by equating $\partial S/\partial R_1$, $\partial S/\partial R_2$, and $\partial S/\partial \gamma$ to constants, where in taking derivatives P_z must be considered to be a function of R_1, R_2 and γ through eqn (3.46). Hence one obtains

$$\int (R_\alpha - A_{\alpha\alpha} x_\alpha^2)^{-1/2}\, \mathrm{d}x_\alpha - z/P_z = C_\alpha \quad \text{(summation on } \alpha \text{ is } \textit{not} \text{ implied here),} \tag{3.47}$$

$$t - (\gamma/P_z)z = C_3. \tag{3.48}$$

The integral of eqn (3.47) evaluates either to an inverse trigonometric or to an inverse hyperbolic function, depending on whether $A_{\alpha\alpha}$ is positive or negative. In the first case the trajectory of the guiding centre is a stable sinusoidal oscillation about the z-axis with period

$$l_{\mathrm{bet},\alpha} = 2\pi P_z / A_{\alpha\alpha}^{1/2}. \tag{3.49}$$

By analogy with the oscillations caused by the focusing magnetic fields of accelerating machines, this motion is named 'betatron oscillation'. In the second case the guiding centre moves along an unstable trajectory which leads the electron away from the axis.

In general, the principal values A_{11} and A_{22} of the tensor $A_{\alpha\beta}$ are different, and the motion could well be stable in one direction and unstable in the other. However, the constraint (3.38) deriving from the field equations implies a constraint upon the trace of $A_{\alpha\beta}$, i.e.

$$A_{11} + A_{22} = -\left\langle \boldsymbol{a}_\perp(0, z) \cdot \frac{\partial^2 \boldsymbol{a}_\perp}{\partial z^2} \right\rangle + \left\langle \frac{\partial a_z}{\partial x_\alpha} \frac{\partial a_z}{\partial x_\alpha} \right\rangle \geqslant k_u^2 A_u^2/2. \tag{3.50}$$

Therefore, assuming that the undulator magnet has been designed to yield a vector potential as close as possible to eqn (3.21), i.e. to minimize transverse field gradients, we expect A_{11} and A_{22} to be both positive and of the order of $k_u^2 A_u^2/4$. Under these conditions the betatron oscillations are stable, i.e. the field gradient has a focusing effect, and the period of oscillation is typically

$$l_{\mathrm{bet,typ}} = 2P_z l_u / A_u \simeq 2\gamma l_u / A_u. \tag{3.51}$$

For large γ, when $P_z \simeq \gamma$, $l_{\mathrm{bet,typ}}$ is much larger than l_u, thus confirming the initial hypothesis of slow variation of the guiding-centre velocity on the scale of the undulator period.

In the presence of betatron oscillations an electron entering at an angle with respect to the undulator axis, instead of departing indefinitely from it, is captured in a betatron oscillation orbit. An easy calculation shows that, provided that $L \geqslant l_{\mathrm{bet}}/4$, an electron entering a betatron orbit on axis at an angle ϑ_i reaches a maximum distance from the axis of $l_{\mathrm{bet}}\vartheta_i/2\pi$, rather than $L\vartheta_i$ as when betatron oscillations are absent. Therefore the limitation (3.41) on electron beam divergence may be replaced by a milder one:

$$\vartheta_e \ll \max(l_u/L, 2\pi l_u/l_{\mathrm{bet}}). \tag{3.52}$$

In other words, the focusing effect of betatron oscillations is beneficial and contributes to containing the beam cross-section growth. It may be noticed that in a purely two-dimensional field, as that of a plane undulator to a good approximation is, betatron focusing only takes place in the plane of the field, because there are no gradients in the transverse direction. It is, however, relatively easy, especially when the undulator magnet has iron poles, to shape the poles so as to create small transverse gradients and redistribute the focusing in both directions (see, e.g., Section 10.4).

Notes

1. The same name has been given to the dimensional and dimensionless form of all quantities except energy. The dimensionless energy has been given the traditional name of γ. Notice, however, that $\mathscr{E}$, and thus γ, denotes the total energy, inclusive of kinetic energy and electrostatic potential energy. Although this is a somewhat non-standard use of the symbol γ, which is more commonly reserved for the sole kinetic energy $(1 - \mathbf{v}^2/c^2)^{-1/2}$, confusion should not arise because in most situations considered in this book the scalar electric potential is zero or negligible.
2. This result is most easily obtained by using the identity, valid for any complex number or vector c, $[\mathrm{Re}(c)]^2 = \mathrm{Re}[c \cdot \mathrm{Re}(c)] = \mathrm{Re}[c \cdot (c^* + c)/2] = [|c|^2 + \mathrm{Re}(c^2)]/2$. For instance, identifying c with $A_u e^{ik_u z}$ one obtains $\boldsymbol{a}_\perp^2 = [A_u^2 + \mathrm{Re}(A_u^2 e^{2ik_u z})]/2$.

References

Abramowitz, M. and Stegun, I. A. (1968). *Handbook of mathematical functions.* Dover, New York.

Courant, E. D. and Snyder, H. S. (1958). Theory of the alternating-gradient synchrotron. *Ann. Phys.* **3**, 1–48.

Kibble, T. W. B., (1966). Mutual reflection of electrons and photons. *Phys. Rev.* **150**, 1060–9.

Landau, L. D. and Lifschitz, E. M. (1975). *The classical theory of fields.* Pergamon Press, Oxford.

Luchini, P. and Solimeno, S. (1982). Effective mass of an electron in an electromagnetic field. *Phys. Lett.* **91A**, 438–40.

4
UNDULATOR RADIATION

In this chapter the electromagnetic radiation generated by an electron beam passing through an undulator will be quantitatively determined in the absence of electromagnetic fields other than the undulator's, and neglecting the effect of the field produced by the electrons upon each other. Radiation obtained in this way bears the name of undulator radiation. ('Spontaneous' undulator radiation when it is to be opposed to the 'stimulated' radiation emitted under the influence of an external electromagnetic wave, which we are going to consider in subsequent chapters.)

The radiation process may be studied in the framework of either classical or quantum field theory. The contrast of classical and quantum descriptions, with their very different formalism, has been particularly felt in the development of the theory of undulators and free-electron lasers, because the basic part of it falls in the applicability domain of both approaches, and research has gone on in parallel along both paths, as in perhaps no other topic. It is the authors' opinion that at the time this book is written the classical approach is more fruitful, because it opens up the way to three-dimensional and nonlinear analyses, which have not been developed to the same degree of refinement along the quantum route, and on the other hand the deviations from classical behaviour foreseen by quantum theory (so-called 'quantum effects') are, in practice, quite marginal. Therefore, the classical theory of radiation has been chosen as the basis upon which to illustrate undulator and FEL phenomena.

4.1 The classical theory of radiation from moving charges

The classical theory of electromagnetic radiation is based upon Maxwell's equations, which can be transformed into wave equations for the scalar and vector potentials. In mks units the latter are

$$\Delta_2 \varphi - \frac{1}{c^2}\frac{\partial^2 \varphi}{\partial t^2} = -\frac{\rho}{\varepsilon_0} \tag{4.1}$$

and

$$\Delta_2 \boldsymbol{a} - \frac{1}{c^2}\frac{\partial^2 \boldsymbol{a}}{\partial t^2} = -\mu_0 \boldsymbol{j}. \tag{4.2}$$

where φ and $\boldsymbol{a}$ are the scalar and vector potentials, such that the electric field is $\boldsymbol{e} = -\mathrm{grad}\,\varphi - \partial \boldsymbol{a}/\partial t$ and the magnetic field is $\boldsymbol{b} = \mu_0 \boldsymbol{h} = \mathrm{curl}\,\boldsymbol{a}$, and ρ and $\boldsymbol{j}$ are the electric charge and current densities. ε_0 and μ_0 are the vacuum dielectric and magnetic permeability constants, tied to the speed of light c by the relation $\varepsilon_0 \mu_0 c^2 = 1$. Equations (4.1) and (4.2) are valid for the potentials in the Lorentz gauge. This means that φ and $\boldsymbol{a}$ are not independent, but are related by the Lorentz gauge equation:

$$\mathrm{div}\,\boldsymbol{a} + c^{-2}\,\partial\varphi/\partial t = 0. \tag{4.3}$$

Similarly, the charge and current densities are not independent, but are related by the charge conservation equation:

$$\mathrm{div}\,\boldsymbol{j} + \partial\rho/\partial t = 0. \tag{4.4}$$

Since eqns (4.1)–(4.4) are not independent, it is possible to dispense with some of them. In particular, in the case of oscillating fields such as the ones we are going to consider it is preferable to obtain $\boldsymbol{a}$ from the solution of eqn (4.2) and then φ from eqn (4.3). It is not necessary to solve eqn (4.1), nor to know the charge density explicitly.

Equation (4.2) can be solved in closed form when the current $\boldsymbol{j}$ is a δ-function of space, centred at a generic point $\boldsymbol{r}'$, and an arbitrary function of time, i.e. $\boldsymbol{j} = \boldsymbol{f}(t)\delta(\boldsymbol{r} - \boldsymbol{r}')$. The well known result is the retarded-potential formula:

$$\boldsymbol{a} = \mu_0 \boldsymbol{f}(t - |\boldsymbol{r} - \boldsymbol{r}'|/c)/4\pi|\boldsymbol{r} - \boldsymbol{r}'|. \tag{4.5}$$

Equation (4.5) represents the vector potential of the field generated by a δ-function current distribution (a current element) with an arbitrary time dependence, under the hypothesis that this current distribution is isolated in an infinite vacuum space. It is the unique solution of eqn (4.2) under the condition, expressing the principle of causality, that if $\boldsymbol{f}(t)$ is zero for all times less than some time t_0, $\boldsymbol{a}$ should also be zero for $t < t_0$.

By representing a generic current distribution $\boldsymbol{j}(\boldsymbol{r}, t)$ as a superposition of δ-functions, through the formula

$$\boldsymbol{j}(\boldsymbol{r}, t) = \int \boldsymbol{j}(\boldsymbol{r}', t)\delta(\boldsymbol{r} - \boldsymbol{r}')\,\mathrm{d}^3\boldsymbol{r}' \tag{4.6}$$

the field radiated by any current distribution can be obtained from eqn (4.5) by superposition, exploiting the linearity of the electromagnetic field equations. In particular, a point charge e, located at point $\boldsymbol{r}'(t)$ and moving with velocity $\boldsymbol{v} = \mathrm{d}\boldsymbol{r}'/\mathrm{d}t$, determines a current density $\boldsymbol{j} = e\boldsymbol{v}(t)\delta[\boldsymbol{r} - \boldsymbol{r}'(t)]$. The corresponding electric field is worked out in any textbook on electromagnetism by means of the Liénard–Wiechert potential. For instance, we quote from Landau and Lifschitz (1975), eqn (63.8) (apart from conversion to mks units):

$$\boldsymbol{e} = -\frac{e}{4\pi\varepsilon_0}\left\{\frac{(1 - v^2/c^2)(\boldsymbol{R} - R\boldsymbol{v}/c)}{(R - \boldsymbol{R}\cdot\boldsymbol{v}/c)^3} + \frac{\boldsymbol{R} \wedge [(\boldsymbol{R} - R\boldsymbol{v}/c) \wedge \mathrm{d}\boldsymbol{v}/\mathrm{d}t]}{c^2(R - \boldsymbol{R}\cdot\boldsymbol{v}/c)^3}\right\}, \tag{4.7}$$

where $\boldsymbol{R} = \boldsymbol{r} - \boldsymbol{r}'(t')$, $\boldsymbol{v}$ and $d\boldsymbol{v}/dt$ are evaluated at time t', and t' is the solution of the equation $t' + R/c = t$.

Obtaining $\boldsymbol{e}$ from eqn (4.7) for a complicated electron trajectory, such as those determined in the previous chapter, is not easy, especially with the implicit definition of time t'. On the other hand, it may not be necessary to know the exact total electromagnetic field at any possible location and time. In practical situations one may only be interested in the radiation field far from the source and at a particular frequency. In these cases it is preferable to represent the given current distribution $\boldsymbol{j}(\boldsymbol{r}, t)$ by a Fourier integral in time and a superposition of δ-functions in space, and to use an approximate form of the radiated field valid at a large distance from the source.

Electromagnetic theory often deals with complex fields and sources in the frequency domain, represented by complex scalar or vectorial functions of spatial co-ordinates. (We shall distinguish complex fields from their real counterparts by denoting them with capital letters.) A complex field, e.g. a current density $\boldsymbol{J}(\boldsymbol{r})$, may be considered either as a representation of the real current density $\boldsymbol{j}(\boldsymbol{r}, t) = \mathrm{Re}(\boldsymbol{J}\mathrm{e}^{\mathrm{i}\omega t})$, time-harmonic at the angular frequency ω, or as a Fourier component of a current density $\boldsymbol{j}(\boldsymbol{r}, t)$ with a general time dependence, in which case the direct and inverse Fourier transform formulae give

$$\boldsymbol{J} = \int_{-\infty}^{\infty} \boldsymbol{j}(\boldsymbol{r}, t)\,\mathrm{e}^{-\mathrm{i}\omega t}\,\mathrm{d}t, \qquad \boldsymbol{j} = \int_{-\infty}^{\infty} \boldsymbol{J}\,\mathrm{e}^{\mathrm{i}\omega t}\,\mathrm{d}\omega/2\pi. \tag{4.8, 4.9}$$

We shall have occasion to use both interpretations. For complex potentials and sources in the frequency domain, eqn (4.2) becomes

$$\Delta_2 \boldsymbol{A} + k^2 \boldsymbol{A} = -\mu_0 \boldsymbol{J}, \tag{4.10}$$

where $k = \omega/c$. If eqn (4.5) is applied to a time-harmonic current element represented by $\boldsymbol{J} = \tilde{\boldsymbol{J}}\delta(\boldsymbol{r} - \boldsymbol{r}')$, with $\tilde{\boldsymbol{J}}$ being a constant complex vector, the expression obtained for the vector potential is:

$$\boldsymbol{A} = \mu_0 \tilde{\boldsymbol{J}} \mathrm{e}^{-\mathrm{i}k|\boldsymbol{r}-\boldsymbol{r}'|}/4\pi|\boldsymbol{r} - \boldsymbol{r}'|. \tag{4.11}$$

Equation (4.11) is not the only solution of eqn (4.10). In order to determine the particular solution (4.11), consistent with eqn (4.5) and with the principle of causality, one must impose the condition that the field should assume for $r \to \infty$ the form of an outgoing locally plane wave, which may be formulated in terms of the electric and magnetic fields $\boldsymbol{E}$ and $\boldsymbol{H}$, as

$$\lim_{r \to \infty} (r\boldsymbol{E} + \zeta \boldsymbol{r} \wedge \boldsymbol{H}) = 0, \tag{4.12}$$

where $\zeta = c\mu_0 \simeq 375\,\Omega$ is the so-called free-space impedance. It may be easily verified that another independent solution of eqn (4.10), for instance the one similar to eqn (4.11) but having the opposite sign before the exponent (which, incidentally, represents an incoming wave), does not satisfy eqn (4.12).

Equation (4.12) is known as Sommerfeld's radiation condition, and expresses for time-harmonic fields the requirement of causality.

The electric field radiated by a time-harmonic current element may be worked out from eqn (4.11) by first calculating the scalar potential from eqn (4.3), or by applying the relation $i\omega \boldsymbol{E} = c^2$ curl curl $\boldsymbol{A}$, which is obtained from one of Maxwell's equations and $\boldsymbol{B} =$ curl $\boldsymbol{A}$. The well known result is:

$$\boldsymbol{E} = -ik\zeta \frac{e^{-ikR}}{4\pi R}\left[\hat{\boldsymbol{R}} \wedge \tilde{\boldsymbol{J}} \wedge \hat{\boldsymbol{R}} + \left(\frac{1}{ikR} - \frac{1}{k^2R^2}\right)[\tilde{\boldsymbol{J}} - 3(\tilde{\boldsymbol{J}}\cdot\hat{\boldsymbol{R}})\hat{\boldsymbol{R}}]\right], \quad (4.13)$$

where $\boldsymbol{R} = \boldsymbol{r} - \boldsymbol{r}'$ and $\hat{\boldsymbol{R}}$ is the unit vector $\boldsymbol{R}/R$. From eqn (4.4) it can be seen that an oscillating δ-function current distribution, represented by $\boldsymbol{J} = \tilde{\boldsymbol{J}}\delta(\boldsymbol{r} - \boldsymbol{r}')$, entails a charge distribution proportional to $\mathrm{div}[\tilde{\boldsymbol{J}}\delta(\boldsymbol{r} - \boldsymbol{r}')] = \tilde{\boldsymbol{J}} \cdot \mathrm{grad}\, \delta(\boldsymbol{r} - \boldsymbol{r}')$, i.e. to the derivative of the δ-function in the direction of $\tilde{\boldsymbol{J}}$. Such a charge distribution characterizes an electric dipole, and therefore eqn (4.13) also gives the electric field generated by an oscillating electric dipole and is generally known under that name.

When the distance from point $\boldsymbol{r}'$ to point $\boldsymbol{r}$ is much larger than the wavelength $\lambda = 2\pi/k$, the second term in eqn (4.13) may be neglected, giving the radiation, or wave-zone, approximation:

$$\boldsymbol{E}_r = -ik\zeta \frac{e^{-ikR}}{4\pi R} \hat{\boldsymbol{R}} \wedge \tilde{\boldsymbol{J}} \wedge \hat{\boldsymbol{R}}. \quad (4.14)$$

By combining eqn (4.14) with eqns (4.6) and (4.8), the component at frequency ω of the radiation electric field generated by a source (current distribution $\boldsymbol{j}(\boldsymbol{r}, t)$) with a general space and time dependence may be written as

$$\boldsymbol{E}_r = -ik\zeta \int \frac{e^{-i(\omega t + kR)}}{4\pi R} \hat{\boldsymbol{R}} \wedge \boldsymbol{j}(\boldsymbol{r}', t) \wedge \hat{\boldsymbol{R}}\, d^3\boldsymbol{r}'\, dt. \quad (4.15)$$

4.2 Far field

When $r \to \infty$, the direction $\hat{\boldsymbol{r}}$ being fixed, the distance $R = |\boldsymbol{r} - \boldsymbol{r}'|$ may be approximated according to the expansion

$$|\boldsymbol{r} - \boldsymbol{r}'| = r(1 - 2\hat{\boldsymbol{r}}\cdot\boldsymbol{r}'/r + r'^2/r^2)^{1/2} = r - \hat{\boldsymbol{r}}\cdot\boldsymbol{r}' + O(r'^2/r). \quad (4.16)$$

Replacing R by $r - \hat{\boldsymbol{r}}\cdot\boldsymbol{r}'$ in the exponent of eqn (4.15) and just by r in the denominator and, at the same time, replacing $\hat{\boldsymbol{R}}$ by $\hat{\boldsymbol{r}}$, one obtains the far-field radiation formula:

$$\boldsymbol{E}_{rff} = -ik\zeta \frac{e^{-ikr}}{4\pi r} \int \hat{\boldsymbol{r}} \wedge \boldsymbol{j}(\boldsymbol{r}', t) \wedge \hat{\boldsymbol{r}}\, e^{-i(\omega t - k\hat{\boldsymbol{r}}\cdot\boldsymbol{r}')}\, d^3\boldsymbol{r}'\, dt. \quad (4.17)$$

According to this equation, the far-field expression of the electric field is proportional to the product of the spherical-wave factor e^{-ikr}/r times a function of $\hat{r}$ (that is, of direction only), independent of absolute distance r. Notice that $\hat{r} \wedge j \wedge \hat{r}$ is simply the component of j in a plane perpendicular to r (as can be obtained, for instance, from the expansion of the double vector product as $\hat{r} \wedge j \wedge \hat{r} = (\hat{r}\cdot\hat{r})j - (\hat{r}\cdot j)\hat{r}$). Therefore, the far-field electric-field vector depends only on the component of j perpendicular to r and is itself always perpendicular to r. By an approximation similar to that leading to eqn (4.17) one finds that the far-field magnetic-field vector is tied to E simply as $H = \zeta^{-1}\hat{r} \wedge E$; that is, by the same relation holding between the magnetic and electric fields of a plane wave travelling in the direction pointed at by $\hat{r}$. H too is perpendicular to $\hat{r}$.

In the case of a point-like moving particle of charge e and co-ordinates $r'(t)$, the current $j(r, t)$ must be replaced by $e(dr'/dt)\delta[r - r'(t)]$. Equation (4.17) then becomes

$$E_{rff} = -ik\zeta e \frac{e^{-ikr}}{4\pi r} \int_{-\infty}^{\infty} \hat{r} \wedge \frac{dr'}{dt} \wedge \hat{r}\, e^{-i[\omega t - k\hat{r}\cdot r'(t)]}\, dt. \tag{4.18}$$

If the motion of the particle is expressed by giving the transverse co-ordinates x and time t as functions of the axial co-ordinate z, rather than by specifying $r'(t)$, eqn (4.18) can also be written in the form of an integral in dz:

$$E_{rff} = -ik\zeta e \frac{e^{-ikr}}{4\pi r} \int_{-\infty}^{\infty} \hat{r} \wedge \frac{dr'}{dz} \wedge \hat{r}\, e^{-i[\omega t(z) - k\hat{r}\cdot r'(z)]}\, dz, \tag{4.19}$$

where $r'(z)$ has transverse and axial components $x(z)$ and z, and dr'/dz has components dx/dz and 1.

Later, we shall want to use eqn (4.19) to calculate the far-field pattern of undulator radiation. Before this is done, however, some words of caution are in order.

The far-field approximation is a standard and proven instrument of electromagnetic theory, used for instance in the design of radio antennae, and, in its quantum version, in the theory of atomic emission. Nevertheless, one must not forget that it is an approximation. Two conditions must be satisfied for this approximation to be acceptable, in addition to the wave-zone requirement that $r \gg \lambda$. First, for R to be replaceable by r in the denominator of eqn (4.15), the maximum modulus of r', that is the overall size of the source, must be small compared to r itself. Secondly, for kR to be replaceable by $k(r - \hat{r}\cdot r')$ in the exponent of eqn (4.15) it is necessary that the neglected terms, which according to eqn (4.16) are of the order of kr'^2/r, be much smaller than unity. A more careful error analysis shows that these terms are actually of order $kr'^2_{\perp}/r$, $r'_{\perp}$ being the modulus of the component of r' in a plane perpendicular to r. Therefore, the second requirement is that the dimensionless Fresnel number $F = s^2/\lambda d$ should be small, s being the size of the projection of the source onto a plane perpendicular to the direction of observation, and d the distance from the observation point.

Somewhat paradoxically, in the case of undulator radiation the most stringent requirement is the first. In fact, since undulator radiation is emitted in a narrow cone around the axis, the size of the electron trajectory as seen from within this cone is pretty small, and the Fresnel number is indeed acceptably low. On the other hand, the overall size of the source is not always small compared to distance. This is a singular condition, because, for instance, in problems related to radio antennae source size is of the order of metres and distances are of the order of kilometres, and in the case of atomic emission, say in a light bulb, source size is of the order of angstroms and distances are of the order of metres. On the contrary, in the case of undulators the size of the source is of the order of metres, and the size of the laboratory in which the emitted radiation is utilized may be not much larger. Therefore, one must not hastily identify the radiation pattern observed in the laboratory with the far-field pattern unless the distance of observation is large enough.

Moreover, applying the far-field approximation to radiation from charges undergoing unbounded motion requires some care from a mathematical viewpoint. In fact, it may happen, as is the case in undulators, that charges are accelerated only in a finite region, which may under appropriate conditions be small compared to the distance from the observation point, but the motion of the charge is, at least theoretically, infinite, the accelerated portion being preceded and followed by uniform motion. In this case the far-field approximation in the form of eqn (4.17) must be applied only to the acceleration region, using a different approximation for the two pieces of uniform motion. It is not possible to use eqns (4.17) and (4.18) as they stand, if only because a non-convergent integral would result, while neglecting the contribution of the uniform regions altogether, on the assumption that charges in uniform motion do not radiate, gives an incorrect result because the transition between uniform and accelerated motion is not accounted for properly. In order to deal with this situation, let us rewrite eqn (4.19) using the far-field approximation only for the part of the z-axis that is inside the undulator (say, $0 < z < L$) and retaining the form of eqn (4.15) outside:

$$\begin{aligned} \boldsymbol{E}_r \simeq -\mathrm{i}k\zeta \mathrm{e}\frac{\mathrm{e}^{-\mathrm{i}kr}}{4\pi r}\Bigg\{&\int_{-\infty}^{0} \hat{\boldsymbol{R}} \wedge \frac{\mathrm{d}\boldsymbol{r}'}{\mathrm{d}z} \wedge \hat{\boldsymbol{R}}\frac{r}{R}\,\mathrm{e}^{-\mathrm{i}[\omega t + k(R-r)]}\,\mathrm{d}z \\ &+ \int_{0}^{L} \hat{\boldsymbol{r}} \wedge \frac{\mathrm{d}\boldsymbol{r}'}{\mathrm{d}z} \wedge \hat{\boldsymbol{r}}\,\mathrm{e}^{-\mathrm{i}(\omega t - k\hat{\boldsymbol{r}}\cdot\boldsymbol{r}')}\,\mathrm{d}z \\ &+ \int_{L}^{\infty} \hat{\boldsymbol{R}} \wedge \frac{\mathrm{d}\boldsymbol{r}'}{\mathrm{d}z} \wedge \hat{\boldsymbol{R}}\frac{r}{R}\,\mathrm{e}^{-\mathrm{i}[\omega t + k(R-r)]}\,\mathrm{d}z\Bigg\}. \end{aligned} \tag{4.20}$$

In the first and third integral $\mathrm{d}\boldsymbol{r}'/\mathrm{d}z$ is a constant as they refer to regions of uniform motion. By evaluating these integrals in the limit for $r \to \infty$ (which can be done by the stationary-phase method, for which see, e.g., Erdelyi, 1956) the far-field expression of radiation from a charge in unbounded motion may eventually be obtained as[1]

$$\boldsymbol{E}_{rff} = -\mathrm{i}k\zeta e\frac{\mathrm{e}^{-\mathrm{i}kr}}{4\pi r}\left\{\hat{\boldsymbol{r}}\wedge\left(\frac{\mathrm{d}\boldsymbol{r}'}{\mathrm{d}z}\right)_i\wedge\hat{\boldsymbol{r}}\frac{\mathrm{e}^{-\mathrm{i}(\omega t_i - k\hat{\boldsymbol{r}}\cdot\boldsymbol{r}_i)}}{-\mathrm{i}[\omega(\mathrm{d}t/\mathrm{d}z)_i - k\hat{\boldsymbol{r}}\cdot(\mathrm{d}\boldsymbol{r}'/\mathrm{d}z)_i]}\right.$$
$$+\int_0^L \hat{\boldsymbol{r}}\wedge\frac{\mathrm{d}\boldsymbol{r}}{\mathrm{d}z}\wedge\hat{\boldsymbol{r}}\,\mathrm{e}^{-\mathrm{i}(\omega t - k\boldsymbol{r}\cdot\boldsymbol{r}')}\,\mathrm{d}z$$
$$\left.+\hat{\boldsymbol{r}}\wedge\left(\frac{\mathrm{d}\boldsymbol{r}'}{\mathrm{d}z}\right)_f\wedge\hat{\boldsymbol{r}}\frac{\mathrm{e}^{-\mathrm{i}(\omega t_f - k\hat{\boldsymbol{r}}\cdot\boldsymbol{r}_f)}}{\mathrm{i}[\omega(\mathrm{d}t/\mathrm{d}z)_f - k\hat{\boldsymbol{r}}\cdot(\mathrm{d}\boldsymbol{r}'/\mathrm{d}z)_f]}\right\}, \qquad (4.21)$$

where t_i, $\boldsymbol{r}_i$ and t_f, $\boldsymbol{r}_f$ are the time and position at which the electron enters and exits the undulator (in particular, $z_i = 0$ and $z_f = L$), and $(\mathrm{d}\boldsymbol{r}'/\mathrm{d}z)_i$ and $(\mathrm{d}\boldsymbol{r}'/\mathrm{d}z)_f$ refer to the constant electron velocity before and after the undulator.

4.3 Paraxial and ultrarelativistic approximation

In optical devices, and undulators and free-electron lasers are no exception, it often happens that electromagnetic waves propagate preferentially in a particular direction (the 'axis'), or at small angles with respect to it, and little interest is attached to radiation going in other directions. When this happens the previous equations may be further simplified, by introducing an approximation which uses the angle between the direction of propagation and the axis as a small parameter. This goes by the name of Fresnel or paraxial approximation.

Starting from the expression of the radiation field of a generic current distribution, eqn (4.15), the paraxial approximation may be obtained as follows. If the vector $\boldsymbol{R} = \boldsymbol{r} - \boldsymbol{r}'$ is assumed to form a small angle with the z-axis, i.e. to have a transverse component $\Delta\boldsymbol{x} = \boldsymbol{x} - \boldsymbol{x}'$ small compared to the axial component $\Delta z = z - z'$ (or, in the case of an extended source, if the integrand in eqn (4.15) is assumed to give non-negligible contributions to the integral only when $\boldsymbol{R}$ satisfies this condition), the modulus R of $\boldsymbol{R}$ may be expanded in inverse powers of Δz as

$$R = (\Delta z^2 + |\Delta\boldsymbol{x}|^2)^{1/2} = \Delta z + |\Delta\boldsymbol{x}|^2/2\Delta z + O(|\Delta\boldsymbol{x}|^4/\Delta z^3). \qquad (4.22)$$

Retaining only the leading term of eqn (4.22) in the denominator and the first two terms in the exponent of eqn (4.15) and, similarly, replacing the unit vector $\hat{\boldsymbol{R}}$ by $\hat{z} + \Delta\boldsymbol{x}/\Delta z$, gives

$$\boldsymbol{E}_r \simeq -\mathrm{i}k\zeta\int\frac{\mathrm{e}^{-\mathrm{i}(\omega t + k\Delta z + k\Delta\boldsymbol{x}^2/2\Delta z)}}{4\pi\Delta z}\left(\hat{z}+\frac{\Delta\boldsymbol{x}}{\Delta z}\right)\wedge\boldsymbol{j}_\perp(\boldsymbol{r}',t)\wedge\left(\hat{z}+\frac{\Delta\boldsymbol{x}}{\Delta z}\right)\mathrm{d}^3\boldsymbol{r}'\,\mathrm{d}t.$$

(4.23)

Equation (4.23) is Fresnel's integral expression of the radiated field.

In particular, when eqn (4.23) is evaluated in the far-field limit, by letting $\boldsymbol{x}$ and z go to infinity with the ratio $\boldsymbol{x}/z$ held fixed, the transverse component of the radiation electric field turns out to be

$$\boldsymbol{E}_{rff\perp} \simeq -\mathrm{i}k\zeta \frac{\mathrm{e}^{-\mathrm{i}k(z+\boldsymbol{x}^2/2z)}}{4\pi z} \int \mathrm{e}^{-\mathrm{i}(\omega t - \boldsymbol{k}_\perp \cdot \boldsymbol{x}' - k_z z')}$$

$$\times \left[\boldsymbol{j}_\perp(\boldsymbol{x}', z', t) - \frac{\boldsymbol{k}_\perp}{k} j_z(\boldsymbol{x}', z', t) \right] \mathrm{d}^2\boldsymbol{x}'\, \mathrm{d}z'\, \mathrm{d}\boldsymbol{t}, \tag{4.24}$$

where we have defined $\boldsymbol{k}_\perp = k\boldsymbol{x}/z$ and $k_z = k - |\boldsymbol{k}_\perp|^2/2k$.

The paraxial approximation may be also obtained directly from the wave equation (4.10). By representing the vector potential in the form $\boldsymbol{A}(\boldsymbol{x}, z) = \tilde{\boldsymbol{A}}(\boldsymbol{x}, z)\mathrm{e}^{-ikz}$, eqn (4.10) may be transformed into

$$\Delta_{2\perp}\tilde{\boldsymbol{A}} + \partial^2\tilde{\boldsymbol{A}}/\partial z^2 - 2ik\partial\tilde{\boldsymbol{A}}/\partial z = -\mu_0\boldsymbol{J}. \tag{4.25}$$

When $\tilde{\boldsymbol{A}}$ and $\partial\tilde{\boldsymbol{A}}/\partial z$ vary little over one wavelength in the z-direction, as they do if $\boldsymbol{A}$ represents a wave propagating almost parallel to the z-axis, $\partial^2\tilde{\boldsymbol{A}}/\partial z^2$ is negligible with respect to $2ik\partial\tilde{\boldsymbol{A}}/\partial z$. Deleting this term transforms eqn (4.25) into Leontovich's parabolic wave equation,

$$\Delta_{2\perp}\tilde{\boldsymbol{A}} - 2ik\partial\tilde{\boldsymbol{A}}/\partial z = -\mu_0\boldsymbol{J}, \tag{4.26}$$

which is the wave equation in the paraxial approximation. As may be verified, eqn (4.23) represents the electric field corresponding to the solution of eqn (4.26) with a δ-function for $\boldsymbol{J}$.

Similarly, letting $\boldsymbol{a} = \mathrm{Re}[\tilde{\boldsymbol{A}}\mathrm{e}^{\mathrm{i}(\omega t - kz)}]$ and $\varphi = \mathrm{Re}[\tilde{\Phi}\mathrm{e}^{\mathrm{i}(\omega t - kz)}]$ in the Lorentz gauge condition (4.3) and neglecting $\partial\tilde{A}_z/\partial z$ gives

$$\mathrm{i}k(\tilde{\Phi}/c - \tilde{A}_z) + \mathrm{div}_\perp\, \tilde{\boldsymbol{A}}_\perp = 0, \tag{4.27}$$

which is the gauge condition appropriate to the paraxial approximation.

Let us point out the similarity between the paraxial approximation of wave propagation and the ultrarelativistic approximation of particle motion. If eqn (4.2) is Fourier-transformed with respect to all three spatial co-ordinates and time it becomes $(\omega^2/c^2 - \boldsymbol{k}^2)\boldsymbol{A} = -\mu_0\boldsymbol{J}$. Free plane wave solutions of this equation, corresponding to zero current density, are therefore characterized by the dispersion relation $\omega^2 = \boldsymbol{k}^2c^2$. Similarly Fourier-transforming eqn (4.26), it can be obtained that free plane wave solutions of this second equation are characterized by the dispersion relation

$$k_z = k - \boldsymbol{k}_\perp^2/2k \tag{4.28}$$

(where k stands for ω/c, not $|\boldsymbol{k}|$). It is known that in the ray-optics limit the dispersion relation takes on the role of a Hamiltonian, with frequency in the place of energy and wave vector in the place of momentum. Within this interpretation, eqn (4.28) may be thought to have been obtained from the exact dispersion equation by a process very similar to that leading to the ultrarelativistic Hamiltonian (eqn (3.15)) from the exact Hamiltonian (eqn (3.6)); that is, a two-term expansion of the axial component of wave vector (momentum) in reciprocal powers of frequency (energy).

Let us now apply the classical theory of radiation to the calculation of radiation emitted by an electron moving in a given electromagnetic field, focusing our attention at first on the far-field formula (4.21). In order to be able to evaluate eqn (4.21) we need the expression of the functions $t(z)$, $\boldsymbol{r}'(z)$ (having transverse component $\boldsymbol{x}(z)$ and axial component z) and of the derivative $\mathrm{d}\boldsymbol{r}'/\mathrm{d}z$. The axial component of $\mathrm{d}\boldsymbol{r}'/\mathrm{d}z$ is 1; the transverse component, $\mathrm{d}\boldsymbol{x}/\mathrm{d}z$, was found in Section 3.4 to be given by

$$\frac{\mathrm{d}\boldsymbol{x}}{\mathrm{d}z} = -\frac{\partial \mathscr{H}}{\partial \boldsymbol{P}_\perp} = \frac{\boldsymbol{P}_\perp - \boldsymbol{a}_\perp}{[(\gamma - \varphi)^2 - (\boldsymbol{P}_\perp - \boldsymbol{a}_\perp)^2 - 1]^{1/2}}, \tag{4.29}$$

where φ and $\boldsymbol{a} = \boldsymbol{a}_\perp + a_z\hat{z}$ are the scalar and vector potentials of the given field, $\boldsymbol{P}_\perp$ is the electron transverse momentum and $\mathscr{H}$ is the Hamiltonian (axial momentum). In deriving eqn (4.29) we used a system of units chosen so as to make $e = m_e = c = 1$. In order to be able to use the results of the previous chapter we need only replace ωt by kt in eqn (4.21), since in the special system of units adopted there spatial co-ordinates are unaltered while t (with the dimensions of a length) represents time multiplied by the speed of light c. In the rest of this chapter the symbols γ, $\boldsymbol{P}$ and $\boldsymbol{a}$ will stand for the dimensionless energy, momentum and vector potential we have found convenient to use. The radiation field, however, will still be expressed by a dimensional (mks) electric field.

Taking the expression of $\mathrm{d}\boldsymbol{x}/\mathrm{d}z$ from eqn (4.29) allows eqn (4.21) to be rewritten as

$$\begin{aligned} \boldsymbol{E}_{rff} = -\mathrm{i}k\zeta e \frac{\mathrm{e}^{-\mathrm{i}kr}}{4\pi r} \hat{\boldsymbol{r}} \wedge \Bigg\{ & \frac{\boldsymbol{P}}{-\mathrm{i}(k\gamma - \boldsymbol{k}\cdot\boldsymbol{P})} [\mathrm{e}^{-\mathrm{i}(kt_i - \boldsymbol{k}\cdot\boldsymbol{r}_i)} - \mathrm{e}^{-\mathrm{i}(kt_f - \boldsymbol{k}\cdot\boldsymbol{r}_f)}] \\ & + \int_0^L \left(\frac{\boldsymbol{P}_\perp - \boldsymbol{a}_\perp}{[(\gamma - \varphi)^2 - (\boldsymbol{P}_\perp - \boldsymbol{a}_\perp)^2 - 1]^{1/2}} + \hat{z} \right) \mathrm{e}^{-\mathrm{i}(kt - \boldsymbol{k}\cdot\boldsymbol{r})}\,\mathrm{d}z \Bigg\} \wedge \hat{\boldsymbol{r}}. \end{aligned} \tag{4.30}$$

Equation (4.30) is the relativistically correct expression of the far-field radiation of an electron moving in the field generated by the potentials φ and $\boldsymbol{a}$. When the electron speed is ultrarelativistic (i.e. $\gamma \gg 1$), eqn (4.30) can be simplified by adopting the ultrarelativistic form of the Hamiltonian, eqn (3.15), which gives, instead of eqn (4.29), $\mathrm{d}\boldsymbol{x}/\mathrm{d}z = -\partial\mathscr{H}/\partial\boldsymbol{P}_\perp = (\boldsymbol{P}_\perp - \boldsymbol{A}_\perp)/\gamma$. Since, as was shown in Chapter 2, in the ultrarelativistic case we expect radiation of significant intensity only at small angles from the axis, it is consistent to adopt at the same time the paraxial approximation of the radiation integral, that is radiation formula (4.24).

When the ultrarelativistic Hamiltonian is used together with the paraxial form of far-field radiation (4.24), the expression of radiation from a moving electron becomes

$$E_{rff} = -ik\zeta e \frac{e^{-ikr}}{4\pi r} \left\{ \frac{P_\perp/\gamma - k_\perp/k}{-ik[\gamma^{-2} + (P_\perp/\gamma - k_\perp/k)^2]/2} [e^{-i(kt_i - k\cdot r_i)} - e^{-i(kt_f - k\cdot r_f)}] + \int_0^L \left(\frac{P_\perp - a_\perp}{\gamma} - \frac{k_\perp}{k} \right) e^{-i(kt - k_z z - k_\perp \cdot x)} \, dz \right\}. \quad (4.31)$$

One of the advantages of introducing the paraxial approximation is that, in combination with the ultrarelativistic approximation of the electron motion, it produces the effect that $P_\perp$ and $k_\perp$ appear in eqn (4.31) only in the combination $(k_\perp/k - P_\perp/\gamma)$. This combination has a simple interpretation. In fact the components of the transverse vector $k_\perp/k$, which by definition equals x/z, represent (in the paraxial approximation, in which these angles may be confused with their sines) the angles that the projections of the direction of observation on the x_1, z and x_2, z co-ordinate planes form with the z-axis, while the components of $P_\perp/\gamma$ are similarly (in the ultrarelativistic approximation, according to which $|P_\perp| \ll |P| \simeq \gamma$, so that again these angles may be confused with their sines) the angles that the projections of the electron trajectory on the co-ordinate planes form with the z-axis (see Fig. 4.1). Since all these angles are small, it is possible to confuse them with their sines or tangents and add them like vectors. Thus the components of $(k_\perp/k - P_\perp/\gamma)$ are the angles that the projections on the co-ordinate planes of the direction of observation and electron trajectory of motion form with each other. For this combination we shall introduce the notation

$$\boldsymbol{\vartheta}_{we} = k_\perp/k - P_\perp/\gamma. \quad (4.32)$$

The absolute angle between the observation and motion directions is $\vartheta_{we} = |\boldsymbol{\vartheta}_{we}|$.

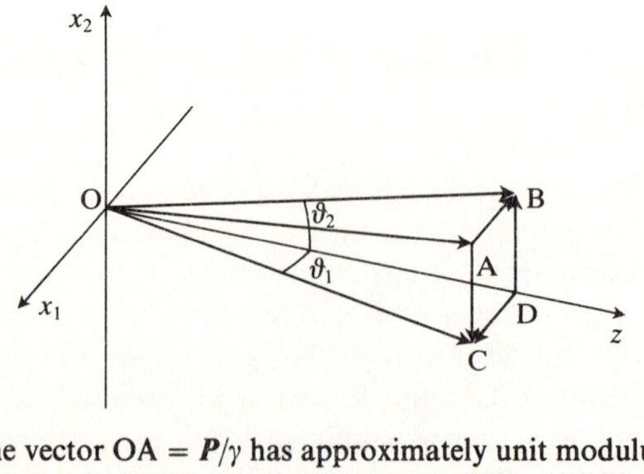

FIG. 4.1. Since the vector OA = P/γ has approximately unit modulus and a direction close to that of the z-axis, its transverse components DC and DB are approximately equal to the angles ϑ_1 and ϑ_2 that the projections OC and OB form with the z-axis.

4.4 Radiation from an ideal sinusoidal undulator

We are now ready to calculate the far-field pattern of radiation from the ideal sinusoidal undulator considered in Section 3.4. Inserting the undulator vector potential (3.21) and the expressions (3.24) of $\boldsymbol{x}(z)$ and $t(z)$ into eqn (4.31) gives[2]

$$\boldsymbol{E}_{rff} = -\mathrm{i}k\zeta e\frac{\mathrm{e}^{-\mathrm{i}kr}}{4\pi r}\mathrm{e}^{-\mathrm{i}(kt_i - \boldsymbol{k}_\perp\cdot\boldsymbol{x}_i)}\left(\frac{\boldsymbol{\vartheta}_{we}}{\mathrm{i}k(1/2\gamma^2 + \vartheta_{we}^2/2)}(\mathrm{e}^{-\mathrm{i}\kappa L} - 1)\right.$$
$$+ \int_0^L [\boldsymbol{\vartheta}_{we} + \mathrm{Re}(\boldsymbol{A}_u\mathrm{e}^{\mathrm{i}k_u z})/\gamma] \exp -\mathrm{i}\left\{\kappa z + \mathrm{Re}\left[\frac{k\boldsymbol{A}_u\cdot\boldsymbol{\vartheta}_{we}}{\mathrm{i}\gamma k_u}(\mathrm{e}^{\mathrm{i}k_u z} - 1)\right.\right.$$
$$\left.\left.\left. + \frac{kA_u^2}{8\mathrm{i}\gamma^2 k_u}(\mathrm{e}^{2\mathrm{i}k_u z} - 1)\right]\right\}\mathrm{d}z\right) \tag{4.33}$$

where

$$\kappa = k\left(1 + \frac{M^2 + \boldsymbol{P}_\perp^2}{2\gamma^2}\right) - \frac{\boldsymbol{k}_\perp\cdot\boldsymbol{P}_\perp}{\gamma} - k_z = (M^2/\gamma^2 + \vartheta_{we}^2)k/2. \tag{4.34}$$

In eqn (4.34) the paraxial dispersion equation (4.28) has been used to eliminate k_z and the angle vector $\boldsymbol{\vartheta}_{we}$ has been introduced according to its definition (4.32). M^2, the square of the electron effective mass (3.26), is $1 + |\boldsymbol{A}_u|^2/2$.

The integrand function in eqn (4.33) has the characteristic feature of being the product of the factor $\mathrm{e}^{-\mathrm{i}\kappa z}/2\gamma$ times a periodic function with period $l_u = 2\pi/k_u$. It is convenient to represent this periodic function by means of a Fourier series as

$$2[\mathrm{Re}(\boldsymbol{A}_u\mathrm{e}^{\mathrm{i}k_u z}) + \gamma\boldsymbol{\vartheta}_{we}]\exp -\mathrm{i}\,\mathrm{Re}\left[\frac{k\boldsymbol{A}_u\cdot\boldsymbol{\vartheta}_{we}}{\mathrm{i}\gamma k_u}(\mathrm{e}^{\mathrm{i}k_u z} - 1) + \frac{kA_u^2}{8\mathrm{i}\gamma^2 k_u}(\mathrm{e}^{2\mathrm{i}k_u z} - 1)\right]$$
$$= \sum_{n=-\infty}^{\infty} \boldsymbol{A}_{ueff,n}(\boldsymbol{\vartheta}_{we})\mathrm{e}^{\mathrm{i}nk_u z}, \tag{4.35}$$

with coefficients

$$\boldsymbol{A}_{ueff,n}(\boldsymbol{\vartheta}_{we}) = \frac{1}{l_u}\int_0^{l_u}(\boldsymbol{A}_u\mathrm{e}^{\mathrm{i}k_u z} + \boldsymbol{A}_u^*\mathrm{e}^{-\mathrm{i}k_u z} + 2\gamma\boldsymbol{\vartheta}_{we}) \exp -\mathrm{i}\left\{nk_u z\right.$$
$$\left. + \mathrm{Re}\left[\frac{k\boldsymbol{A}_u\cdot\boldsymbol{\vartheta}_{we}}{\mathrm{i}\gamma k_u}(\mathrm{e}^{\mathrm{i}k_u z} - 1) + \frac{kA_u^2}{8\mathrm{i}\gamma^2 k_u}(\mathrm{e}^{2\mathrm{i}k_u z} - 1)\right]\right\}\mathrm{d}z. \tag{4.36}$$

(The real part of $\boldsymbol{A}_u\mathrm{e}^{\mathrm{i}k_u z}$ has been written as half the sum of this vector and its complex conjugate.) The reason for the choice of the name $\boldsymbol{A}_{ueff,n}$ for the coefficients will be clear later.

Inserting eqn (4.35) into eqn (4.33) gives

$$\boldsymbol{E}_{rff} = -\mathrm{i}k\zeta e \frac{\mathrm{e}^{-\mathrm{i}kr}}{4\pi r} \mathrm{e}^{-\mathrm{i}(kt_i - \boldsymbol{k}_\perp \cdot \boldsymbol{x}_i)} \left[\frac{\boldsymbol{\vartheta}_{we}}{\mathrm{i}k(1/2\gamma^2 + \vartheta_{we}^2/2)} (\mathrm{e}^{-\mathrm{i}\kappa L} - 1) + \frac{1}{2\gamma} \int_0^L \sum_{n=-\infty}^{\infty} \boldsymbol{A}_{ueff,n}(\boldsymbol{\vartheta}_{we}) \mathrm{e}^{\mathrm{i}(nk_u - \kappa)z}\, \mathrm{d}z \right], \tag{4.37}$$

i.e.

$$\boldsymbol{E}_{rff} = \frac{-\mathrm{i}kL\zeta e}{2\gamma} \frac{\mathrm{e}^{-\mathrm{i}kr}}{4\pi r} \sum_{n=-\infty}^{\infty} \boldsymbol{A}_{ueff,n}\, \mathrm{e}^{-\mathrm{i}(\nu_n + kt_i - \boldsymbol{k}_\perp \cdot \boldsymbol{x}_i)} \frac{\sin \nu_n}{\nu_n}, \tag{4.38}$$

where ν_n is defined as

$$\nu_n = (\kappa - nk_u)L/2 = (M^2/\gamma^2 + \vartheta_{we}^2)kL/4 - nk_u L/2, \tag{4.39}$$

and in eqn (4.38) the coefficient $\boldsymbol{A}_{ueff,0}$ has been redefined as

$$\boldsymbol{A}_{ueff,0} = \frac{1}{l_u} \int_0^{l_u} (\boldsymbol{A}_u \mathrm{e}^{\mathrm{i}k_u z} + \boldsymbol{A}_u^* \mathrm{e}^{-\mathrm{i}k_u z} + 2\gamma \boldsymbol{\vartheta}_{we}) \times \exp -\mathrm{i}\left\{ \mathrm{Re}\left[\frac{k\boldsymbol{A}_u \cdot \boldsymbol{\vartheta}_{we}}{\mathrm{i}\gamma k_u} (\mathrm{e}^{\mathrm{i}k_u z} - 1) + \frac{kA_u^2}{8\mathrm{i}\gamma^2 k_u} (\mathrm{e}^{2\mathrm{i}k_u z} - 1) \right] \right\} \mathrm{d}z - 2\gamma \boldsymbol{\vartheta}_{we}(M^2/\gamma^2 + \vartheta_{we}^2)/(1/\gamma^2 + \vartheta_{we}^2) \tag{4.40}$$

to include the first term of eqn (4.37).

κ and ν_n are parameters we shall meet often in the theory of free-electron lasers. We shall call them 'reduced wavenumber' and 'detuning parameter'.

Equation (4.38) deserves careful study. We preliminarily observe that each term of the sum is the product of a rather smooth function of wavenumber and angle, the coefficient $\boldsymbol{A}_{ueff,n}$, and a sharp-peaked one, the factor $\sin(\nu)/\nu$. Even without a detailed analysis of the wavenumber and angle dependence of the coefficients $\boldsymbol{A}_{ueff,n}$, which will be carried out in the next section, we can already understand many properties of undulator radiation. In particular, we observe that the definition of the detuning parameter, eqn (4.39), may be rewritten as

$$\nu_n = k_u L(k - nk_r)/2k_r = \pi N(k - nk_r)/k_r, \tag{4.41}$$

where $N = L/l_u = k_u L/2\pi$ is the number of periods of the undulator, and k_r is defined as

$$k_r = 2k_u/(M^2/\gamma^2 + \vartheta_{we}^2). \tag{4.42}$$

From eqns (4.38) and (4.41) one can see that the spectral structure of undulator radiation is characterized by evenly spaced lines, each having a $(\sin \nu)/\nu$ shape, centred about the harmonics of the resonance wavenumber k_r. The full width at half maximum power (*fwhm*) of each line may be deduced from the *fwhm* of the $\sin^2 v/v^2$ function, which is $\delta v = 2.782$ (from $v = -1.391$ to 1.391). From eqn (4.41) it then appears that $\delta k = \delta v k_r/\pi N$. Dividing this by the central wavenumber of the nth harmonic, nk_r, gives the relative linewidth as

$$\delta\lambda_{fwhm}/\lambda \simeq \delta k_{fwhm}/k = 0.89/nN. \tag{4.43}$$

Since the ratio between line width and line separation is of the order of $1/N$, the lines are quite well separated, and the more so the larger N is. Therefore it makes sense to call them 'lines'. The parameter v_n gives a dimensionless measure of the displacement of the radiation wavenumber from the line centre, and therefore has been given the name of detuning parameter.

As is clearly seen from eqn (4.42), the resonance wavenumber k_r, and with it the central wavenumber of each harmonic line, is a decreasing function of the angle ϑ_{we}. Hence we can gain some knowledge of the angular dependence of undulator radiation even before we know the behaviour of $A_{ueff,n}(\vartheta_{we})$. In fact if the radiation is observed at a fixed wavenumber coinciding with the centre of one line, it will be reduced to half maximum power at the angles at which the central wavenumber deviation makes $v_n = \pm 1.391$. The value of this angle can be obtained from eqn (4.39) by letting $v_n = \pm 1.391$ and $k = nk_r = 2nk_u\gamma^2/M^2$, remembering also that $k_u L/2\pi$ is the number of periods N. The resulting *fwhm* angular aperture of the radiation beam is

$$\delta\vartheta_{we,fwhm} = 0.94(nN)^{-1/2}M/\gamma \tag{4.44a}$$

which, for emission at the wavenumber $k \simeq nk_r = 2n(\gamma/M)^2 k_u$, may also be written

$$\delta\vartheta_{we,fwhm} = (1.78\lambda/L)^{1/2} \tag{4.44b}$$

since $M/\gamma = (2k_u/k_r)^{1/2} = (2n\lambda/l_u)^{1/2}$ and $Nl_u = L$. Although radiation of non-negligible intensity is present at angles much larger than $\delta\vartheta_{we,fwhm}/2$, roughly up to the angle $\vartheta_r \simeq 1/\gamma_z = M/\gamma$ considered in Chapter 2, all single monochromatic components of the radiation extend only over an angle range of the order given by eqn (4.44).

Let us now consider what happens when all the electrons in a perfectly collimated and monoenergetic electron beam radiate simultaneously inside the undulator (the effects of electron momentum and energy spread will be dealt with later). It can be seen from eqn (4.38) that the fields radiated by different electrons differ only by the phase factor $e^{-i(kt_i - \boldsymbol{k}_\perp \cdot \boldsymbol{x}_i)}$. Since in a uniform beam the phase $kt_i - \boldsymbol{k}_\perp \cdot \boldsymbol{x}_i$ is uniformly distributed, the fields of different electrons sum up incoherently; that is, the total field produced has zero mean and a power equal to the sum of the individually radiated powers. Therefore, the power S_ω per unit frequency, beam area and solid angle emitted by the whole beam may be obtained by multiplying the number of electrons traversing the undulator per unit time and transverse area, that is the beam current density divided by e, times the power radiated by each electron per unit frequency and solid angle or, in other words, the Poynting vector[3] $2|\boldsymbol{E}|^2/\zeta$ multiplied by the area occupied by a unit solid angle over a spherical surface of radius r, i.e. r^2. Thus one has

$$S_\omega(\omega, \vartheta_{we}) = \frac{2r^2 j_b}{\zeta e} |\boldsymbol{E}_{rff}|^2, \tag{4.45}$$

j_b being the beam current density. Since single terms of eqn (4.38) are substantially different from zero in non-overlapping wavenumber ranges, it is possible to neglect cross-products between terms when squaring the electric field. Within the same approximation, terms with negative n may be omitted altogether because they are never significantly different from zero for any positive ω. With this simplification the power per unit frequency, beam area and solid angle of undulator radiation from an electron beam of current density j_b turns out to be, in mks units,

$$S_\omega(\omega, \vartheta_{we}) \simeq \frac{k^2 L^2 \zeta e j_b}{32\pi^2\gamma^2} \sum_{n=0}^{\infty} |\boldsymbol{A}_{ueff,n}(\vartheta_{we})|^2 \left[\frac{\sin(v_n)}{v_n}\right]^2. \tag{4.46}$$

A three-dimensional plot of the function $[(\sin v)/v]^2$ in the k, ϑ_{we}-plane, for v given as a function of k and ϑ_{we} by eqn (4.39), is shown in Fig. 4.2. As the figure shows, at the resonance wavenumber $k_r = (2\gamma^2/M^2)k_u$ radiation is peaked on the axis of the incoming electron beam ($\vartheta_{we} = 0$), and extends laterally up to an angle of the order of $\delta\vartheta_{we,fwhm}/2$ of eqn (4.44), whereas at slightly lower wavenumbers it splits up into two beams going laterally at an angle such that eqn (4.42) is satisfied. These two beams retain a roughly constant peak amplitude as long as $\boldsymbol{A}_{ueff,n}$ does not vary significantly, i.e., as will be seen in the next section, up to an angle of the order of M/γ, and then fade off. This is the typical wavenumber and angle intensity distribution of any single line of spontaneous undulator radiation.

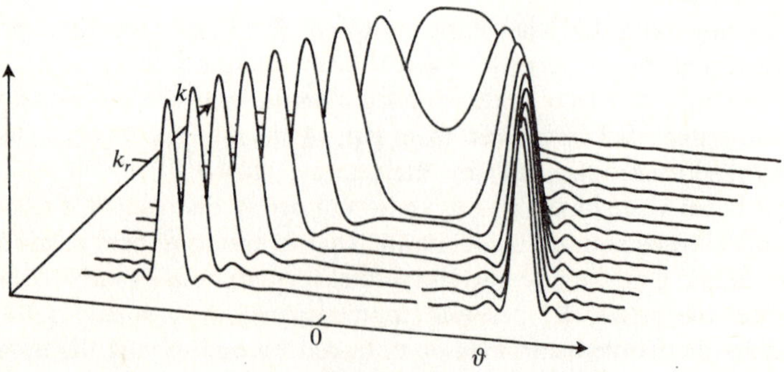

FIG. 4.2. 3-D plot in the k, ϑ_{we}-plane of the $\sin^2 v/v^2$ factor characterizing the undulator-radiation spectrum.

4.5 Intensity of undulator emission lines

In order to obtain the intensity of each harmonic and further information about its angular dependence the coefficient $A_{ueff,n}(k, \vartheta_{we})$, which will be seen to play the role of an effective undulator vector potential, must be studied in detail. We notice preliminarily that it is sufficient to evaluate $A_{ueff,n}$ for $k \simeq nk_r$, for only then is the factor $[(\sin v_n)/v_n]^2$ in eqn (4.46) significantly different from zero. Therefore we can substitute nk_r for k in the expression of $A_{ueff,n}(k, \vartheta_{we})$ and treat it as independent of wavenumber.

Let us begin by studying the radiation on axis, at $\vartheta_{we} = 0$. A result which can be immediately obtained by inspection of eqn (4.36) is that on axis only the odd harmonics are present. In fact for n even the integrand is an odd periodic function of z and its integral is zero. Therefore all the $A_{ueff,n}$ with n even are zero on axis. An expression of $A_{ueff,n}$ with n odd can be found with the aid of the mutually inverse formulas (Abramowitz and Stegun 1968, eqns 9.1.43–44)

$$\exp(iv \sin z) = \sum_{p=-\infty}^{\infty} J_p(v) e^{ipz}, \tag{4.47}$$

$$\frac{1}{2\pi} \int_0^{2\pi} e^{i(v \sin z - pz)} \, dz = J_p(v), \tag{4.48}$$

where $J_p(v)$ denotes a Bessel function of the first kind. The integral (4.36) can be reduced to two integrals of the form (4.48), when $\vartheta_{we} = 0$, by separating the terms containing A_u and A_u^*, changing the integration variable to $2k_u z$, and moving the origin of the z-axis until the coefficient of $e^{2ik_u z}$ becomes purely imaginary. On doing so, $A_{ueff,2p+1}$, with $2p + 1$ denoting an odd number, may be expressed in terms of Bessel functions, of orders p and $p + 1$, of the argument $v = k|A_u^2|/8\gamma^2 k_u$. The following result is obtained:

$$A_{ueff,2p+1}(0) = (-1)^p \left[A_u J_p\left(\frac{k|A_u^2|}{8\gamma^2 k_u}\right) - \frac{A_u^2 A_u^*}{|A_u^2|} J_{p+1}\left(\frac{k|A_u^2|}{8\gamma^2 k_u}\right) \right]$$
$$\times \exp i\left[p \arg(A_u^2) - \mathrm{Im}\left(\frac{kA_u^2}{8\gamma^2 k_u}\right) \right]. \tag{4.49}$$

It must be remembered that A_u^2 is the square of the complex vector A_u, different from the square of its modulus. $|A_u^2| = |A_u|^2$ only for linear polarization, whereas for circular polarization $A_u^2 = 0$. Since $J_0(0) = 1$ and $J_p(0) = 0$ for $p > 0$, circularly polarized undulators emit only the first harmonic on axis. On the contrary, the emission from a linearly polarized undulator, in which $|A_u^2|$ is maximum, has the largest harmonic contents. It is also interesting to observe that when the polarization is linear or circular the vector $A_{ueff,n}$ is parallel to A_u (because in the first case $J_{p+1} = 0$ and in the second $A_u^2 A_u^*/|A_u^2| = A_u$), while in all other cases the polarization of the vector $A_{ueff,n}$, and therefore the polarization of the radiated field, according to eqn (4.38), is different from the polarization of the undulator field A_u.

The expression (4.49) may be further simplified if k is replaced by nk_r. Taking the modulus of eqn (4.49), as only the modulus appears in eqn (4.46), and replacing k by nk_r with k_r given by eqn (4.42), where $M^2 = 1 + |A_u|^2/2$, one has

$$|A_{ueff,2p+1}(0)| \simeq \left| A_u \mathrm{J}_p\left[\frac{(2p+1)|A_u^2|}{4+2|A_u|^2}\right] - \frac{A_u^2 A_u^*}{|A_u^2|}\mathrm{J}_{p+1}\left[\frac{(2p+1)|A_u^2|}{4+2|A_u|^2}\right]\right|. \quad (4.50)$$

In particular, for circular polarization, $|A_{ueff,1}| = A_u$ and $|A_{ueff,n}| = 0$ for $n > 1$, as remarked above, while for linear polarization, when $|A_u^2| = |A_u|^2$, eqn (4.50) reduces to

$$|A_{ueff,2p+1}(0)| \simeq \left\{\mathrm{J}_p\left[\frac{(2p+1)A_u^2}{4+2A_u^2}\right] - \mathrm{J}_{p+1}\left[\frac{(2p+1)A_u^2}{4+2A_u^2}\right]\right\}A_u. \quad (4.51)$$

Bounds on the intensity of emission lines may be obtained by observing that, since $0 \leqslant |A_u^2| \leqslant |A_u|^2$, the argument of the Bessel functions in eqn (4.50) always lies between 0 and $p + \frac{1}{2}$, the upper limit being approached for linear polarization and $|A_u| \to \infty$. As a consequence, the modulus of the first-harmonic coefficient $|A_{ueff,1}|$ is found to lie between $[\mathrm{J}_0(\frac{1}{2}) - \mathrm{J}_1(\frac{1}{2})]A_u \simeq 0.696A_u$ and A_u, while for $p > 0$ $|A_{ueff,2p+1}|$ lies between zero and $[\mathrm{J}_p(p+\frac{1}{2}) - \mathrm{J}_{p+1}(p+\frac{1}{2})]A_u$. Notice that the coefficient $A_{ueff,1}$ is never considerably different from A_u, thus justifying the name of effective undulator vector potential.

A particular interest is attached to the higher-harmonic contents of undulator radiation because of the possibility to emit and, as we shall see later, amplify short wavelength radiation. We shall consider the case of linear polarization, when $|A_u^2|$ is maximum and equals $|A_u|^2$, and therefore the harmonic contents is highest. Let us rewrite the argument of the Bessel functions in eqn (4.51) as $p + \frac{1}{2} - (p+\frac{1}{2})/(1 + A_u^2/2)$. Defining s such that

$$p^{1/3}s = (p+\tfrac{1}{2})/(1 + A_u^2/2) - \tfrac{1}{2}, \quad (4.52)$$

we may then exploit, for large p and A_u, the asymptotic expression of $\mathrm{J}_p(p - p^{1/3}s)$ for $p \to \infty$ (Abramowitz and Stegun 1968, eqn 9.3.4):

$$\mathrm{J}_p(p - p^{1/3}s) \simeq (2/p)^{1/3}\mathrm{Ai}[2^{1/3}s] + O(p^{-1}). \quad (4.53)$$

In eqn (4.53), Ai(x) denotes the Airy function, which has an oscillating behaviour for $x < 0$ and decays exponentially as $\mathrm{e}^{-(2/5)x^{3/2}}$ for $x \to +\infty$, becoming less than 0.01 for $x > 2.7$. According to its definition (4.52), the argument s of the Airy function is an increasing function of p which, provided that $A_u \gg 1$, becomes of order unity when $p^{2/3}/A_u^2$ is of order unity; that is, when $p \simeq A_u^3$. Hence it can be seen that the amplitude of the $(2p+1)$th harmonic is of the order of $p^{-1/3}A_u$ up to $p \simeq A_u^3$, and decreases exponentially with p afterwards. For reasonably high A_u hundreds of harmonics can, in principle, be generated with non-negligible intensity.

Equation (4.53) gives an acceptable approximation of the Bessel functions appearing in eqn (4.51) for all qualitative purposes. When a better approximation is desired, e.g. in computer programs, it may be noticed that, since in eqn (4.51) the argument of the Bessel functions is smaller than their index, a straightforward power series expansion gives a very good approximation using only the first few terms.

Having examined the relative amplitude of the coefficients $A_{ueff,n}$ at $\vartheta_{we} = 0$, we may now proceed to study their behaviour off axis, with the aim of understanding from an order-of-magnitude analysis when and how $A_{ueff,n}(\boldsymbol{\vartheta}_{we})$ can be expected to become noticeably different from $A_{ueff,n}(0)$. For this purpose we go back to eqn (4.36) and observe, first of all, that the undulator length L or number of periods N do not appear in this equation in any way. Thus, whatever the extent of the region in which $A_{ueff,n}(\boldsymbol{\vartheta}_{we})$ may be confused with $A_{ueff,n}(0)$, for high enough N the angular range $\delta\vartheta_{we,fwhm}$ of eqn (4.44) will eventually fall in this region.

In further detail, it may also be noticed that for n even $A_{ueff,n}(\boldsymbol{\vartheta}_{we})$ is an odd function, while for n odd it is an even function. Therefore, for even harmonics the Taylor series expansion of $A_{ueff,n}$ must begin with the linear term, and for odd ones the expansion of the difference $A_{ueff,n}(\boldsymbol{\vartheta}_{we}) - A_{ueff,n}(0)$ must begin with the quadratic term.

Studying odd harmonics first, one can obtain the expression of the quadratic term by expanding eqn (4.36) in powers of $\boldsymbol{\vartheta}_{we}$. Having done so, we find that this term becomes comparable to $A_{ueff,n}(0)$ when $k\vartheta_{we}^2/k_u \simeq 1$; that is, when $\vartheta_{we} \simeq n^{-1/2}M/\gamma$. Notice that this value is $N^{1/2}$ times greater than the angular aperture of eqn (4.44). Therefore, for angles less than $N^{1/2}$ times $\delta\vartheta_{we,fwhm}$ of eqn (4.44) the amplitude of the $(2p + 1)$th harmonic is approximated well by its value on axis. Since the next term in the expansion of $A_{ueff,n}$ is quadratic in $\boldsymbol{\vartheta}_{we}$, at $\vartheta_{we} = \delta\vartheta_{we,fwhm}/2$ the error involved in identifying $A_{ueff,n}(\boldsymbol{\vartheta}_{we})$ with $A_{ueff,n}(0)$ is of the order of N^{-1}.

For the even harmonics it is not possible to make a similar comparison of off-axis with respect to on-axis amplitude because the latter is zero. On comparing their amplitude with that of neighbouring odd harmonics, it can be observed that the first (linear) term of the expansion of an even harmonic in powers of $\boldsymbol{\vartheta}_{we}$ becomes comparable to the first (constant) term of the expansion of a nearby odd harmonic for $\vartheta_{we} \sim |A_u|/\gamma = (nN)^{1/2}\delta\vartheta_{we,fwhm}$. Moreover, the following (third-power) term of the expansion of an even harmonic coefficient becomes comparable to the first for the same values of $\vartheta_{we} \simeq N^{1/2}\delta\vartheta_{we,fwhm}$ for which this happens for an odd harmonic. Therefore it turns out that for angles less than $N^{1/2}\delta\vartheta_{we,fwhm}$ the amplitude of even harmonics is less than that of neighbouring odd harmonics. At $\vartheta_{we} \simeq \delta\vartheta_{we,fwhm}$ the intensity of an even harmonic is $1/(nN)^{1/2}$ times that of the neighbouring odd harmonics.

Notice that the even values of n also include 0, which corresponds to a low-wavenumber component of undulator radiation extending from $k = 0$ up to $k \simeq k_r/N$. For this low-wavenumber portion the above estimates are not completely valid, and in particular it is not possible to consider the coefficient $A_{ueff,0}$ as independent of wavenumber. We shall not undertake further study of this component because it has no practical applications. In any case, its properties are all contained in eqn (4.40).

Summarizing, for angles less than $\delta\vartheta_{we,fwhm}/2$ of eqn (4.44) the effective undulator vector potential $A_{ueff,n}(\vartheta_{we})$ (with $n > 0$) is approximated by its value at $\vartheta_{we} = 0$ within an error of the order of $1/N$. This is sufficient for the great majority of purposes. If a greater precision is needed, the r.h.s. of eqn (4.36) must be evaluated numerically.

4.6 Brightness, monochromaticity, and coherence of undulator radiation

Brightness of a beam of radiation impinging on a given surface is the power falling on a unit area from a unit solid angle. This quantity is important because, as opposed to intensity, which is power per unit area integrated over all solid angles, it cannot be increased by an optical lens, and therefore source brightness determines the maximum intensity of radiation that can be obtained upon focusing. Monochromaticity is the property of emitting in a narrow wavelength range. It is measured by the linewidth, that is by the frequency or wavelength range in which most of the radiation power is emitted, or by the spectral brightness, that is brightness per unit frequency or wavelength. As far as orders of magnitude are concerned, and as long as the spectral structure is simple, spectral brightness can be expected to be of the order of total brightness divided by the linewidth.

Coherence is a statistical property, concerning the correlation existing between the phase of the field at different times and locations. It is measured by the correlation function, defined as the statistical average of the product of two values of the wave field (say, the electric field) taken at different times and locations. Two global parameters that characterize the correlation function are the coherence time and length, which are the maximum time difference and spatial distance, respectively, between the two field samples, for which the correlation function is still significantly different from zero. A coherence length may be defined either in the direction of propagation or in the direction normal to it. The longitudinal coherence length is simply related to the coherence time, being nothing other than the coherence time multiplied by c, while the transverse coherence length is an independent parameter.

Coherence parameters are related to the concepts of monochromaticity and brightness because of the Wiener–Khinchin theorem, according to which the power spectrum of a signal is the Fourier transform of its correlation function. It follows from this theorem and from the delocalization properties of Fourier transforms that the coherence time of a radiation beam is equal to or greater than the reciprocal of its frequency bandwidth, and its transverse coherence length is equal to or greater than the reciprocal of its $\boldsymbol{k}_\perp$ bandwidth (i.e. its angular spread divided by the wavelength). Therefore, highly monochromatic and collimated radiation is also coherent. The ultimate limit to the coherence time and length of a radiation beam is given by its temporal duration and spatial extension, since the correlation function, by its very definition, must vanish for times and distances larger than these. Correspondingly, the frequency bandwidth cannot be reduced below the reciprocal of the beam life, and the bandwidth in $\boldsymbol{k}_\perp$ space (angular spread divided by the wavelength) cannot be reduced below the reciprocal of its width. A radiation beam which attains these limits is said to be temporally or spatially fully coherent.

Notice that, just as monochromaticity may be improved at the expense of power by filtering through suitable selective absorbers, transverse coherence may be improved at the expense of power by spatial filtering through pin-holes. In fact, this was the only means of obtaining spatially coherent optical radiation before the development of lasers, and spatial filtering is still routinely adopted in the use of lasers when a very good beam quality is needed. The operation of spatial filtering preserves brightness, so that the radiation intensity that can be obtained by a given source upon filtering through a pin-hole is determined by its brightness. Similarly, the radiation intensity that can be obtained upon frequency filtering is determined by spectral brightness.

The above discussion of the significance of coherence parameters was given as a preparation to the unusual fact that undulator radiation presents a fairly good transverse spatial coherence together with a low (as compared to lasers) temporal coherence. Because of this, some confusion may arise as to whether or not undulator radiation deserves the attribute of 'coherent'. This is only a semantic problem, of course.

That temporal coherence of undulator radiation is far from approaching that of lasers (including free-electron lasers) may be easily recognized by comparing the relative frequency bandwidth of undulator radiation, which is given by the reciprocal of the number of undulator periods, typically of the order of 100, with that of lasers, which ranges from 10^{-5}–10^{-6} without particular precautions, to 10^{-8}–10^{-9} and beyond with the use of a diffraction grating or a Fabry–Perot interferometer.

Transverse coherence of undulator radiation may be estimated from its beam width and divergence. According to eqn (4.44b), beam divergence $\delta\vartheta$ is of the order of $(\lambda/L)^{1/2}$. Beam width, on the other hand, is given either by $(\lambda L)^{1/2}$, the Fresnel diffraction limit, or by the width σ of the electron beam, whichever the largest. If we measure the degree of coherence by the ratio of actual beam divergence, $(\lambda/L)^{1/2}$, to the divergence of a perfectly coherent (Gaussian) beam of the same cross-section size σ, i.e. λ/σ, it is easily seen that this ratio is given, as far as orders of magnitude are concerned, by $\sigma/(\lambda L)^{1/2}$, the square root of a parameter which may be called the e-beam Fresnel number, $F = \sigma^2/\lambda L$.

For the purpose of comparison with lasers, it may be said that radiation emitted from an undulator of Fresnel number of order unity or less is as coherent (transversely) as radiation from a single-transverse-mode laser or from a pin-hole of diameter λ illuminated by completely incoherent light; while when the Fresnel number is greater than unity, the degree of coherence is the same as for a laser oscillating on F transverse modes or a pin-hole of diameter $\lambda\sqrt{F}$. In a typical 2-metre long undulator emitting in the visible spectrum with an electron beam 3 mm thick, the Fresnel number is of the order of 10.

The situation changes radically when the undulator is made into a free-electron laser, as monochromaticity then reaches the levels of conventional lasers while brightness is itself improved, with a dramatic increase in spectral brightness.

4.7 The electromagnetic reciprocity theorem

When undulators are used to generate millimetre- or submillimetre-wave radiation, it is convenient to put a waveguide between the undulator poles and let the electrons travel inside the waveguide. One is then interested in calculating the intensity of radiation coupled into each waveguide mode. When, in the next chapter, we consider stimulated radiation from an undulator, we shall want to calculate the work done by several field configurations (in particular plane waves) upon the electrons. We shall find it useful to be able to relate this work to spontaneous undulator radiation. These two seemingly unrelated problems can be solved by means of the property of reciprocity with which the electromagnetic field is endowed. This is, incidentally, the usual way in which radiation into waveguides of more conventional radiators, say coupling wires, is calculated.

The basic statement of the reciprocity theorem is that, given two time-harmonic electromagnetic sources (represented by their complex current density distributions) and the corresponding radiated fields, the integral of the scalar product of the current density of the first source with the field generated by the second source, and the integral of the scalar product of the current density of the second source with the field generated by the first one are equal. Put algebraically:

$$\int \boldsymbol{E}_{r2} \cdot \boldsymbol{J}_{s1}\, \mathrm{d}V_1 = \int \boldsymbol{E}_{r1} \cdot \boldsymbol{J}_{s2}\, \mathrm{d}V_2, \tag{4.54}$$

where each integral is extended over all space, or alternatively over the volume occupied by the corresponding source. Equation (4.54) is typically used, for instance, to prove that the transmission and reception characteristics of an antenna are simply related to each other. In free space, eqn (4.54) is an immediate consequence of the symmetry of eqn (4.14) with respect to the source point $\boldsymbol{r}'$ and the observation point $\boldsymbol{r}$ (in German, *Aufpunkt*). In fact, only the difference $\boldsymbol{r} - \boldsymbol{r}'$ appears in eqn (4.14). It has the effect that a similar expression represents the electric field generated at point $\boldsymbol{r}$ by a current element located at point $\boldsymbol{r}'$ and the electric field generated at point $\boldsymbol{r}'$ by a current element located at point $\boldsymbol{r}$. The real usefulness of eqn (4.54) comes, however, from the fact that it does not hold only in free space, but also in the presence of loss-free metallic walls and refractive media of arbitrary geometrical configuration under some mild limitations.

We shall not prove eqn (4.54) in this context, because we are more interested in the limiting form that eqn (4.54) assumes when one of the sources is pushed away to infinity. In this case the corresponding radiated field becomes, in the limit, a source-free field, that is a solution of the homogeneous Maxwell equations, and the reciprocity theorem acquires the form of an equality between the work done by this source-free field upon the source of the other field and a certain bilinear integral combination of the two fields. Let us derive this limiting form directly.

The electromagnetic energy balance equation, also known as Poynting's theorem, in the version which is valid for time-harmonics fields, states that

$$\int_V \mathrm{Re}(\boldsymbol{E}^* \cdot \boldsymbol{J}_s)\,\mathrm{d}V = -\int_{\partial V} \mathrm{Re}(\boldsymbol{E}^* \wedge \boldsymbol{H}) \cdot \hat{\boldsymbol{n}}\,\mathrm{d}S, \qquad (4.55)$$

where the surface ∂V, the outward normal of which is indicated by $\hat{\boldsymbol{n}}$, is the boundary of the (arbitrary) volume V. The l.h.s. of eqn (4.55) is proportional to the work done per unit time by the electric field upon the source current density $\boldsymbol{J}_s$, while the r.h.s. is proportional to the electromagnetic energy flux flowing into the surface ∂V. Poynting's theorem, which is a consequence of Maxwell's equations, states that these two quantities are equal; that is, that electromagnetic fields enforce energy conservation.

Equation (4.55) is valid with $\boldsymbol{E}(\boldsymbol{r})$ and $\boldsymbol{H}(\boldsymbol{r})$ being any complex electric and magnetic fields satisfying the time-harmonic Maxwell equations with the source term $\boldsymbol{J}_s(\boldsymbol{r})$. They are not necessarily the fields radiated by this source, in the sense that they are not required to obey Sommerfeld's radiation condition (4.12), and in general $\boldsymbol{E}$, $\boldsymbol{H}$ may represent the superposition of the fields radiated by the current density $\boldsymbol{J}_s$ and other, source-free, fields. Let $\boldsymbol{E} = \boldsymbol{E}_r + \boldsymbol{E}_f$, $\boldsymbol{H} = \boldsymbol{H}_r + \boldsymbol{H}_f$, where $\boldsymbol{E}_r$, $\boldsymbol{H}_r$ are the fields radiated by source $\boldsymbol{J}_s$, which satisfy Sommerfeld's condition, and $\boldsymbol{E}_f$, $\boldsymbol{H}_f$ are unspecified source-free fields. Equation (4.55) also holds for fields $\boldsymbol{E}_r$, $\boldsymbol{H}_r$ and, with a zero l.h.s., for fields $\boldsymbol{E}_f$, $\boldsymbol{H}_f$. Subtracting from eqn (4.55) the same equation written once for

$\boldsymbol{E}_r$, $\boldsymbol{H}_r$ and then for $\boldsymbol{E}_f$, $\boldsymbol{H}_f$ yields

$$\int_V \mathrm{Re}(\boldsymbol{E}_f^* \cdot \boldsymbol{J}_s)\,\mathrm{d}V = -\int_{\partial V} \mathrm{Re}(\boldsymbol{E}_r \wedge \boldsymbol{H}_f^* + \boldsymbol{E}_f^* \wedge \boldsymbol{H}_r)\cdot \hat{\boldsymbol{n}}\,\mathrm{d}S. \tag{4.56}$$

Equation (4.56) is the result we are seeking. It states the equality of the work done by an external source-free field upon the given source current density to a bilinear integral of the free and radiated fields extended over a surface enclosing the source. When the source has a non-time-harmonic current density $\boldsymbol{j}_s(\boldsymbol{r}, t)$, eqn (4.56) may be combined with the Fourier transform formula, eqn (4.8), to give

$$-\int_{\partial V} \mathrm{Re}(\boldsymbol{E}_r \wedge \boldsymbol{H}_f^* + \boldsymbol{E}_f^* \wedge \boldsymbol{H}_r)\cdot \hat{\boldsymbol{n}}\,\mathrm{d}S = \int_{-\infty}^{\infty} \mathrm{d}t \int_V \mathrm{Re}(\boldsymbol{E}_f^* \cdot \boldsymbol{j}_s \mathrm{e}^{-\mathrm{i}\omega t})\,\mathrm{d}V$$

$$= \int_{-\infty}^{\infty} \mathrm{d}t \int_V \mathrm{Re}(\boldsymbol{E}_f \mathrm{e}^{\mathrm{i}\omega t})\cdot \boldsymbol{j}_s(\boldsymbol{r}, t)\,\mathrm{d}V. \tag{4.57}$$

(We have used the possibility of taking a real factor out of, and conjugating a complex number which appears under, the real-part operator.) The r.h.s. of eqn (4.57) may be read to represent the total work done over all time by the time-harmonic free field $\boldsymbol{e}_f = \mathrm{Re}(\boldsymbol{E}_f \mathrm{e}^{\mathrm{i}\omega t})$ upon the source current density $\boldsymbol{j}_s(\boldsymbol{r}, t)$. This equals, according to eqn (4.57), a bilinear integral combination of the given complex free field and the Fourier component at frequency ω of the radiated field. When $\boldsymbol{j}_s$ is the current density engendered by a moving point charge, $\boldsymbol{j}_s = e(\mathrm{d}\boldsymbol{r}'/\mathrm{d}t)\delta[\boldsymbol{r} - \boldsymbol{r}'(t)]$, eqn (4.57) becomes

$$-\int_{\partial V} \mathrm{Re}(\boldsymbol{E}_r \wedge \boldsymbol{H}_f^* + \boldsymbol{E}_f^* \wedge \boldsymbol{H}_r)\cdot \hat{\boldsymbol{n}}\,\mathrm{d}S$$

$$= \int_{-\infty}^{\infty} e\frac{\mathrm{d}\boldsymbol{r}'}{\mathrm{d}t}\cdot \mathrm{Re}(\boldsymbol{E}_f \mathrm{e}^{\mathrm{i}\omega t})\,\mathrm{d}t = \int_{-\infty}^{\infty} e\frac{\mathrm{d}\boldsymbol{r}'}{\mathrm{d}z}\cdot \mathrm{Re}(\boldsymbol{E}_f \mathrm{e}^{\mathrm{i}\omega t})\,\mathrm{d}z, \tag{4.58}$$

where the second form of the integral is useful when the motion of the particle is expressed using the axial co-ordinate z as the independent variable.

The significance of the bilinear integral combination of fields appearing in eqns (4.56)–(4.58) arises in connection with modal decompositions of the electromagnetic field. In fact, as we shall show for instance for waveguide modes, this integral vanishes when the two fields are replaced by two different members of a mode set, but not when they are the same. Thus, this integral has the properties of an orthogonal projection operator, and can be used to find the components in the mode set of any given field. Equations (4.56)–(4.58) then allow these components to be calculated through the work done by the mode field upon the source of radiation.

4.8 Radiation from charges moving inside a waveguide

Without going through the whole of waveguide theory (for which see, e.g., Motz 1950), let us introduce waveguide modes as a set of fields $\boldsymbol{E}_m$, $\boldsymbol{H}_m$, numbered by the index m $(0 < m < \infty)$, of the form

$$\boldsymbol{E}_m^{\pm} = C_m \boldsymbol{F}_m(\boldsymbol{x}) \mathrm{e}^{\mp \mathrm{i}\beta_m z}, \qquad \boldsymbol{H}_m^{\pm} = \pm \zeta^{-1} C_m \boldsymbol{G}_m(\boldsymbol{x}) \mathrm{e}^{\mp \mathrm{i}\beta_m z}, \tag{4.59}$$

where

$$\beta_m = (k^2 - k_{cm}^2)^{1/2} \tag{4.60}$$

(the positive root is implied), C_m is a constant with the dimensions of an electric field, and $\boldsymbol{F}$ and $\boldsymbol{G}$ are dimensionless vector functions of transverse co-ordinates.[4] Superscript '+' denotes modes propagating in the increasing z direction, and superscript '−' denotes modes propagating in the decreasing z direction. Attention should be drawn to the different sign of the magnetic field in the two cases. The vacuum impedance ζ has been included in the definition of the magnetic field in order to let $\boldsymbol{F}$ and $\boldsymbol{G}$ be dimensionless quantities. The usual distinction between transverse-electric, TE, and transverse-magnetic, TM, modes is not relevant here; eqn (4.59) represents the union of both mode sets. The constant k_{cm}, characteristic of each mode, is named the cut-off wavenumber, because for $k < k_{cm}$ the propagation constant β_m becomes imaginary, and the corresponding mode cannot propagate in the waveguide. We shall generally assume the modes to be set out in order of cut-off wavenumber increasing with m. For any given free-space wavenumber only a finite number or modes are above cut-off and can propagate. When the free-space wavenumber falls in the range between the cut-off wavenumbers of the first (fundamental) and second mode only one mode can propagate, and the waveguide is said to be single-mode. For larger free-space wavenumbers more than one mode can propagate in the waveguide, which is then said to be overmoded. Overmoded waveguides are more difficult to operate because energy may be lost through mode coupling caused by bends, joints and imperfections.

Waveguide modes have, among others, the following orthogonality property:

$$\int_{ws} \boldsymbol{F}_n^* \wedge \boldsymbol{G}_m \cdot \hat{\boldsymbol{z}} \, \mathrm{d}S = \delta_{nm}, \tag{4.61}$$

where $\delta_{nm} = 0$ for $n \neq m$ and $\delta_{nm} = 1$ for $n = m$, and the integral is extended over the waveguide cross-section. Equation (4.61) can be shown to be a consequence of Maxwell's equations and the boundary conditions at the waveguide walls (assumed to be perfectly conducting) when $k_{cn} \neq k_{cm}$; if different modes have the same cut-off wavenumber (which is usually indicated by saying that these modes are degenerate), eqn (4.61) can be made to hold by redefining the degenerate modes as suitable linear combinations of themselves. Equation (4.61) also implies a particular normalization of the functions $\boldsymbol{F}$ and $\boldsymbol{G}$, which makes them uniquely determined up to a phase factor.

Now let the volume V in eqn (4.57) be a portion of the waveguide delimited by two planes $z = z_1$ and $z = z_2$, situated one to the left and one to the right of the source, and let the free field be one of the modes expressed by eqn (4.59), say the mth mode propagating in the increasing z direction,[5] with an arbitrary initial phase φ_0:

$$\boldsymbol{E}_f = \boldsymbol{F}_m(\boldsymbol{x})\mathrm{e}^{\mathrm{i}(\varphi_0 - \beta_m z)}, \qquad \boldsymbol{H}_f = \zeta^{-1}\boldsymbol{G}_m(\boldsymbol{x})\mathrm{e}^{\mathrm{i}(\varphi_0 - \beta_m z)}. \tag{4.62}$$

The radiated field is given, to the right of the source, by a superposition of modes propagating in the increasing z direction with coefficients C_n^+,

$$\boldsymbol{E}_r^+ = \sum_{n=0}^{\infty} C_n^+ \boldsymbol{F}_n(\boldsymbol{x})\mathrm{e}^{-\mathrm{i}\beta_n z}, \qquad \boldsymbol{H}_r^+ = \sum_{n=0}^{\infty} C_n^+ \zeta^{-1}\boldsymbol{G}_n(\boldsymbol{x})\mathrm{e}^{-\mathrm{i}\beta_n z}, \tag{4.63}$$

and, to the left, by a superposition of modes propagating in the decreasing z direction with coefficients C_n^-,

$$\boldsymbol{E}_r^- = \sum_{n=0}^{\infty} C_n^- \boldsymbol{F}_n(\boldsymbol{x})\mathrm{e}^{+\mathrm{i}\beta_n z}, \qquad \boldsymbol{H}_r^- = -\sum_{n=0}^{\infty} C_n^- \zeta^{-1}\boldsymbol{G}_n(\boldsymbol{x})\mathrm{e}^{+\mathrm{i}\beta_n z}. \tag{4.64}$$

Now put the free and radiated fields of eqns (4.62)–(4.64) into eqn (4.57). The surface integral on the l.h.s. of this equation is extended over the boundary of the volume V, i.e. over the two sections of the waveguide with the planes $z = z_1$ and $z = z_2$ and over the portion of waveguide lateral surface included between these two planes. The integral extended over the lateral surface is zero because the tangential component of the electric field is zero there (in fact, e.g., $\boldsymbol{E}_r \wedge \boldsymbol{H}_f^* \cdot \hat{\boldsymbol{n}} = \hat{\boldsymbol{n}} \wedge \boldsymbol{E}_r \cdot \boldsymbol{H}_f^*$, and $\hat{\boldsymbol{n}} \wedge \boldsymbol{E}_r$ is zero when the tangential component of $\boldsymbol{E}_r$ is). The integral extended over the section $z = z_1$ contains $\boldsymbol{E}_r^-$ as the real field (the abscissa z_1 is on the left of the source), and is

$$\zeta^{-1}\,\mathrm{Re} \sum_{n=0}^{\infty} C_n^- \mathrm{e}^{\mathrm{i}[(\beta_n + \beta_m)z_1 - \varphi_0]} \int_{ws} (\boldsymbol{F}_n \wedge \boldsymbol{G}_m^* - \boldsymbol{F}_m^* \wedge \boldsymbol{G}_n)\cdot\hat{\boldsymbol{z}}\,\mathrm{d}S. \tag{4.65}$$

This integral is also zero. In fact all the terms of the sum with $n \neq m$ are zero because of eqn (4.61), and the one with $n = m$ is zero because it is given by the difference of two equal quantities. (Since the mode is a right-going wave while the real field is left-going, their magnetic fields have opposite signs.) The integral extended over the section z_2 is similar to (4.65) but contains the radiated field $\boldsymbol{E}_r^+$:

$$\zeta^{-1}\,\mathrm{Re}\left[\sum_{n=0}^{\infty} C_n^+ \mathrm{e}^{\mathrm{i}[(\beta_m - \beta_n)z_2 + \varphi_0]} \int_{ws} (\boldsymbol{F}_n \wedge \boldsymbol{G}_m^* + \boldsymbol{F}_m^* \wedge \boldsymbol{G}_n)\cdot\hat{\boldsymbol{z}}\,\mathrm{d}S\right]. \tag{4.66}$$

All terms with $n \neq m$ of this sum are zero, but the one with $n = m$ is not. Because of eqn (4.61), this term equals $2\zeta^{-1}\,\mathrm{Re}(C_m^+ e^{i\varphi_0})$. Since this is the only non-zero contribution to the l.h.s. of eqn (4.57), we obtain

$$\mathrm{Re}(C_m^+ e^{i\varphi_0}) = \frac{\zeta}{2}\int_{-\infty}^{\infty} dt \int_V \mathrm{Re}(\boldsymbol{E}_f e^{i\omega t})\cdot \boldsymbol{j}_s \, dV. \qquad (4.67)$$

In particular, when the current density $\boldsymbol{j}_s$ is generated by a single moving point charge, i.e. $\boldsymbol{j}_s = e(d\boldsymbol{r}'/dt)\delta(\boldsymbol{r} - \boldsymbol{r}')$, eqn (4.67) becomes

$$\mathrm{Re}(C_m^+ e^{i\varphi_0}) = \frac{e\zeta}{2}\int_{-\infty}^{\infty} \frac{d\boldsymbol{r}'}{dz'}\cdot \mathrm{Re}[\boldsymbol{E}_f(\boldsymbol{r}')e^{i\omega t}]\, dz'$$

$$= \frac{e\zeta}{2}\int_{-\infty}^{\infty} \frac{d\boldsymbol{r}'}{dz'}\cdot \mathrm{Re}\{\boldsymbol{F}_m[\boldsymbol{x}'(z')]e^{i[\varphi_0 + \omega t(z') - \beta_m z']}\}\, dz', \qquad (4.68)$$

where $\boldsymbol{E}_f$ has been taken from eqn (4.62) and $\boldsymbol{x}'(z')$ and $t(z')$ represent the electron motion, using z' as the independent variable. By applying eqns (4.67) and (4.68) twice, first with $\varphi_0 = 0$ and then with $\varphi_0 = \pi/2$, both the real and imaginary parts of C_m^+ can be determined. Thus the amplitude of the mth radiated mode can be obtained from the work done upon the source by that same mode. Of course, the argument applies to modes travelling to the left as well as to those travelling to the right.

Equations (4.67) and (4.68) can be applied in quite a similar way to free-space modes, as we shall have occasion to do in Chapter 5.

4.9 Undulator radiation in a waveguide

The evaluation of eqn (4.68) for an electron moving in an undulator proceeds along the same lines as the previous analysis of free-space radiation. Notice, however, that in an infinite waveguide the electron momentum $\boldsymbol{P}$ can only be parallel to the z-axis, as otherwise the electron would hit the walls.[6] After some transformations which we omit, and the introduction of a Fourier series similar to (4.35), eqn (4.68) may be cast into a form similar to eqn (4.38):

$$C_m^+ = -\frac{e\zeta L}{4\gamma}\sum_{n=-\infty}^{\infty} B_{nm} e^{-i(\nu_{nm} + kt_i)}\frac{\sin \nu_{nm}}{\nu_{nm}}, \qquad (4.69)$$

where ν_{nm} is defined analogously to ν_n of eqn (4.39) but with the propagation constant β_m of eqn (4.60) in the place of k_z, i.e. as

$$\nu_{nm} = [(1 + M^2/2\gamma^2)k - \beta_m - nk_u]L/2, \qquad (4.70)$$

and the coefficient B_{nm}, describing the coupling of the nth undulator harmonic with the mth waveguide mode, is

$$B_{nm} = \frac{1}{l_u}\int_0^{l_u} F_m^* \left\{ x_i - \mathrm{Re}\left[\frac{A_u}{\mathrm{i}\gamma k_u}(\mathrm{e}^{\mathrm{i}k_u z} - 1)\right]\right\} \cdot \left(A_u \mathrm{e}^{\mathrm{i}k_u z} + A_u^* \mathrm{e}^{-\mathrm{i}k_u z}\right)$$

$$\times \exp -\mathrm{i}\left\{ nk_u z + \mathrm{Re}\left[\frac{kA_u^2}{8\mathrm{i}\gamma^2 k_u}(\mathrm{e}^{2\mathrm{i}k_u z} - 1)\right]\right\} \mathrm{d}z. \tag{4.71}$$

Corresponding to eqn (4.69), the spectral intensity (power per unit ω in the interval $0 < \omega < \infty$) of radiation transmitted to the mth mode travelling in the increasing z direction by a uniform electron beam of current I_b, making I_b/e electrons per unit time pass through the undulator, is

$$S_m^+ = 2\frac{I_b}{e\zeta}|C_m^+|^2 \simeq \frac{e\zeta I_b L^2}{8\gamma^2}\sum_{n=0}^{\infty}|B_{nm}|^2\left[\frac{\sin(\nu_{nm})}{\nu_{nm}}\right]^2. \tag{4.72}$$

Again, as in the free-space case, the most significant information is in the parameter ν_{nm}, rather than in the coefficient B_{nm}, because of the sharp-peakedness of the $\sin(\nu_{nm})/\nu_{nm}$ factor which causes a significant radiation intensity to be emitted only when $\nu_{nm} \simeq 0$. Contrarily to the free-space case, however, ν_{nm} is not a linear function of frequency, and may vanish for more than one value of ω. The root structure of the equation $\nu_{nm} = 0$ is illustrated by Fig. 1.3 of Chapter 1. There, the dispersion relation of the mth waveguide mode, eqn (4.60), is qualitatively traced in the ω, β-plane as the line AB, together with the straight lines $\beta = [1 + M^2/2\gamma^2]\omega/c - nk_u$, lines OE, DF, and GH, so that their intersections represent the points where $\nu_{nm} = 0$. It readily appears that there can be either two or no roots depending on n and m. Moreover, for some n both roots may correspond to positive β, whereas for high enough n one intersection is with the positive and the other with the negative branch of the dispersion curve, corresponding to waves propagating in the decreasing z direction. (This is the normal situation for free-space propagation, where the dispersion equation is $\beta = k = \pm\omega/c$. In the previous sections, however, only the positive branch was considered because we were concerned with forward waves only.) The condition separating the no-root from the two-root region can be obtained by equating to zero $\mathrm{d}\nu_{nm}/\mathrm{d}\omega$, and is

$$\frac{nk_u}{k_{cm}} = \left(\frac{M^2/2\gamma^2}{2 + M^2/2\gamma^2}\right)^{1/2} \simeq \frac{M}{2\gamma}, \tag{4.73}$$

where the last simplification applies in the ultrarelativistic limit ($\gamma \gg M$). When condition (4.73) is satisfied the central free-space wavenumber of the emitted line is

$$k \simeq nk_u\gamma^2/M^2 \simeq k_{cm}\gamma/2M. \tag{4.74}$$

The point beyond which one root begins to correspond to negative β is obtained when $k = k_{cm}$ is a root, i.e. when

$$\frac{nk_u}{k_{cm}} = 1 + M^2/2\gamma^2 \simeq 1. \qquad (4.75)$$

From eqns (4.73)–(4.75) one can draw the following conclusions. First, an undulator radiating in a waveguide emits each harmonic line in only those modes having a low enough k_{cm}, such that nk_u/k_{cm} exceeds the value given by eqn (4.73). In particular, if one wants to obtain excitation of a single mode from the first harmonic, the required k_{c0} is of the order of $(\gamma/M)k_u$, which implies a size of the waveguide cross-section approximately γ/M times smaller than the undulator period. The emitted radiation will have a wavelength another γ/M times smaller than the waveguide size, so that the waveguide is strongly overmoded. A single-mode waveguide cannot be excited at all by an ultrarelativistic electron beam travelling in an undulator. Secondly, radiation will be emitted in each mode having a cut-off wavenumber lying between nk_u and $(2\gamma/M)nk_u$ in the form of two separate lines, one of which falls in the wavenumber interval from $nk_r/2$ to nk_r, k_r being the free-space resonance wavenumber $(2\gamma^2/M^2)k_u$, and the other in the interval from k_{cm} to $nk_r/2$. The *fwhm* of each line is approximately $2.782/|\mathrm{d}\nu_{nm}/\mathrm{d}k|$, with ν_{nm} given by eqn (4.69), and is smaller for the lower-wavenumber line. When the value of eqn (4.73) is approached the two lines coalesce and linewidth is maximum.

One might ask, since the waveguide must be strongly overmoded (that is, wide compared to the wavelength), why not to dispense with the waveguide at all. The decision in practice is, as always, determined by a balance between the wall losses of the waveguide, which decrease with wavelength, and the dispersion losses of free-space propagation, which increase with wavelength. Moreover, use of a waveguide is convenient, whenever it is possible, because the constant cross-section of its mode allows a tighter geometrical fit to the undulator magnet, but is offset at the smaller wavelengths by the exceedingly small mechanical tolerances required in its fabrication. The crossover between these two tendencies is, for today's technology, in the submillimetre range, where some experimenters have found convenient a mixed arrangement in which propagation is limited in the direction of the magnetic field by two plane metallic walls and in the direction orthogonal to it by a cylindrical-mirror open resonator (see, e.g., Section 11.5).

4.10 Effects of electron momentum and energy spread

Until now we have considered the electron beam to be perfectly collimated and monochromatic, i.e. composed of electrons of zero transverse momentum and one and the same energy. When the momentum and energy spread of a real beam is taken into account, the expression (4.46) of radiation intensity must be averaged using the momentum and energy distribution of the electrons in the beam. It is then interesting to find, first, the maximum spread that can be tolerated without a substantial change in radiation intensity and, second, what changes are to be expected when this maximum is exceeded.

We shall use the approximation, justified for a large number of periods N, of neglecting all but one term of the series (4.46) and treating the coefficient $|A_{ueff,n}(\vartheta_{we})|^2$ as a constant.

Under this approximation it is evident that a noticeable spread in momentum (or, equivalently, direction) and/or energy is one that produces a change in v_n of order unity. Moreover, changes in momentum or energy that produce a similar effect upon v_n will be undistinguishable. Therefore we are led to define an 'equivalent detuning spread' δv which, by linearizing the expression (4.39) of v_n, can be related to the angular spread $\delta\vartheta_i$ and energy spread $\delta\gamma_i$ of the electron beam as

$$\delta v = (kL/4)\delta\vartheta_i^2 + (kLM^2/2\gamma^3)\delta\gamma_i \tag{4.76}$$

(for free-space emission). It is easily deduced from eqn (4.76) that the order of magnitude of the angle and energy spread that produce a noticeable effect (i.e. a detuning spread δv varying from -1.391 to 1.391, as in Section 4.4) is

$$\delta\vartheta_{i,fwhm} = (11.13/kL)^{1/2} = (1.78\lambda/L)^{1/2} \simeq (2\lambda/L)^{1/2}, \tag{4.77}$$

$$\delta\gamma_{i,fwhm} = 5.564\gamma^3/kLM^2. \tag{4.78}$$

Not surprisingly, the allowable electron beam angular spread $\delta\vartheta_{i,fwhm}$, in the evaluation of which it is also necessary to include the effect of betatron oscillations caused by the undulator itself, coincides with the angular spread of radiation given by eqn (4.44). This is a direct consequence of the fact that in the ultrarelativistic approximation the emitted radiation depends only on the relative angle ϑ_{we} between the direction of emission and the direction of motion of the electron.

It is interesting to compare the limit on e-beam angular spread imposed by the angular spread of the emitted radiation with that found in Section 3.7 to be necessary for the assumption of a 1-D undulator field. The first limit, $N^{-1/2}M/\gamma$ for the fundamental harmonic according to eqn (4.44), is stronger than the second, N^{-1} according to eqn (3.41), whenever

$$N \ll \gamma^2/M^2. \tag{4.79}$$

Since this inequality is verified for most practical values of N, M, and γ, we see that if the angular spread is contained below the limit imposed by the requirement not to increase the angular divergence of the emitted radiation, one need not be concerned about the validity of a 1-D undulator model.

On the other hand, the relative energy spread $\delta\gamma_{i,fwhm}/\gamma$ acquires a particularly simple and easily memorable form when the product kL is again replaced by $4\pi N\gamma^2/M^2$ (as is always possible near resonance):

$$\delta\gamma_{i,fwhm}/\gamma = 0.443/nN \simeq 1/2nN. \tag{4.80}$$

Notice that $\delta\gamma_{i,fwhm}/\gamma$ coincides with half the relative emission bandwidth which, according to eqn (4.43), is of the order of $1/nN$.

Equation (4.80) is a very important one in the theory of undulators and free-electron lasers because it sets an upper technological limit on the usable undulator length (and on the nice properties of the undulator that improve with increasing N, such as brightness, monochromaticity and, as we are going to see in the next chapter, gain) tied to the minimum electron-beam energy spread that is achievable. For this reason the improvements that today are continuously being obtained in free-electron laser technology are in large part the result of electron energy spread and emittance reductions.

It is also important to observe the appearance in eqn (4.80) of the harmonic number n. It was remarked in section 4.5 that when the dimensionless undulator field strength A_u is moderately large, a considerable number of harmonics can be generated with a relative intensity of the order of $n^{-1/3}$ with respect to the fundamental. However, we now see that with increasing n the limit set by eqn (4.80) will be eventually exceeded, and harmonics will be damped by the effect of energy spread.

So much for electron momentum and energy spread limitations. Let us now investigate what happens when these limitations are exceeded. We shall consider two situations; when the limitations are strongly exceeded, and when the equivalent detuning spread has a Lorentzian distribution. In both cases simplifications may be attained.

In general, the frequency and angle intensity distribution of undulator radiation is

$$S_\omega(\omega, \vartheta_w) = \int S_{\omega se}(\omega, \gamma, \vartheta_w - \vartheta_e) f(\gamma, \vartheta_e)\, \mathrm{d}\gamma\, \mathrm{d}\vartheta_e, \tag{4.81}$$

where $S_{\omega se}$ is the frequency and angle intensity distribution of radiation from electrons of a single energy γ and direction ϑ_e, given by eqn (4.46), and $f(\gamma, \vartheta_e)$ is the probability distribution of beam electrons in γ, ϑ_e-space. In the limit of a very wide electron distribution, when the bounds of eqns (4.77) and (4.78) are strongly exceeded, the $\sin^2(v_n)/v_n^2$ function in eqn (4.46) may be approximated by $\pi\delta(v_n)$. Upon inserting eqn (4.46) into eqn (4.81) and eliminating the integration in $\mathrm{d}\gamma$ by means of the δ-function, the following expression is obtained:

$$S_\omega = \frac{e\zeta j_b kL\gamma}{16\pi M^2} |A_{ueff,n}(0)|^2 \int f[M/(2nk_r/k - \vartheta_{we}^2)^{1/2}, \vartheta_e]\, \mathrm{d}\vartheta_e. \tag{4.82}$$

In particular, if energy spread is large but angular spread is negligible, the radiation intensity is proportional to the energy distribution function $f(\gamma)$.

Rather than approximating the integral of eqn (4.80) by introducing a δ-function, which may only be done for an energy and/or angular spread which is large compared to the values given in eqns (4.77) and (4.78), it is possible to calculate it exactly when the equivalent ν distribution has a Lorentzian shape. In order to do so, let us set up a very simple device which will also be useful later (the idea of which is due to Moore 1985).

Consider an integral of the form

$$\int_{-\infty}^{\infty} f(\nu)g(\nu)\,\mathrm{d}\nu, \tag{4.83}$$

where $f(\nu)$ is the normalized Lorentzian distribution function $\delta\nu/\pi[\delta\nu^2 + (\nu - \bar{\nu})^2]$, with mean $\bar{\nu}$ and width $\delta\nu$. Suppose that $g(\nu)$ is an analytic function of the complex variable ν, regular and vanishing at infinity in the half-plane $\mathrm{Im}(\nu) > 0$. We want to show that

$$\int_{-\infty}^{\infty} f(\nu)g(\nu)\,\mathrm{d}\nu = \int_{-\infty}^{\infty} \frac{\delta\nu g(\nu)}{\pi[\delta\nu^2 + (\nu - \bar{\nu})^2]}\,\mathrm{d}\nu = g(\bar{\nu} + \mathrm{i}\delta\nu). \tag{4.84}$$

In fact, the product $f(\nu)g(\nu)$ has a pole at the point $\bar{\nu} + \mathrm{i}\delta\nu$, with a residue $2\pi i g(\bar{\nu} + \mathrm{i}\delta\nu)$ and, because of the assumed regularity of $g(\nu)$, it has no other poles in the half-plane $\mathrm{Im}(\nu) > 0$ and vanishes at infinity faster than $1/\nu^2$ in the same half-plane. Under these conditions the residue theorem may be applied and eqn (4.84) follows.

In order to apply eqn (4.84) to the integral of eqn (4.81), define $g(\nu)$ as $g(\nu) = (1 + 2\mathrm{i}\nu - \mathrm{e}^{2\mathrm{i}\nu})/2\nu^2$. Then, $\sin^2(\nu)/\nu^2 = \mathrm{Re}[g(\nu)]$, and $g(\nu)$ has the required regularity properties. Therefore eqn (4.84) holds and the integral of eqn (4.81) is simply proportional to the real part of $g(\bar{\nu} + \mathrm{i}\delta\nu)$:

$$S_\omega = \frac{\mathrm{e}\zeta j_b k^2 L^2}{32\pi^2\gamma^2} |A_{ueff,n}(0)|^2 \,\mathrm{Re}[g(\bar{\nu} + \mathrm{i}\delta\nu)]. \tag{4.85}$$

This result may be stated by saying that a Lorentzian detuning distribution (representing energy and/or momentum spread) with mean $\bar{\nu}$ and width $\delta\nu$ can be accounted for by giving the complex value $\bar{\nu} + \mathrm{i}\delta\nu$ to ν in the no-spread results. This is a very effective way to understand the effects of energy spread, and the results can be considered to be qualitatively valid even when the real distribution is not exactly Lorentzian in shape. The function $\mathrm{Re}[(1 + 2\mathrm{i}\nu - \mathrm{e}^{2\mathrm{i}\nu})/2\nu^2]$, representative of the frequency spectrum of a single harmonic of free-space undulator radiation, is plotted in Fig. 4.3 for a number of values of $\delta\nu$. Notice how, for $\delta\nu = 1$, the radiation spectrum already tends to approach the shape of the Lorentzian detuning distribution, in accordance with the approximate formula (4.82).

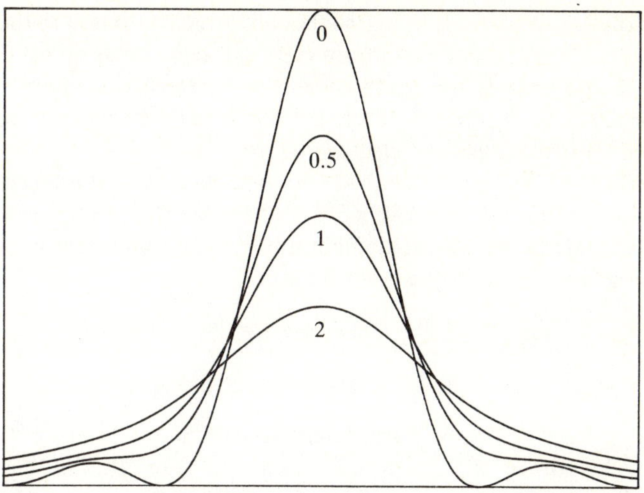

FIG. 4.3. Effect of energy spread upon the undulator-radiation spectrum. Each curve is labelled with the corresponding value of the detuning-spread parameter $\delta v = \mathrm{Im}(v)$.

4.11 Quantum recoil

Within the ultrarelativistic approximation the undulator field is equivalent in the electron frame of reference to an electromagnetic wave running across the electron in the direction opposite to its motion. This equivalence opens the way to a quantum description of the phenomenon of undulator emission as Compton scattering in the electron frame. Doppler shift accounts for a wavelength reduction by a factor of $2\gamma^2$, as shown in Chapter 2.

According to the usual theory of Compton scattering (e.g. Heitler 1954) which, incidentally, is based upon perturbation theory and requires modification to be applied to undulators when the dimensionless undulator field strength A_u is not small, the quantum-mechanical scattering cross-section does not differ significantly from the classical one as long as the photon energy $\hbar\omega = hc/\lambda$ is small compared to the electron rest energy mc^2, that is down to wavelengths (in the electron frame) of the order of the electron Compton wavelength $\lambda_C = h/mc \simeq 2.4 \times 10^{-12}$ m. Since the wavelength of radiation seen in the electron frame is approximately $1/\gamma$ times the undulator period, it can be easily calculated that even with an undulator period in the range of hundreds of micrometres, as could be obtained with an electromagnetic undulator, breaking the Compton wavelength barrier requires $\gamma \simeq 10^8$; that is, an unthinkable electron energy of 10^5 GeV. Moreover, the output wavelength of the undulator would be of the order of 10^{-20} m, corresponding to a photon energy of the order of 10^5 GeV as well. The hugeness of these numbers explains why classical mechanics is adequate, for the great majority of purposes, to describe undulator phenomena.

Nevertheless, even at wavelengths that are large compared to the Compton wavelength, there is a deviation from classical theory that might be observed in undulators. This is the small difference between the input and output frequencies of a Compton scattering event due to the non-zero momentum of the photon which causes the electron to 'recoil'.

This effect can be easily described with reference to a single elastic collision between an electron of energy $mc^2\gamma$ and momentum $mc(\gamma^2 - 1)^{1/2}$ and a photon of energy hc/λ and momentum h/λ. The momentum and energy balance equations for such a collision are:

$$mc(\gamma_i^2 - 1)^{1/2} - h/\lambda_i = mc(\gamma_f^2 - 1)^{1/2} + h/\lambda_f, \qquad (4.86a)$$

$$mc^2\gamma_i + 0 = mc^2\gamma_f + hc/\lambda_f, \qquad (4.86b)$$

where γ_i, γ_f, λ_i, and λ_f are, respectively, the initial and final γ of the electron and wavelength of the photon. Notice that the photon has non-zero momentum but zero energy before the collision (it is one of the virtual photons composing the static magnetic field), and reverses its direction of motion after the collision. From eqns (4.86) it is easy to obtain the properties of the scattered particles as functions of those of the incoming ones:

$$2\gamma_f = \gamma_i - (\gamma_i^2 - 1)^{1/2} + \frac{h}{mc\lambda_i} + \frac{1}{\gamma_i - (\gamma_i^2 - 1)^{1/2} + h/mc\lambda_i}, \qquad (4.87a)$$

$$h/mc\lambda_f = \gamma_f - \gamma_i. \qquad (4.87b)$$

Under the ultrarelativistic condition of $\gamma \gg 1$ one may set $\gamma - (\gamma^2 - 1)^{1/2} \simeq 1/2\gamma$, and eqns (4.87) become

$$1/\gamma_f = 1/\gamma_i + 2h/mc\lambda_i, \qquad \lambda_f = \lambda_i/2\gamma_i^2 + h/mc\gamma_i. \qquad (4.88a,b)$$

With respect to the classical expression for the emission wavelength in an infinitely long undulator, $\lambda_f = \lambda_i/2\gamma_i^2$, eqn (4.88b) differs by the additional term $h/mc\gamma_i = \lambda_C/\gamma_i$, λ_C being the Compton wavelength of the electron, which causes a slight red-shift. This red-shift is due, according to the above derivation, to the non-zero momentum of the photon which forces the electron to change its momentum in the emission, that is, to 'recoil': hence the name quantum recoil.

Let us now try to identify the situations in which quantum recoil might be observed. A condition for this is that the red-shift caused by quantum recoil should be comparable to or greater than the linewidth of undulator spontaneous emission arising from the finite length of the undulator. According to eqn (4.46), the relative linewidth of the first harmonic is $1/N$, so that the wavelength linewidth is $\delta\lambda = \lambda/N$. Comparing this to the quantum-recoil red-shift one obtains the condition:

$$\lambda \lesssim N\lambda_C/\gamma. \qquad (4.89)$$

Equation (4.89) may be recognized to be equivalent to conditions (2.15) and (2.16), derived in Chapter 2 from delocalization considerations, as those under which quantum effects would become important. We see now that these quantum effects simply consist of the quantum-recoil red-shift, without a noticeable change in the line shape or intensity of emission. The operating conditions under which eqn (4.89) holds might perhaps be reached in the future in a very-high-N, high-γ, X-ray undulator, using an infrared laser wave as an undulator field, but are not even approached in today's undulator and FEL devices.

Notes

1. Actually, the limiting form of eqn (4.20) contains an additional term, not included in eqn (4.21), coming from the contribution of the stationary-phase point where the line-of-sight vector $\boldsymbol{R} = \boldsymbol{r} - \boldsymbol{r}'$ is perpendicular to the electron trajectory. This term, which represents the impact upon the observer of the 'wake field' that the electron carries along because of its charge, has not been considered because it is assumed that in reality the electron will not go on indefinitely along a rectilinear path, but will be deflected by a guiding magnet or dumped by some other means before going past the observer.
2. The difference $e^{i(kt_f - \boldsymbol{k}\cdot\boldsymbol{r}_f)} - e^{i(kt_i - \boldsymbol{k}\cdot\boldsymbol{r}_i)}$ appearing in eqn (4.31) is written as $e^{-i(kt_i - \boldsymbol{k}_\perp\cdot\boldsymbol{x}_i)}(e^{-i\kappa L} - 1)$ in eqn (4.32). This is only allowed if the undulator length L is an integral multiple of the period l_u, as we shall assume it to be for the sake of simplicity. At any rate, this term affects only the low-frequency tail of the undulator radiation spectrum (the coefficient $A_{ueff,0}$) and is not important for emission near the resonance lines identified below.
3. According to a standard convention, S_ω is defined as radiated power per unit beam area, solid angle, and frequency in the interval $0 < \omega < \infty$. This is the reason for the factor of 2 in the expression of the Poynting vector, which would not be there had S_ω been defined as an even function in the interval $-\infty < \omega < \infty$.
4. We continue to denote by the symbol k the free-space wavenumber ω/c. This should not be confused with the propagation constant inside the waveguide, which is β.
5. The coefficient C_m is set equal to 1 in eqn (4.62). As often happens with modal decompositions, the resulting field $\boldsymbol{E}_f$ turns out not to have the right dimensions, but this causes no harm because, for all purposes it will be used for, this field is defined only up to an arbitrary multiplicative constant.
6. More precisely, an electron in a waveguide is never in free motion, even in the absence of external magnetic fields, because of the electrostatic interaction between the electron and the metallic walls, and thus transverse momentum is not conserved. In order to make statements about electron momentum in a waveguide we must assume that the undulator is shorter than the distance over which the effect of electrostatic interaction on the electron path becomes significant. Even if the necessity of considering a finite distance makes a transverse momentum $\boldsymbol{P}_\perp$ slightly different from zero admissible, for the scope of this section we shall take $\boldsymbol{P}_\perp$ to be exactly zero.

References

Abramowitz, M. and Stegun, I. A. (1968). *Handbook of mathematical functions.* Dover, New York.

Erdélyi, A. (1956). *Asymptotic expansions.* Dover, New York.

Heitler, W. (1954). *The quantum theory of radiation.* Oxford University Press, Oxford.

Landau, L. D. and Lifschitz, E. M. (1975). *The classical theory of fields.* Pergamon Press, Oxford.

Moore, G. T. (1985). High-gain, small-signal modes of the free-electron laser. II. Effect of electron energy spread. *Optics Communs* **54**, 121–5.

Motz, H. (1950). *Electromagnetic problems of microwave theory.* Methuen, London and John Wiley, New York.

5

SMALL-SIGNAL, SMALL-GAIN FREE-ELECTRON LASER

When an electromagnetic wave of approximately the same frequency as the spontaneous undulator emission is driven through the undulator simultaneously with the electron beam, an interaction between the beam and the wave takes place which can lead to a noticeable exchange of energy. In this case the undulator becomes a coherent electromagnetic amplifier capable of working from microwaves to visible and shorter than visible wavelengths, and takes on the name of Free-Electron Laser (FEL).

To study the interaction between the electromagnetic wave and the electron beam several points of view can be, and have been, taken, which lead to completely different pictures of the phenomenon. What distinguishes one model from the other is its operating regime, corresponding to a small or large input-field amplitude (leading to linear or non-linear amplification) and a small or large amplification (leading to negligible or substantial modification of the input wave in traversing the undulator).

In this chapter, after an introduction to the differences between the spontaneous-emission and coherent-amplification phenomena, we begin the study of free-electron lasers from the simplest operating regime, the small-signal, small-gain one. One of the following chapters is devoted to each of the other regimes.

5.1 Stimulated versus spontaneous emission

The radiation electric field produced by an electron moving under the action of an undulator magnet has a phase which, according to eqn (4.38), depends on the time, t_i, of arrival of the electron at the undulator entrance. For this reason the fields produced by different electrons in a uniform input beam have a random phase relation to each other and sum up incoherently. In analogy with thermal emission from heated matter, in which the fields of different atoms or molecules have a random phase relation to each other and sum up incoherently, this radiation may be named 'spontaneous'.

On the other hand, just like the atoms or molecules of heated matter, the electrons moving in an undulator can be driven by an external electromagnetic wave in a synchronous oscillation, the phase of which is no longer random but locked to the phase of the external wave. The radiated fields of different electrons then sum up coherently to each other and to the external wave, leading either to a decrease or to an increase of power, depending on whether destructive or constructive interference is realized. As a result the external wave is either absorbed or amplified. Using the language of atomic and molecular lasers, we may call an amplification process of this kind 'stimulated emission'. Notice, however, that while it can be shown that in heated matter in thermal equilibrium the interference between the coherent radiated field and the external wave can only be destructive, that is only absorption is possible, and in order to make a laser amplifier the medium has to be driven far from thermodynamic equilibrium, the distribution of electrons in an electron beam is already anything but 'thermal', and both absorption and amplification are possible.

In fact, amplification of electromagnetic waves by interaction with an electron beam is the basis of most classical microwave amplifiers, with the notable exception of the maser, the only difference between one and the other being the particular interaction mechanism. The free-electron laser could well be described as an optical-wavelength member of this class of devices, rather than as a laser, but since the atomic laser has for some time been the only kind of coherent amplifier operating at optical wavelength, the free-electron laser has inherited the name 'laser', as well as some of the related terminology.

5.2 Free-electron-laser regimes

The equations of motion of an electron in an electromagnetic field are strongly non-linear, and thus the process through which the oscillations of different electrons inside an undulator may be influenced and synchronized by an external field is also basically non-linear. Nevertheless, when the external field is weak the equations may be linearized, and it makes sense to speak of a linear amplification process in which the output field is proportional to the input field. The ratio of the output to the input field in this small-input[1] linear regime is called, as in all amplifiers, small-signal gain. The process by which linearity is lost with increasing input field is called saturation, and the non-linear operating regime which is attained at saturation is called, in opposition to the small-signal regime, the large-signal regime. The description of the small-signal and saturation regimes of a free-electron laser requires a completely different approach, and in this book separate chapters are devoted to each.

In addition to the distinction between small-signal and large-signal regimes (which we shall only later be able to trace quantitatively), another important distinction must be made between a regime in which the interaction of electrons with each other is negligible, so that it is possible to calculate separately the amplification of the input field provided by each electron and

Table 5.1. FEL regimes

	Small-gain	Large-gain
Small-signal (linear)	Linearized electron motion subject only to incident field. Madey's theorem yields the gain	Linearized electron motion subject to the evolving laser field. Analysis similar to TWT
Large-signal (non-linear)	Pendulum analogy in uniform undulator. Synchrotron oscillations lead to saturation	Moving staircase analogy in tapered undulator. Similar to LINAC. Can function as accelerator

then just sum up, and a regime in which the amplification process is collective, and the electron distribution must be considered as a whole. The first regime is the one in which the gain of the free-electron laser is small (compared to unity), as the electromagnetic field seen by each electron then coincides substantially with the input field, and the radiation field of the other electrons is negligible with respect to it; the second is when gain is of order unity or larger, and the modifications produced by the other electrons on the field seen by each one of them during its motion along the undulator cannot be neglected.

By combining the above conditions in all possible ways, we are led to define four basic free-electron laser operating regimes, indicated in Table 5.1. Different schematizations and approximations are appropriate to each of these situations.

Among these the small-signal, small-gain regime is peculiar for the number of different ways in which it may be approached, which include classical and quantum-mechanical perturbation analysis and suitably approximated versions of the techniques appropriate to the large-gain and large-signal regimes. Different authors have different personal preferences among these different approaches. We have chosen to devote this chapter to the techniques based on classical perturbation analysis, because they have a much wider scope; for instance, allowing a relatively easy account of three-dimensional effects. We shall, however, indicate in other chapters how some of the same results may also be obtained in a different way. In the following, we shall take as our working assumption that the equations of motion can be linearized in the amplitude of the radiation field and that the latter can be identified with the input one, thereby neglecting the effect on electron motion of the field radiated by the electrons themselves. Under these circumstances the motion and radiation processes do not interact; the motion may be determined first, using mechanical perturbation theory and considering the electromagnetic field as completely known, and the radiated field determined thereafter by suitable radiation formulae.

For the mechanical part of this analysis we need to determine perturbatively the motion of an electron in the superposition of the undulator magnetic field and the electromagnetic field of the input wave. A fundamental tool will be Madey's theorem, derived in Section 5.7, which will enable us to reduce the problem of determining FEL gain to the already solved one of determining its spontaneous emission. Before doing that, however, we shall give an example of free-electron laser amplification with reference to the simplest situation: one-dimensional laser and electron beams in a one-dimensional circularly polarized undulator. This example will enable us to determine the basic FEL gain formula, which will be generalized later with the help of Madey's theorem.

5.3 The definition and calculation of gain

The input and output of a circuital amplifier are scalar signals represented by single voltages or currents, and it is immediately possible to define gain as the ratio of output to input. The only distinction that must be made is whether the ratio of amplitudes or powers is considered. If the amplifier is linear, knowledge of the gain at all interesting frequencies is a quite complete external characterization of the amplifier itself.

The input and output of an electromagnetic wave amplifier are spatial field configurations, and a complete characterization of the linear amplification properties of such a device can only be obtained through a Green function, or scattering operator, which transforms the input in the output wave. In addition, an exact definition of what is meant by 'input' and 'output' can only be given at an infinite distance from the device, where incoming and outgoing components may be clearly distinguished in the wave field. It is then possible to impose the input field configuration and calculate, or measure, the output one. For free wave amplifiers, amplitude gain is no longer a number but an operator (transition matrix).

Power gain may be defined, just as for any amplifier, as the ratio of output to input power. In order to apply this definition it is necessary, again, to separate the power flows associated with outgoing and incoming waves.

Let the incident field $\boldsymbol{e}_{\mathrm{w}} = \mathrm{Re}[\boldsymbol{E}_{\mathrm{w}}(\boldsymbol{r})\mathrm{e}^{\mathrm{i}kt}]$ be assigned in the form of an arbitrary source-free harmonic solution of Maxwell's equations. The energy flux associated with this wave is expressed by its Poynting vector $\frac{1}{2}\,\mathrm{Re}(\boldsymbol{E}_{\mathrm{w}}^{*} \wedge \boldsymbol{H}_{\mathrm{w}})$. The total incident power P_{in} is therefore given by the integral of the Poynting vector over any surface crossed by the wave, which in all the cases we shall consider may be assumed to be a plane perpendicular to the z-axis:

$$P_{\mathrm{in}} = \int_{z=\bar{z}} \tfrac{1}{2}\,\mathrm{Re}(\boldsymbol{E}_{\mathrm{w}}^{*} \wedge \boldsymbol{H}_{\mathrm{w}}) \cdot \hat{\boldsymbol{z}}\,\mathrm{d}^2\boldsymbol{x}. \tag{5.1}$$

On the other hand, the integral of this Poynting vector over any closed surface is zero because, according to Poynting's theorem, when the field is source-free its Poynting vector has identically zero divergence, implying that the power P_{in} associated with waves crossing that surface inwards exactly balances the power P_{out} associated with waves travelling outwards. By the same token the power P_{in} of eqn (5.1) is independent of the choice of the point $\bar{z}$.

If, now, an amplifier (actually, any object capable of interacting with electromagnetic radiation) is inserted in space, the configuration of outgoing waves will be modified and the integral of the Poynting vector over a surface enclosing the amplifier will no longer balance, but will equal the power increment generated by the amplifier, $P_{\text{inc}} = P_{\text{out}} - P_{\text{in}}$. Denoting the total electromagnetic fields existing in the presence of the amplifier by $\boldsymbol{E}_{\text{tot}}$ and $\boldsymbol{H}_{\text{tot}}$, one has

$$P_{\text{inc}} = \int_{\partial V} \tfrac{1}{2}\,\text{Re}(\boldsymbol{E}^*_{\text{tot}} \wedge \boldsymbol{H}_{\text{tot}})\cdot\hat{\boldsymbol{n}}\,\text{d}S, \tag{5.2}$$

where ∂V is the boundary of any volume enclosing the amplifier and $\hat{\boldsymbol{n}}$ the outward normal to this boundary.

The power gain of an amplifier is usually defined as $G = P_{\text{out}}/P_{\text{in}}$. Since, however, under the small-gain assumption the field radiated by the amplifier is small and $P_{\text{out}} \simeq P_{\text{in}}$, it has become customary to give a different definition of gain in connection with small-gain free-electron lasers, as the ratio of power increment $P_{\text{inc}} = P_{\text{out}} - P_{\text{in}}$ to input power. Let us name this ratio incremental gain g. Therefore,

$$g = (P_{\text{out}} - P_{\text{in}})/P_{\text{in}} = P_{\text{inc}}/P_{\text{in}}. \tag{5.3}$$

Obviously, $G = g + 1$. When g is a small number (which is what we mean by small-gain regime) it is more convenient to work with g than with G.

A useful expression of P_{inc}, and therefore of g, may be obtained in the small-gain regime from reciprocity relations. Let us write the total electric and magnetic fields existing around the FEL as the sum of the given input fields $\boldsymbol{E}_w$, $\boldsymbol{H}_w$, which are a source-free solution of Maxwell's equations, and unknown radiated fields $\boldsymbol{E}_r$, $\boldsymbol{H}_r$. According to the small-gain assumption, we may consider $\boldsymbol{E}_r$, $\boldsymbol{H}_r$ as small compared to $\boldsymbol{E}_w$, $\boldsymbol{H}_w$, and linearize eqn (5.2) with respect to the radiated fields. This linearization yields

$$P_{\text{inc}} \simeq \int_{\partial V} \tfrac{1}{2}\,\text{Re}(\boldsymbol{E}^*_w \wedge \boldsymbol{H}_w + \boldsymbol{E}^*_w \wedge \boldsymbol{H}_r + \boldsymbol{E}^*_r \wedge \boldsymbol{H}_w)\cdot\hat{\boldsymbol{n}}\,\text{d}S. \tag{5.4}$$

Since, however, the integral of the Poynting vector $\tfrac{1}{2}\,\text{Re}(\boldsymbol{E}^*_w \wedge \boldsymbol{H}_w)$ over any closed surface is zero, the first term may be dropped out, leaving

$$P_{\text{inc}} \simeq \int_{\partial V} \tfrac{1}{2}\,\text{Re}(\boldsymbol{E}^*_w \wedge \boldsymbol{H}_r + \boldsymbol{E}^*_r \wedge \boldsymbol{H}_w)\cdot\hat{\boldsymbol{n}}\,\text{d}S. \tag{5.5}$$

Equation (5.5) leads us back to reciprocity relations, in particular to eqn (4.56), which allows us to write immediately

$$P_{\text{inc}} \simeq -\int_V \tfrac{1}{2}\,\text{Re}(\boldsymbol{E}_w^* \cdot \boldsymbol{J}_r)\,\text{d}V = -W_{wr}, \tag{5.6}$$

i.e. the wave power increment equals the opposite of the work W_{wr} done per unit time by the input wave upon the current density existing in the radiator.

Notice that when the input field is a member of a mode set (e.g. in a waveguide) the integral in eqn (5.5) may be also interpreted as a projection operator giving the projection of the radiated field onto the input mode. Therefore eqn (5.6) automatically calculates the 'interesting' power radiated in the same mode where power is input. It is a consequence of the small-gain assumption that this coincides at first order with the total power increment.

5.4 One-dimensional amplification in a circularly polarized ideal undulator

The motion of an electron in the magnetic field of an ideal sinusoidal undulator was determined in Section 3.4 in the absence of additional external fields. We are now going to consider the motion of electrons in the field corresponding to the dimensionless vector potential:

$$\boldsymbol{a}_\perp = \text{Re}[\boldsymbol{A}_u \text{e}^{\text{i}k_u z} + \boldsymbol{A}_w \text{e}^{\text{i}k(t-z)}], \tag{5.7}$$

where the first term represents an ideal sinusoidal undulator and the second term the incident electromagnetic plane wave to be amplified, the amplitude of which is assumed to be small enough to be treated as a perturbation. $\boldsymbol{A}_u$ and $\boldsymbol{A}_w$ are circular-polarization complex vectors, such that $\boldsymbol{A}_u^2 = \boldsymbol{A}_w^2 = \boldsymbol{A}_u \cdot \boldsymbol{A}_w = 0$. In this section we want to determine the amplification caused by a spatially uniform, perfectly collimated and monoenergetic beam of electrons moving in such a field.

The radiation power increment may be obtained from eqn (5.6) after calculating the work per unit transverse area and time done by the input wave upon the electron beam moving in the wave itself. (Work and power must be calculated *per unit transverse area* because a 1-D model necessarily deals with plane waves and electron beams of infinite cross-section and the total power is infinite.)

In turn, the total work per unit transverse area and time, W_{wr}, done by the input field upon a uniform electron beam in which n_e electrons per unit transverse area and time pass through the undulator is n_e times the average work, W_{we}, done upon a single electron. The work done by the electric field, $\boldsymbol{e}_w = \text{Re}(\boldsymbol{E}_w \text{e}^{\text{i}kt})$, upon a single electron may be expressed as

$$W_{we} = \int_{-\infty}^{\infty} e\frac{\text{d}\boldsymbol{r}}{\text{d}t} \cdot \boldsymbol{e}_w\,\text{d}t = \int_{-\infty}^{\infty} e\frac{\text{d}\boldsymbol{x}(z)}{\text{d}z} \cdot \text{Re}\{\boldsymbol{E}_w \text{e}^{\text{i}k[t(z)-z]}\}\,\text{d}z, \tag{5.8}$$

in the second form of which z has been chosen as the independent variable rather than t. Notice that $\boldsymbol{E}_w$ and W_{we} in eqn (5.8) have their usual mks dimensions, whereas t denotes time multiplied by c, just as in the previous chapters. The relation between the dimensional electric field $\boldsymbol{E}_w$ and the dimensionless vector potential $\boldsymbol{A}_w$ is $\boldsymbol{E}_w = -ikmc^2\boldsymbol{A}_w/e$.

Let us now introduce a perturbative expansion of $t(z)$ and $\boldsymbol{x}(z)$ of the form $t = t_0 + t_1$, $\boldsymbol{x} = \boldsymbol{x}_0 + \boldsymbol{x}_1$, where $t_0(z)$, $\boldsymbol{x}_0(z)$ correspond to the motion of the electron in the undulator field alone and t_1, $\boldsymbol{x}_1$ represent the first terms of corresponding expansions in powers of the input wave amplitude $\boldsymbol{A}_w$. To first order in the perturbation eqn (5.8) becomes

$$W_{we} \simeq \mathrm{Re}\int_0^L e\left(\frac{\mathrm{d}\boldsymbol{x}_0}{\mathrm{d}z} + \frac{\mathrm{d}\boldsymbol{x}_1}{\mathrm{d}z} + \mathrm{i}kt_1\frac{\mathrm{d}\boldsymbol{x}_0}{\mathrm{d}z}\right)\cdot \boldsymbol{E}_w \mathrm{e}^{\mathrm{i}k(t_0 - z)}\,\mathrm{d}z, \tag{5.9}$$

where the integral has been extended over the undulator length, $0 < z < L$, rather than from $-\infty$ to ∞, because before and after the undulator $\mathrm{d}\boldsymbol{x}/\mathrm{d}z$ is zero. The unperturbed motion represented by $\boldsymbol{x}_0(z)$ and $t_0(z)$ was determined in Section 3.4. In the particular case of circular polarization (such that $\boldsymbol{A}_u^2 = 0$) and zero transverse momentum ($\boldsymbol{P}_\perp = 0$), eqns (3.24) give

$$\mathrm{d}\boldsymbol{x}_0/\mathrm{d}z = -\gamma_i^{-1}\,\mathrm{Re}(\boldsymbol{A}_u \mathrm{e}^{\mathrm{i}k_u z}), \qquad \gamma_0 = \gamma_i, \qquad t_0 = t_i + (1 + M^2/2\gamma_i^2)z, \tag{5.10a-c}$$

where $M^2 = 1 + A_u^2/2$, as in chapter 3, and subscript i denotes initial values at the entrance of the undulator, where $z = 0$.

In order to calculate the first-order quantities $\mathrm{d}\boldsymbol{x}_1/\mathrm{d}z$ and t_1 appearing in eqn (5.9) we have to consider the generalized Hamilton equations (3.4) obtained from the ultrarelativistic Hamiltonian (3.15), i.e.

$$\frac{\mathrm{d}\boldsymbol{x}}{\mathrm{d}z} = -\frac{\partial\mathscr{H}}{\partial\boldsymbol{P}_\perp} = -\frac{\boldsymbol{a}}{\gamma}, \qquad \frac{\mathrm{d}\gamma}{\mathrm{d}z} = -\frac{\partial\mathscr{H}}{\partial t} = \frac{\boldsymbol{a}}{\gamma}\cdot\frac{\partial\boldsymbol{a}}{\partial t}, \qquad \frac{\mathrm{d}t}{\mathrm{d}z} = \frac{\partial\mathscr{H}}{\partial\gamma} = 1 + \frac{1 + \boldsymbol{a}_\perp^2}{2\gamma^2}. \tag{5.11a-c}$$

Inserting the expression (5.7) of the vector potential into eqns (5.11) and linearizing with respect to the input wave amplitude $\boldsymbol{A}_w$ gives[2]

$$\frac{\mathrm{d}\boldsymbol{x}_1}{\mathrm{d}z} = -\frac{1}{\gamma_0}\mathrm{Re}[\boldsymbol{A}_w \mathrm{e}^{\mathrm{i}k(t_0 - z)}] + \frac{\gamma_1}{\gamma_0^2}\mathrm{Re}(\boldsymbol{A}_u \mathrm{e}^{\mathrm{i}k_u z}), \tag{5.12a}$$

$$\frac{\mathrm{d}\gamma_1}{\mathrm{d}z} = \frac{1}{2\gamma_0}\mathrm{Re}\{\mathrm{i}k\boldsymbol{A}_u^* \cdot \boldsymbol{A}_w \mathrm{e}^{\mathrm{i}[kt_0 - (k + k_u)z]}\}, \tag{5.12b}$$

$$\frac{\mathrm{d}t_1}{\mathrm{d}z} = \frac{1}{2\gamma_0^2}\mathrm{Re}\{\boldsymbol{A}_u^* \cdot \boldsymbol{A}_w \mathrm{e}^{\mathrm{i}[kt_0 - (k + k_u)z]}\} - M^2\frac{\gamma_1}{\gamma_0^3}. \tag{5.12c}$$

Equations (5.12) are differential equations that must be solved with the initial conditions $\boldsymbol{x}_1(0) = 0$, $\gamma_1(0) = t_1(0) = 0$. First, integrating the equation for γ_1 with the aid of the expression (5.10) of $t_0(z)$ yields

$$\gamma_1 = \frac{k}{2\mu\gamma_i} \operatorname{Re}[\boldsymbol{A}_u^* \cdot \boldsymbol{A}_w \mathrm{e}^{\mathrm{i}kt_i}(\mathrm{e}^{\mathrm{i}\mu z} - 1)], \tag{5.13}$$

where the name μ has been given to the coefficient of z in the expression of $kt_0(z) - (k + k_u)z$, i.e.

$$\mu = kM^2/2\gamma_i^2 - k_u. \tag{5.14}$$

One may notice that μ is the difference between the reduced wavenumber κ, which was defined in eqn (4.34) (for $\vartheta_{we} = 0$) and the undulator wavenumber k_u. We shall call μ the 'beat wavenumber'. Then, integrating the equation for t_1 gives

$$t_1 = \operatorname{Re}\left[\frac{1}{2\mathrm{i}\mu\gamma_i^2} \boldsymbol{A}_u^* \cdot \boldsymbol{A}_w \mathrm{e}^{\mathrm{i}kt_i}(\mathrm{e}^{\mathrm{i}\mu z} - 1) - \frac{kM^2}{2\mathrm{i}\mu^2\gamma_i^4} \boldsymbol{A}_u^* \cdot \boldsymbol{A}_w \mathrm{e}^{\mathrm{i}kt_i}(\mathrm{e}^{\mathrm{i}\mu z} - 1 - \mathrm{i}\mu z)\right]. \tag{5.15}$$

Finally, inserting the results of eqns (5.10)–(5.15) into eqn (5.9) we obtain

$$\begin{aligned} W_{we} = \operatorname{Re} \int_0^L e\Bigg\{ & \operatorname{Re}\left[- \frac{\boldsymbol{A}_u}{\gamma_i} \mathrm{e}^{\mathrm{i}k_u z} - \frac{\boldsymbol{A}_w}{\gamma_i} \mathrm{e}^{\mathrm{i}kt_i + \mathrm{i}(\mu + k_u)z} \right. \\ & \left. + \frac{k}{2\mu\gamma_i^3} \boldsymbol{A}_u^* \cdot \boldsymbol{A}_w \mathrm{e}^{\mathrm{i}kt_i}(\mathrm{e}^{\mathrm{i}\mu z} - 1) \operatorname{Re}(\boldsymbol{A}_u e^{\mathrm{i}k_u z}) \right] \\ & + \mathrm{i} \operatorname{Re}\left[- \frac{k}{2\mathrm{i}\mu\gamma_i^3} \boldsymbol{A}_u^* \cdot \boldsymbol{A}_w \mathrm{e}^{\mathrm{i}kt_i}(\mathrm{e}^{\mathrm{i}\mu z} - 1) \right. \\ & \left. + \frac{k^2M^2}{2\mathrm{i}\mu^2\gamma_i^5} \boldsymbol{A}_u^* \cdot \boldsymbol{A}_w \mathrm{e}^{\mathrm{i}kt_i}(\mathrm{e}^{\mathrm{i}\mu z} - 1 - \mathrm{i}\mu z) \right] \\ & \times \operatorname{Re}(\boldsymbol{A}_u \mathrm{e}^{\mathrm{i}k_u z}) \Bigg\} \cdot \boldsymbol{E}_w \mathrm{e}^{\mathrm{i}[kt_i + (\mu + k_u)z]} \, \mathrm{d}z. \end{aligned} \tag{5.16}$$

Equation (5.16) represents the work done by the input wave upon an electron of arrival time t_i, the trajectory of which is modified at first order by the input wave itself. In order to calculate the total work W_{wr} done per unit time and transverse area by the input wave upon the whole beam, eqn (5.16) must be summed over all the electrons in the beam. Since the input beam is assumed to be uniform and the electrons are uniformly distributed in arrival time, the sum may be obtained simply by multiplying the average over t_i of eqn (5.16) by the number n_e of electrons per unit time and transverse area in the beam.

The expression of W_{we} contained in eqn (5.16) may be written as the sum of several terms multiplied by different powers of $\mathrm{e}^{\mathrm{i}kt_i}$, by replacing the real part operator with half the sum of its argument and its complex conjugate. The average, $\langle W_{we} \rangle$, is then simply given by the sum of those terms which turn out to be independent of e^{ikt_i}:

$$\langle W_{we} \rangle = \mathrm{Re} \int_0^L -\frac{eA_w^* \cdot E_w}{2\gamma_i} + eA_u \cdot A_w^* \,\mathrm{Re}(A_u e^{ik_u z}) \cdot E_w e^{ik_u z}$$

$$\times \left\{ \frac{k}{2\mu\gamma_i^3}(1 - e^{i\mu z}) - \frac{k^2 M^2}{4\mu^2 \gamma_i^5}[1 - (1 - i\mu z)e^{i\mu z}] \right\} dz. \quad (5.17)$$

On taking into account that $eE_w = -ikmc^2 A_w$ and using the properties of circularly polarized vectors, namely that $A_u \cdot A_w = 0$ and $|A_u^* \cdot A_w| = A_u A_w$, eqn (5.17) may be transformed into

$$\langle W_{we} \rangle = \mathrm{Re} \int_0^L A_u^2 A_w^2 \left\{ -\frac{ikmc^2}{4\mu\gamma_i^3}(1 - e^{i\mu z}) + \frac{ik^3 M^2 mc^2}{8\mu^2 \gamma_i^5}[1 - (1 - i\mu z)e^{i\mu z}] \right\} dz$$

$$= \frac{k^3 L^3 M^2 mc^2 A_u^2 A_w^2}{8\gamma_i^5} \left(\frac{2\gamma_i^2}{M^2 kL} \frac{\cos \mu L - 1}{\mu^2 L^2} - \frac{2\cos \mu L - 2 + \mu L \sin \mu L}{\mu^3 L^3} \right). \quad (5.18)$$

It is easy now to calculate from eqn (5.6) the power per unit area added to the input wave, $P_{\mathrm{inc}} = -n_e \langle W_{we} \rangle$, and dividing this by the input power per unit area, $P_{\mathrm{in}} = E_w^2/2\zeta = (mc^2 kA_w)^2/2\zeta e^2$, gives the gain as

$$g = -\frac{2\zeta e^2 n_e}{k^2 (mc^2)^2} \frac{\langle W_{we} \rangle}{A_w^2}$$

$$= \frac{\zeta e^2 k L^3 M^2 A_u^2 n_e}{4mc^2 \gamma_i^5} \left[\frac{2\cos \mu L - 2 + \mu L \sin \mu L}{\mu^3 L^3} + \frac{2\gamma_i^2}{M^2 kL} \frac{1 - \cos \mu L}{\mu^2 L^2} \right]. \quad (5.19)$$

5.5 The basic free-electron laser gain formula

Equation (5.19) is one of the most important results of this chapter, since it gives a reasonably good estimate of the gain of a free-electron laser under much more general conditions than those under which it was derived. We now want to put this gain formula in a simpler form which will allow us to point out the role of the various undulator parameters.

The grouping of natural constants appearing in eqn (5.19) may be expressed through the Alfvén current, I_A, defined as $I_A = 4\pi mc/\mu_0 e = 4\pi mc^2/\zeta e$. I_A is a natural constant itself, with the dimensions of a current, and its approximate value is 17.03 kA. On writing the number of electrons per unit transverse area and time, n_e, as j_b/e, j_b being the e-beam current density and introducing, in addition, the detuning parameter $\nu = \mu L/2$, eqn (5.19) may be recast[3] as

$$g = \frac{\pi k L^3 M^2 A_u^2 j_b}{4\gamma_i^5 I_A}\left[\frac{\cos 2\nu - 1 + \nu \sin 2\nu}{\nu^3} + \frac{2\gamma_i^2}{M^2 k L}\frac{1 - \cos 2\nu}{\nu^2}\right]. \quad (5.20)$$

The detuning parameter ν may be recognized to be the same which was already introduced in Chapter 4 (eqn 4.39, for $\vartheta_{we} = 0$ and $n = 1$) and, according to eqn (4.41), may be written as

$$\nu = N(k - k_r)/\pi k_r, \quad (5.21)$$

N being the number of periods L/l_u of the undulator and k_r the undulator resonance wavenumber of eqn (4.42),

$$k_r = 2\gamma_i^2 k_u/M^2. \quad (5.22)$$

From the appearance of ν in eqn (5.20) it may be deduced that FEL amplification occurs near the same wavenumber k_r near which spontaneous radiation is emitted, and within a bandwidth of about the same extension ($1/N$, as in eqn 4.43). Since only wavenumbers $k \simeq k_r$ are of interest, it is possible to replace k by k_r in eqn (5.20). Using eqn (5.22) one may then observe that the coefficient of the second term in eqn (5.20), $2\gamma_i^2/M^2kL$, becomes $1/k_uL = 2\pi/N$ and may be neglected when N is large. Equation (5.20) then simplifies into what we shall call the basic FEL gain formula in the small-gain regime:

$$g_0 = N^3 A_u^2 \gamma_i^{-3} \frac{\pi^2 l_u^2 j_b}{I_A} \frac{\cos 2\nu - 1 + \nu \sin 2\nu}{\nu^3}. \quad (5.23)$$

In eqn (5.23) the undulator length L has been replaced by the product Nl_u, N being the number of periods in the undulator and $l_u = 2\pi/k_u$ the period, and the coefficients have been purposely collected into separate dimensionless groupings. According to this formula the gain of a free-electron laser in the small-gain regime turns out to be proportional to the cube of the number of periods, to the square of the dimensionless undulator vector potential and to the power -3 of the electron energy factor γ; it depends on e-beam current through the ratio of the current flowing across a disc with radius l_u and the Alfvén current, and on radiation wavelength through the universal function $(\cos 2\nu - 1 + \nu \sin 2\nu)/\nu^3$ of the detuning parameter. A plot of this function is shown in Fig. 5.1, superposed on one of the function $[(\sin \nu)/\nu]^2$ which, as explained in the previous chapter, represents the most rapidly varying term in the expression (4.46) of the spontaneous-radiation spectrum.

As may be seen from the figure, the spectral distribution of gain of an undulator is quite different from that of spontaneous radiation and, in particular, maximum amplification does not occur at the same wavenumber

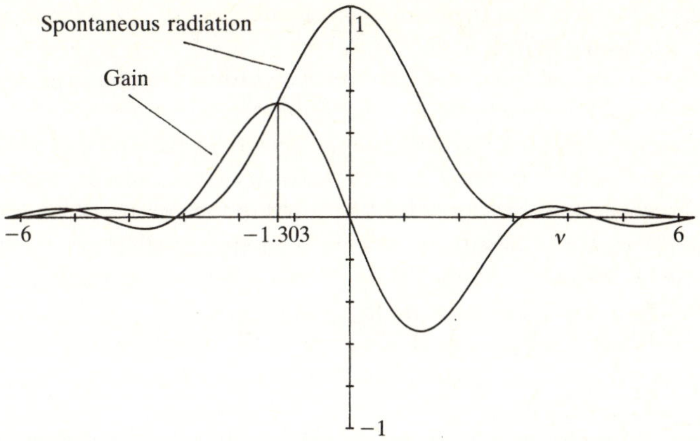

FIG. 5.1. Superposed plots of the gain and spontaneous-radiation spectra of an undulator.

as maximum spontaneous emission. In terms of detuning parameter, maximum spontaneous emission occurs at $\nu = 0$, whereas maximum amplification occurs at $\nu \simeq -1.303$. In fact, the spectral gain curve is antisymmetric about $\nu = 0$ and presents regions of positive as well as negative amplification (i.e. attenuation). Gain at the fundamental wavenumber k_r, corresponding to $\nu = 0$, is zero. The point of maximum gain, $\nu \simeq -1.303$, corresponds in terms of wavenumber to $k \simeq (1 - 0.414/N)k_r$. This maximum gain is given by

$$g_{0\max} = 0.54N^3A_u^2\gamma_i^{-3}\pi^2l_u^2\, j_b/I_A. \tag{5.24}$$

Another useful form of the basic FEL gain formula is obtained by writing the coefficient of eqn (5.20) as the cubed ratio of two lengths, the undulator length L and a quantity we shall name gain length L_g, which we shall define as

$$L_g = (2\gamma_i^5 I_A/\pi M^2 A_u^2 k j_b)^{1/3} \tag{5.25}$$

or, on replacing k by k_r, also as

$$L_g = \gamma_i(I_A/\pi A_u^2 k_u j_b)^{1/3}. \tag{5.26}$$

In terms of L_g, eqns (5.23) and (5.24) may be rewritten as

$$g_0 = \frac{1}{2}\left(\frac{L}{L_g}\right)^3 \frac{\cos 2\nu - 1 + \nu \sin 2\nu}{\nu^3}, \qquad g_{0\max} = 0.27(L/L_g)^3. \qquad (5.27, 5.28)$$

The gain length allows one to estimate how long the undulator must be in order to achieve a set maximum gain. If, for instance, a gain of 1 per cent is desired, it results immediately from eqn (5.28) that L must be $0.33L_g$, whereas a gain of 10 per cent necessitates $L = 0.7L_g$. L_g has the meaning of the typical undulator length at which gain g attains order unity (whence the name 'gain length'), and the present small-gain approximation loses validity. In the next chapter we shall find that the behaviour of gain for $L \gtrsim L_g$ (in the 'large-gain' regime) is considerably different.

The dependence of the gain length on the various undulator parameters is most readily seen from eqn (5.26). The undulator length necessary to achieve a set gain increases, for a given undulator period, as the first power of γ_i, and may be decreased by increasing current density or undulator field strength. Notice, however, that current density is raised to the power $-\frac{1}{3}$, and therefore large increases are necessary to achieve a moderate effect on L_g, while an increase in A_u also modifies the resonance wavenumber, through the effective mass M, unless a parallel increase is given to γ_i.

In the following sections we shall re-derive the gain formulae (5.20)–(5.28) using more powerful analytical instruments which will enable us to remove the assumption of circular polarization and eventually that of one-dimensionality. It is instructive, however, to emphasize now the elementary steps that the phenomenon of FEL amplification may be ideally subdivided into, for these will be hidden later. Analysing the straightforward perturbative calculation performed in the previous section one may identify three main steps. First, electron energy is modified at first order due to interaction with the input wave (eqn (5.13)); then, the time of arrival at a given position $t(z)$ is modified due both to the input wave and to the change in energy (eqn (5.15)); finally, radiation is emitted by the electrons because of the change in their time of arrival and transverse velocity (eqn (5.16)). In the following sections we shall bypass these three steps and derive the gain by more powerful, and more abstract, analytical techniques, but remembering the above sequence helps us to connect the operation mechanism of the FEL with that of related microwave devices, and to understand the non-linear analysis we shall present in Chapters 7 and 8, where electron beam modulation ('bunching') and radiation emission will have to be separated again.

5.6 Mechanical perturbation analysis

This section is devoted to preparing the analytical mechanical instruments by which Madey's theorem will be proved. The required elements of Hamiltonian mechanics are summarized in Appendix A.

Consider a general mechanical system governed by the Hamiltonian $\mathscr{H} = \mathscr{H}_0(\boldsymbol{r}, \boldsymbol{P}, t) + V(\boldsymbol{r}, \boldsymbol{P}, t)$, where V is a small perturbation, and suppose that the equations of motion corresponding to the Hamiltonian $\mathscr{H}_0$ can be solved completely, giving the solution $\boldsymbol{r} = \boldsymbol{r}_0(\boldsymbol{r}_i, \boldsymbol{P}_i; t)$, $\boldsymbol{P} = \boldsymbol{P}_0(\boldsymbol{r}_i, \boldsymbol{P}_i; t)$ for any choice of the initial conditions $\boldsymbol{r}_i$, $\boldsymbol{P}_i$. In order to find a perturbative expansion of the motion corresponding to the Hamiltonian $\mathscr{H}$ it is useful to first perform a change of variables which leads to a simpler Hamiltonian. In particular, it is a result of classical Hamiltonian mechanics (which may be found in Appendix A) that the change of variables $\boldsymbol{r} = \boldsymbol{r}_0(\boldsymbol{r}', \boldsymbol{P}'; t)$, $\boldsymbol{P} =$

$\boldsymbol{P}_0(\boldsymbol{r}', \boldsymbol{P}'; t)$, obtained by substituting the new variables $\boldsymbol{r}'$ and $\boldsymbol{P}'$ for $\boldsymbol{r}_i$ and $\boldsymbol{P}_i$ in the expression of the unperturbed motion, is a canonical transformation. Therefore the equations determining the time evolution of the variables $\boldsymbol{r}'$, $\boldsymbol{P}'$ keep the form of Hamilton equations. The Hamiltonian corresponding to the variables $\boldsymbol{r}'$, $\boldsymbol{P}'$ is V (as in eqn (A.27)), and the relevant equations of motion are

$$\mathrm{d}\boldsymbol{r}'/\mathrm{d}t = \partial V/\partial \boldsymbol{P}', \qquad \mathrm{d}\boldsymbol{P}'/\mathrm{d}t = -\partial V/\partial \boldsymbol{r}'. \tag{5.29}$$

These equations must be solved with the initial conditions, at $t = 0$, $\boldsymbol{r}'(0) = \boldsymbol{r}(0) = \boldsymbol{r}_i$, $\boldsymbol{P}'(0) = \boldsymbol{P}(0) = \boldsymbol{P}_i$.

Now expand $\boldsymbol{r}'$ and $\boldsymbol{P}'$ in a perturbative series with respect to V. At zero and first order one obtains

$$\mathrm{d}\boldsymbol{r}'_0/\mathrm{d}t = 0, \qquad \mathrm{d}\boldsymbol{P}'_0/\mathrm{d}t = 0; \tag{5.30a}$$

$$\mathrm{d}\boldsymbol{r}'_1/\mathrm{d}t = \partial V(\boldsymbol{r}'_0, \boldsymbol{P}'_0, t)/\partial \boldsymbol{P}'_0, \qquad \mathrm{d}\boldsymbol{P}'_1/\mathrm{d}t = -\partial V(\boldsymbol{r}'_0, \boldsymbol{P}'_0, t)/\partial \boldsymbol{r}'_0. \tag{5.30b}$$

The corresponding initial conditions are $\boldsymbol{r}'_0(0) = \boldsymbol{r}_i$, $\boldsymbol{P}'_0(0) = \boldsymbol{P}_i$, $\boldsymbol{r}'_n(0) = \boldsymbol{P}'_n(0) = 0$ for $n > 0$. According to eqn (5.30a), the zero-order solution is constant, and $\boldsymbol{r}'_0$ and $\boldsymbol{P}'_0$ equal $\boldsymbol{r}_i$ and $\boldsymbol{P}_i$ for all t. Substituting $\boldsymbol{r}_i$ and $\boldsymbol{P}_i$ for $\boldsymbol{r}'_0$ and $\boldsymbol{P}'_0$ in eqn (5.30b), we can write the first-order perturbations as

$$\boldsymbol{r}'_1 = \int_0^t \frac{\partial V(\boldsymbol{r}_i, \boldsymbol{P}_i; \tau)}{\partial \boldsymbol{P}_i}\,\mathrm{d}\tau, \qquad \boldsymbol{P}'_1 = -\int_0^t \frac{\partial V(\boldsymbol{r}_i, \boldsymbol{P}_i; \tau)}{\partial \boldsymbol{r}_i}\,\mathrm{d}\tau. \tag{5.31}$$

Equations (5.31) give the correction to $\boldsymbol{r}'$ and $\boldsymbol{P}'$ due to the perturbation.

More generally, let $F(\boldsymbol{r}, \boldsymbol{P}; t)$ be any function of the state of the mechanical system, the alteration of which in the presence of the perturbation is desired. Expressing F as a function of $\boldsymbol{r}'$, $\boldsymbol{P}'$, t and expanding it in a perturbation series gives, to first order,

$$F \simeq F(\boldsymbol{r}'_0, \boldsymbol{P}'_0; t) + \frac{\partial F}{\partial \boldsymbol{r}'}\cdot \boldsymbol{r}'_1 + \frac{\partial F}{\partial \boldsymbol{P}'}\cdot \boldsymbol{P}'_1. \tag{5.32}$$

Substituting again $\boldsymbol{r}_i$ and $\boldsymbol{P}_i$ for $\boldsymbol{r}'_0$ and $\boldsymbol{P}'_0$, and taking the values of $\boldsymbol{r}'_1$, $\boldsymbol{P}'_1$ from eqn (5.31) then yields

$$F \simeq F_0(\boldsymbol{r}_i, \boldsymbol{P}_i; t) + \int_0^t \left[\frac{\partial F_0(\boldsymbol{r}_i, \boldsymbol{P}_i; t)}{\partial \boldsymbol{r}_i}\cdot\frac{\partial V_0(\boldsymbol{r}_i, \boldsymbol{P}_i; \tau)}{\partial \boldsymbol{P}_i} - \frac{\partial F_0(\boldsymbol{r}_i, \boldsymbol{P}_i; t)}{\partial \boldsymbol{P}_i}\cdot\frac{\partial V_0(\boldsymbol{r}_i, \boldsymbol{P}_i; \tau)}{\partial \boldsymbol{r}_i}\right]\mathrm{d}\tau, \tag{5.33}$$

where $F_0(\boldsymbol{r}_i, \boldsymbol{P}_i; t)$ and $V_0(\boldsymbol{r}_i, \boldsymbol{P}_i; t)$ are the functions $F(\boldsymbol{r}, \boldsymbol{P}; t)$ and $V(\boldsymbol{r}, \boldsymbol{P}; t)$ evaluated along the unperturbed motion $\boldsymbol{r} = \boldsymbol{r}_0(\boldsymbol{r}_i, \boldsymbol{P}_i; t)$, $\boldsymbol{P} = \boldsymbol{P}_0(\boldsymbol{r}_i, \boldsymbol{P}_i; t)$ corresponding to initial conditions $\boldsymbol{r} = \boldsymbol{r}_i$ and $\boldsymbol{P} = \boldsymbol{P}_i$ at $t = 0$.

Equation (5.33) gives the first-order change of a property F of the mechanical system due to the perturbation, and has important applications in Hamiltonian mechanics and, in particular, in the proof of thermodynamic fluctuation–dissipation relations (see, e.g., Krinsky 1985).

The above discussion was carried out for a general mechanical system, the motion of which is specified by spatial co-ordinates $\boldsymbol{r}$ varying in time t. In order to apply it to the problem of FEL amplification we need to extend the results to the generalized Hamiltonian formalism in which z is the independent variable, since we have found it convenient in describing the electron motion. This extension may be performed simply by renaming variables. In fact, on changing z for t as the independent variable and $-\mathscr{K}$ (negative axial momentum) for $\mathscr{H}$ as the Hamiltonian, and considering $\boldsymbol{x}$ and t as co-ordinates, and $\boldsymbol{P}_\perp$ and $-\mathscr{E}$ as momenta, the standard Hamilton equations (3.2) become the generalized Hamilton equations (3.4).

Performing the substitution of z for t, t and $\boldsymbol{x}$ for $\boldsymbol{r}$, and $\boldsymbol{P}_\perp$ and $-\mathscr{E}$ for $\boldsymbol{P}$ in eqns (5.31) and (5.33) one obtains the generalized results:

$$\boldsymbol{x}_1' = -\int_0^z \frac{\partial V_0(\zeta)}{\partial \boldsymbol{P}_{\perp i}}\,\mathrm{d}\zeta, \qquad t_1' = \int_0^z \frac{\partial V_0(\zeta)}{\partial \mathscr{E}_i}\,\mathrm{d}z; \tag{5.34}$$

$$\boldsymbol{P}_{\perp 1}' = \int_0^z \frac{\partial V_0(\zeta)}{\partial \boldsymbol{x}_i}\,\mathrm{d}\zeta, \qquad \mathscr{E}_1' = -\int_0^z \frac{\partial V_0(\zeta)}{\partial t_i}\,\mathrm{d}z; \tag{5.35}$$

$$F \simeq F_0(z) + \int_0^z \left[\frac{\partial F_0(z)}{\partial \boldsymbol{P}_{\perp i}}\cdot\frac{\partial V_0(\zeta)}{\partial \boldsymbol{x}_i} - \frac{\partial F_0(z)}{\partial \boldsymbol{x}_i}\cdot\frac{\partial V_0(\zeta)}{\partial \boldsymbol{P}_{\perp i}} - \frac{\partial F_0(z)}{\partial \mathscr{E}_i}\frac{\partial V_0(\zeta)}{\partial t_i} + \frac{\partial F_0(z)}{\partial t_i}\frac{\partial V_0(\zeta)}{\partial \mathscr{E}_i}\right]\mathrm{d}\zeta; \tag{5.36}$$

where $V = \mathscr{K} - \mathscr{K}_0$, and the subscript 0 attached to V and F reminds us that these functions are evaluated along the unperturbed motion generated by the Hamiltonian $\mathscr{K}_0$ (and therefore are functions of $\boldsymbol{x}_i$, $\boldsymbol{P}_{\perp i}$, t_i, $\mathscr{E}_i$, and z).

5.7 Madey's theorem

It might have been noticed that the gain curve which was shown in Fig. 5.1 looks like the derivative of the spontaneous radiation curve plotted in the same figure. In fact it is exactly so: it may be easily checked analytically that the expression $(\cos 2v - 1 + v \sin 2v)/v^3$ appearing in eqns (5.23) and (5.27) is the derivative of the factor $[(\sin v)/v]^2$ which determines the spectral distribution of spontaneous radiation.

That the spectral distribution of gain is proportional to the derivative of the spectral distribution of spontaneous radiation is more or less the contents of Madey's theorem, and is true under much more general conditions than those assumed in Section 5.4. In this section we are going to state precisely and prove this theorem, which will serve us to extend the validity of the one-dimensional gain formula previously found for circular polarization. We shall consider a completely general three-dimensional wave of any polarization and an undulator magnetic field having an unspecified (not necessarily sinusoidal) variation in space, giving a generalized version of Madey's theorem similar to that of Luchini and Solimeno (1985). The original formulation given by Madey (1979) was quite different and was limited to one-dimensional free-electron lasers.

The basic step in the previous calculation of gain was the evaluation of the average work $\langle W_{we} \rangle$ done by the input wave upon a single electron, the trajectory of which is modified at first order by the wave itself. We shall now repeat this calculation using the perturbation formula (5.36), for an electron moving under the electromagnetic field generated by the dimensionless potentials

$$\boldsymbol{a} = \boldsymbol{a}_u(\boldsymbol{x}, z) + \boldsymbol{a}_w(\boldsymbol{x}, t, z), \qquad \varphi = \varphi_w(\boldsymbol{x}, t, z), \tag{5.37}$$

where $\boldsymbol{a}_u(\boldsymbol{x}, z)$ is the undulator vector potential, independent of time but having an unspecified dependence on spatial co-ordinates, and $\boldsymbol{a}_w$, φ_w represent a completely general electromagnetic wave, the amplitude of which will be assumed to be small enough that perturbation theory may be applied. The Hamiltonian $\mathscr{H}$ for an electron moving in this field may be written as $\mathscr{H} = \mathscr{H}_0(\boldsymbol{x}, \boldsymbol{P}_\perp, \mathscr{E}, z) + V(\boldsymbol{x}, \boldsymbol{P}_\perp, t, \mathscr{E}, z)$, where $\mathscr{H}_0$ is the time-independent Hamiltonian corresponding to the undulator field alone, and V is the time-dependent perturbation caused by the addition of the input wave field.

In particular, inserting the expression (5.37) of the potentials into the ultrarelativistic expression (3.15) of the Hamiltonian $\mathscr{H}$ for the motion of an electron in an electromagnetic field, and expanding the result in powers of the electromagnetic wave amplitude, one finds the unperturbed Hamiltonian,

$$\mathscr{H}_0 = \gamma + a_{uz} - \frac{1 + (\boldsymbol{P}_\perp - \boldsymbol{a}_{u\perp})^2}{2\gamma}, \tag{5.38}$$

and the first-order perturbation,

$$V = a_{wz} - \varphi_w + \boldsymbol{a}_{w\perp} \cdot (\boldsymbol{P}_\perp - \boldsymbol{a}_u)/\gamma. \tag{5.39}$$

In order to calculate the work W_{we} done by the input field upon a single electron let us start from energy conservation: the work done by the electric field acting upon the electron equals the energy change of the latter. Since the undulator field is a static magnetic field and does no work, the work done by the incident electromagnetic wave upon the electron equals its energy change. Using the equation $\mathrm{d}\mathscr{E}/\mathrm{d}z = -\partial\mathscr{H}/\partial t$ (one of the generalized Hamilton equations (3.4b)) one may write

$$W_{we} = -\int_0^L \frac{\partial \mathscr{H}}{\partial t}\, \mathrm{d}z. \tag{5.40}$$

In order to calculate the small-signal gain of the free-electron laser, we now want to evaluate eqn (5.40) along the electron trajectory perturbed at first order by the wave. In the primed co-ordinates introduced in the previous section, in terms of which the physical ones are expressed by the equations of the unperturbed motion corresponding to the Hamiltonian $\mathscr{H}_0$, the role of Hamiltonian is taken by the perturbation V, and since $\mathscr{E}' = \mathscr{E}$, because in the unperturbed motion energy is constant, one has $\mathrm{d}\mathscr{E}/\mathrm{d}z = \mathrm{d}\mathscr{E}'/\mathrm{d}z = -\partial V/\partial t'$. Equation (5.40) may then be rewritten as

$$W_{we} = -\int_0^L \frac{\partial V}{\partial t'}\,\mathrm{d}z. \tag{5.41}$$

Let us now apply eqn (5.36) with $F = \partial V/\partial t'$. Equation (5.41) becomes, up to first order,

$$\begin{aligned} W_{we} \simeq -\int_0^L \frac{\partial V_0(z)}{\partial t_i}\,\mathrm{d}z - \int_0^L \mathrm{d}z \int_0^z \Bigg[& \frac{\partial^2 V_0(z)}{\partial t_i\,\partial \boldsymbol{P}_{\perp i}}\cdot\frac{\partial V_0(\zeta)}{\partial \boldsymbol{x}_i} \\ & - \frac{\partial^2 V_0(z)}{\partial t_i\,\partial \boldsymbol{x}_i}\cdot\frac{\partial V_0(\zeta)}{\partial \boldsymbol{P}_{\perp i}} - \frac{\partial^2 V_0(z)}{\partial t_i\,\partial \mathscr{E}_i}\frac{\partial V_0(\zeta)}{\partial t_i} + \frac{\partial^2 V_0(z)}{\partial t_i^2}\frac{\partial V_0(\zeta)}{\partial \mathscr{E}_i}\Bigg]\mathrm{d}\zeta \end{aligned} \tag{5.42}$$

where, just as in eqn (5.36), $\partial V/\partial t'$ evaluated at $t' = t_i$ has been written simply as $\partial V_0/\partial t_i$, i.e. as V evaluated along the unperturbed motion corresponding to the Hamiltonian $\mathscr{H}_0$ and differentiated with respect to initial time t_i.

Equation (5.42) formally solves the problem of calculating the work W_{we} up to first order. However, what one really needs in order to calculate the gain of a free-electron laser is, as in Section 5.4, the average of W_{we} over a uniform initial-time distribution. We shall calculate this average as

$$\langle W_{we}\rangle = \lim_{T\to\infty} \frac{1}{2T}\int_{-T}^{T} W_{we}(t_i)\,\mathrm{d}t_i. \tag{5.43}$$

A well known property of this kind of average is that if W_{we} is the derivative with respect to t_i of a bounded function, i.e. $W_{we} = \mathrm{d}F/\mathrm{d}t_i$ with $|F(t_i)| < M$ for some fixed number M and $-\infty < t_i < \infty$, its average is zero. For, in this case,

$$\left|\int_{-T}^{T} W_{we}(t_i)\,\mathrm{d}t_i\right| = |F(T) - F(-T)| < 2M \tag{5.44}$$

and $|\langle W_{we}\rangle| \leqslant \lim_{T\to\infty} M/T = 0$.

In order to simplify the average, the work W_{we} of eqn (5.42) may be rewritten as

$$\begin{aligned} & -\frac{\partial}{\partial t_i}\int_0^L V_0\,\mathrm{d}z + \frac{\partial}{\partial t_i}\int_0^L \mathrm{d}z\int_0^z \left[\frac{\partial V_0(z)}{\partial \boldsymbol{x}_i}\cdot\frac{\partial V_0(\zeta)}{\partial \boldsymbol{P}_{\perp i}} - \frac{\partial V_0(z)}{\partial t_i}\frac{\partial V_0(\zeta)}{\partial \mathscr{E}_i}\right]\mathrm{d}\zeta \\ & - \int_0^L \mathrm{d}z\int_0^z \Bigg[\frac{\partial^2 V_0(z)}{\partial t_i\,\partial \boldsymbol{P}_{\perp i}}\cdot\frac{\partial V_0(z)}{\partial \boldsymbol{x}_i}\cdot\frac{\partial V_0(z)}{\partial \boldsymbol{x}_i}\frac{\partial^2 V_0(\zeta)}{\partial t_i\,\partial \boldsymbol{P}_{\perp i}} \\ & - \frac{\partial^2 V_0(z)}{\partial t_i\,\partial \mathscr{E}_i}\frac{\partial V_0(\zeta)}{\partial t_i} - \frac{\partial V_0(z)}{\partial t_i}\frac{\partial^2 V_0(\zeta)}{\partial t_i\,\partial \mathscr{E}_i}\Bigg]\mathrm{d}\zeta. \end{aligned} \tag{5.45}$$

As the first two terms of this expression are derivatives with respect to t_i of bounded quantities, $\langle W_{we}\rangle$ equals the average of the last term alone. Using the expressions (5.35) of $P'_{\perp 1}$ and $\mathscr{E}'_1$, where the prime on $\mathscr{E}$ may be omitted because, being $\mathscr{E}$ constant in the unperturbed motion, $\mathscr{E}' = \mathscr{E}$, one finally obtains

$$\langle W_{we}\rangle = \left\langle \int_0^L \left(\frac{\partial^2 \mathscr{E}_1}{\partial \boldsymbol{P}_{\perp i}\,\partial z} \cdot \boldsymbol{P}'_{\perp 1} + \frac{\partial \boldsymbol{P}'_{\perp 1}}{\partial z} \cdot \frac{\partial \mathscr{E}_1}{\partial \boldsymbol{P}_{\perp i}} + \frac{\partial^2 \mathscr{E}_1}{\partial \mathscr{E}_i\,\partial z} \mathscr{E}_1 + \frac{\partial \mathscr{E}_1}{\partial z} \frac{\partial \mathscr{E}_1}{\partial \mathscr{E}_i} \right) \mathrm{d}z \right\rangle, \tag{5.46}$$

i.e.

$$\langle W_{we}\rangle = \left\langle \frac{\partial \mathscr{E}_1}{\partial \boldsymbol{P}_{\perp i}} \cdot \boldsymbol{P}'_{\perp 1} + \frac{\partial \mathscr{E}_1}{\partial \mathscr{E}_i} \mathscr{E}_1 \right\rangle, \tag{5.47}$$

where

$$\mathscr{E}_1 = -\int_0^L \frac{\partial V_0}{\partial t_i}\,\mathrm{d}z, \qquad \boldsymbol{P}'_{\perp 1} = \int_0^L \frac{\partial V_0}{\partial \boldsymbol{x}_i}\,\mathrm{d}z. \tag{5.48}$$

Equation (5.47) is the result we were seeking, the generalized Madey theorem. In the one-dimensional case, when the undulator and wave fields are independent of $\boldsymbol{x}$, $\boldsymbol{P}'_{\perp 1} = 0$ and eqn (5.47) becomes even simpler:

$$\langle W_{we}\rangle = \left\langle \mathscr{E}_1 \frac{\partial \mathscr{E}_1}{\mathscr{E}_i} \right\rangle = \frac{1}{2} \frac{\partial}{\partial \mathscr{E}_i} \langle \mathscr{E}_1^2 \rangle. \tag{5.49}$$

Equation (5.49) is the original form of Madey's theorem (Madey 1979). Notice that the first-order work W_{we} equals the second-order energy perturbation $\mathscr{E}_2$, as may be verified by carrying on the expansion (5.30) to higher orders. Therefore, an alternative statement of Madey's theorem is that the average second-order perturbation of the electron energy equals half the derivative with respect to initial energy of the average squared first-order perturbation. By the same token, the zero-order work W_{we}, calculated along the unperturbed electron motion, equals the first-order energy perturbation $\mathscr{E}_1$. This equivalence will be useful for calculating $\mathscr{E}_1$ in the next section.

Since $\langle W_{we}\rangle$ is proportional to gain and $\langle \mathscr{E}_1^2\rangle$ is proportional in the 1-D case to spontaneous emission, as will appear more clearly in the next section, Madey's theorem may also be stated by saying that free-electron laser gain is proportional to the derivative with respect to electron energy of spontaneous emission. As spontaneous emission depends on electron energy mainly through the detuning parameter, it is also approximately true that gain is proportional to the derivative of spontaneous emission with respect to detuning parameter or to wavenumber, as observed at the beginning of this section.

In order to proceed from eqns (5.47)–(5.49) to the small-signal gain of a free-electron laser one need only multiply the average work $\langle W_{we} \rangle$ by the number of electrons entering the undulator per unit time and transverse area, integrating over the beam cross-section if the beam is finite, and divide by the input wave power. This will be done in the next sections. We only want to stress here that the validity of eqns (5.47)–(5.49) is not limited by any other assumption than those of small signal and small gain. In particular, neither a particular shape of the undulator and/or wave field nor a particular form of the Hamiltonian has to be assumed, even though we shall always consider a sinusoidal undulator and use the ultrarelativistic Hamiltonian.

A further generalization of Madey's theorem exists which allows the calculation of the complete transition matrix of a three-dimensional free-electron laser. For this generalization the interested reader is referred to Luchini and Solimeno (1985). The connections existing between Madey's theorem and similar theorems holding in the theory of stochastic processes and in statistical mechanics have been discussed by Krinsky, Wang, and Luchini (1982) and Krinsky (1985).

5.8 Gain of a transversely infinite free-electron laser for input plane waves of any direction and polarization

In this section we want to particularize the previous results to the case of an electromagnetic field composed of a sinusoidal undulator magnetic field, the dimensionless vector potential of which is given by (3.21) as

$$\boldsymbol{a}_{u\perp} = \mathrm{Re}(\boldsymbol{A}_u \mathrm{e}^{\mathrm{i}k_u z}), \qquad \boldsymbol{a}_{uz} = 0, \tag{5.50}$$

and a plane electromagnetic wave, represented by the potentials

$$\boldsymbol{a}_{w\perp} = \mathrm{Re}[\boldsymbol{A}_w \mathrm{e}^{\mathrm{i}(kt - k_z z - \boldsymbol{k}_\perp \cdot \boldsymbol{x})}], \qquad a_{wz} = -\boldsymbol{k}_\perp \cdot \boldsymbol{a}_{w\perp}/k_z, \qquad \varphi_w = 0. \tag{5.51}$$

$\boldsymbol{A}_u$ and $\boldsymbol{A}_w$ are complex transverse-plane vectors of arbitrary polarization. The direction of the wave will be assumed to be close to that of the z-axis, in the sense that $|k_\perp| \ll k$, as it will result that only in this case is gain non-negligible. Therefore, the paraxial approximation of eqn (4.28) may be used to express k_z.

In order to apply Madey's theorem we must first calculate the first-order perturbations $\mathscr{E}_1$ and $\boldsymbol{P}'_{\perp 1}$ according to eqn (5.48). Inserting the potentials of eqns (5.50) and (5.51) into the perturbation Hamiltonian V of eqn (5.39), one may observe that the derivatives with respect to t and $\boldsymbol{x}$ of V, like those of $\boldsymbol{a}_w$ itself, are linearly related. Since $\partial V/\partial \boldsymbol{x}_i = \partial V/\partial \boldsymbol{x} = -(\boldsymbol{k}_\perp/k)\partial V/\partial t_i$, it follows from eqns (5.48) that $\boldsymbol{P}_{\perp 1} = (\boldsymbol{k}_\perp/k)\mathscr{E}_1$. Therefore Madey's theorem (5.47) may be written in this case as

$$\langle W_{we}\rangle = \left\langle \mathscr{E}_1 \frac{\partial \mathscr{E}_1}{\partial \mathscr{E}_i} + \mathscr{E}_1 \frac{\boldsymbol{k}_\perp}{k} \cdot \frac{\partial \mathscr{E}_1}{\partial \boldsymbol{P}_{\perp i}} \right\rangle = \tfrac{1}{2}\left(\frac{\partial \langle \mathscr{E}_1^2\rangle}{\partial \mathscr{E}_i} + \frac{\boldsymbol{k}_\perp}{k} \cdot \frac{\partial \langle \mathscr{E}_1^2\rangle}{\partial \boldsymbol{P}_{\perp i}} \right). \tag{5.52}$$

$\mathscr{E}_1$ itself may be calculated as the work done along the unperturbed electron trajectory by the electric field corresponding to eqn (5.51), i.e.

$$\boldsymbol{e}_w = \mathrm{Re}[\boldsymbol{E}_w \mathrm{e}^{\mathrm{i}(kt - k_z z - \boldsymbol{k}_\perp \cdot \boldsymbol{x})}], \tag{5.53}$$

with $\boldsymbol{E}_w = ikmc^2 \boldsymbol{A}_w/e$. Thus

$$\mathscr{E}_1 = \mathrm{Re} \int_{-\infty}^{\infty} e \frac{\mathrm{d}\boldsymbol{r}_0}{\mathrm{d}z} \cdot \boldsymbol{E}_w \mathrm{e}^{\mathrm{i}(kt_0 - k_z z - \boldsymbol{k}_\perp \cdot \boldsymbol{x}_0)}\, \mathrm{d}z, \tag{5.54}$$

where $\boldsymbol{r}_0$ is the vector of components $\boldsymbol{x}_0(z)$ and z.

We have already met a similar integral in Chapter 4, namely in the far-field radiation formula (4.19).[4] In fact, there is a reciprocity relation between the far field generated by a moving particle (or, for that matter, by any current distribution) and the work done by a plane wave on it. This relation is

$$\mathrm{Re}\left(\frac{4\pi r}{-\mathrm{i}k\zeta} \mathrm{e}^{\mathrm{i}kr} \boldsymbol{E}_{rff}(k, \hat{\boldsymbol{k}}r) \cdot \boldsymbol{E}_w^* \right) = \int_{-\infty}^{\infty} e \frac{\mathrm{d}\boldsymbol{r}}{\mathrm{d}z} \cdot \boldsymbol{e}_w \, \mathrm{d}z = \mathscr{E}_1, \tag{5.55}$$

where $\boldsymbol{E}_{rff}(k, \hat{\boldsymbol{k}}r)$ denotes the far field radiated at the same wavenumber k as the impinging plane wave (5.53) and at the position $\hat{\boldsymbol{k}}r$, $\hat{\boldsymbol{k}}$ being a unit vector in the direction of propagation of the wave and r an arbitrary distance. Equation (5.55) may be shown to be a particular case of the general reciprocity relation (4.58), with $\boldsymbol{E}_{rff}$ as the radiated field $\boldsymbol{E}_r$ and $\boldsymbol{E}_w$ as the free field $\boldsymbol{E}_f$, by evaluating the integral on the l.h.s. of eqn (4.58) in the limit for $r \to \infty$ with the aid of the stationary-phase method, but it is actually easier to verify it directly by comparison of eqn (4.19) with eqn (5.54) (eqn (5.53) may also be written as $\boldsymbol{e}_w = \mathrm{Re}[\boldsymbol{E}_w^* \mathrm{e}^{-\mathrm{i}(kt - k_z z - \boldsymbol{k}_\perp \cdot \boldsymbol{x})}]$).

Having connected the first-order energy perturbation $\mathscr{E}_1$ with the far field of spontaneous undulator radiation, we may now use the expression for $\boldsymbol{E}_{rff}$ found in the previous chapter. In particular, putting the expression (4.38) for $\boldsymbol{E}_{rff}$ into eqn (5.55), squaring, and averaging over t_i gives

$$\langle \mathscr{E}_1^2 \rangle = \frac{e^2 L^2}{8\gamma_i^2} \sum_{n=0}^{\infty} |\boldsymbol{A}_{ueff,n}(\boldsymbol{\vartheta}_{we}) \cdot \boldsymbol{E}_w^*|^2 \left(\frac{\sin \nu_n}{\nu_n} \right)^2, \tag{5.56}$$

where the detuning parameter ν_n and the effective undulator vector potential $\boldsymbol{A}_{ueff,n}$ are those which were introduced and discussed in Sections 4.4 and 4.5. The vector $\boldsymbol{\vartheta}_{we}$ is defined by (4.32) as

$$\boldsymbol{\vartheta}_{we} = \boldsymbol{k}_\perp/k - \boldsymbol{P}_{\perp i}/\gamma_i, \tag{5.57}$$

and its components are the angles formed by the wave and electron directions of motion in the x_1, z- and x_2, z-planes.

As appears immediately, $\langle \mathscr{E}_1^2 \rangle$ of eqn (5.56) is proportional to the spectral power density of spontaneous radiation of eqn (4.46), and therefore has been subjected to the same approximations, i.e. neglecting cross-products of terms with different indices and all terms with negative n.

Since $\partial/\partial \mathscr{E}_i = (mc^2)^{-1}\partial/\partial\gamma_i$, eqn (5.52) now gives

$$\langle W_{we} \rangle = \frac{e^2L^2}{16mc^2} \sum_{n=0}^{\infty} \left(\frac{\partial}{\partial \gamma_i} + \frac{\boldsymbol{k}_\perp}{k} \cdot \frac{\partial}{\partial \boldsymbol{P}_{\perp i}} \right) \left[|\boldsymbol{A}^*_{ueff,n}(\vartheta_{we}) \cdot \boldsymbol{E}_w|^2 \frac{\sin^2 \nu_n}{\gamma_i^2 \nu_n^2} \right], \tag{5.58}$$

and multiplying $\langle W_{we} \rangle$ by the number of electrons per unit time and transverse area $n_e = j_b/e$, and dividing by the input power $E_w^2/2\zeta$, yields the gain, according to eqns (5.3) and (5.6), as

$$g = -\frac{2\zeta j_b \langle W_{we} \rangle}{eE_w^2}$$

$$= -\frac{\zeta e j_b L^2}{8mc^2} \sum_{n=0}^{\infty} \left(\frac{\partial}{\partial \gamma_i} + \frac{\boldsymbol{k}_\perp}{k} \cdot \frac{\partial}{\partial \boldsymbol{P}_{\perp i}} \right) \left[|\boldsymbol{A}_{ueff,n}(\vartheta_{we}) \cdot \hat{\boldsymbol{E}}^*_w|^2 \frac{\sin^2 \nu_n}{\gamma_i^2 \nu_n^2} \right], \tag{5.59}$$

where $\hat{\boldsymbol{E}}_w$ denotes the unit vector $\boldsymbol{E}_w/E_w$, which represents the polarization of the input wave.

Equation (5.59) may be further simplified, when the number N of periods of the undulator is large, by introducing approximations similar to those already adopted in the description of spontaneous radiation. In particular, it appears that amplification, just as spontaneous emission, mainly takes place at harmonic wavenumbers close to integral multiples of the resonance wavenumber, $k_r = 2\gamma_i^2 k_u/M^2$. Since the terms of the sum appearing in eqn (5.59) are sensibly different from zero in disjoint ranges of wavenumber k, it is permissible to neglect all but the term corresponding to the harmonic that falls closest to the input wavenumber. In addition, in evaluating the derivatives with respect to γ_i and $\boldsymbol{P}_{\perp i}$ one may neglect the explicit dependence on these variables that appears from eqn (5.59) with respect to the much stronger dependence that is created through the detuning parameter ν_n. The error connected with both these approximations is of order $1/N$. Since, according to eqn (4.39),

$$\left(\frac{\partial}{\partial \gamma_i} + \frac{\boldsymbol{k}_\perp}{k} \cdot \frac{\partial}{\partial \boldsymbol{P}_{\perp i}} \right) \nu_n = -\frac{M^2 kL}{2\gamma_i^3} - \frac{\vartheta_{we}^2 kL}{2\gamma_i}, \tag{5.60}$$

eqn (5.59) may be reduced to

$$g \simeq \frac{\zeta e j_b k L^3 M^2}{16mc^2\gamma_i^5} \left(1 + \frac{\gamma_i^2 \vartheta_{we}^2}{M^2} \right) \frac{|\boldsymbol{A}_{ueff,n}(\vartheta_{we}) \cdot \hat{\boldsymbol{E}}^*_w|^2}{A_u^2} \frac{\partial}{\partial \nu_n} \frac{\sin^2 \nu_n}{\nu_n^2}. \tag{5.61}$$

The derivative with respect to v of $(\sin^2 v)/v^2$ is exactly the factor $(\cos 2v - 1 + v \sin 2v)/v^3$ appearing in eqn (5.25), so that eqn (5.61) may be expressed in terms of the basic FEL gain formula as

$$g = \left(1 + \frac{\gamma_i^2 \vartheta_{we}^2}{M^2}\right) \frac{|\boldsymbol{A}_{ueff,n}(\vartheta_{we}) \cdot \hat{\boldsymbol{E}}_w^*|^2}{A_u^2} g_0. \tag{5.62}$$

Notice that the value of v to be inserted into the expression (5.25) of g_0 must now be calculated by putting the complete expression (4.42) of k_r into eqn (5.21), and not the expression (5.22) which is only valid on axis. When $\vartheta_{we} = 0$ and $n = 1$, and both $\boldsymbol{A}_u$ and $\boldsymbol{E}_w$ are circularly polarized so that $\boldsymbol{A}_{ueff,1} = \boldsymbol{A}_u$ and $|\boldsymbol{A}_{ueff,1} \cdot \hat{\boldsymbol{E}}_w^*| = A_u$, gain reduces to just g_0, as it should. In the general case, eqn (5.62) gives the correction to g_0 resulting from different directions and/or polarizations.

With regard to the polarization of the input wave, gain is proportional to the square of the scalar product $\boldsymbol{A}_{ueff,n} \cdot \hat{\boldsymbol{E}}_w^*$, and is maximum when the polarization of $\boldsymbol{E}_w$ coincides with that of the effective undulator vector potential $\boldsymbol{A}_{ueff,n}$.

With regard to the dependence on direction, eqn (5.62) may give the impression that gain increases with ϑ_{we} owing to the factor $(1 + \gamma_i^2 \vartheta_{we}^2/M^2)$, but this impression is wrong because $\boldsymbol{A}_{ueff,n}$ also depends on ϑ_{we}. Quite often, anyway, the dependence of gain on direction is in practice negligible. In fact, in studying the magnitude of $\boldsymbol{A}_{ueff,n}$ in Section 4.5, it was observed that if

$$\vartheta_{we} \lesssim \Delta\vartheta_{we} = (nN)^{-1/2} M/\gamma_i, \tag{5.63}$$

$\Delta\vartheta_{we}$ being the value of ϑ_{we} at which the variation of v_n induced by ϑ_{we} reaches π, $\boldsymbol{A}_{ueff,n}(\vartheta_{we})$ may be replaced by $\boldsymbol{A}_{ueff,n}(0)$ within an error $O(1/N)$. Equation (5.63) turns out to be a meaningful limitation in the case of stimulated as well as spontaneous emission, so that one is most often interested in conditions such that eqn (5.63) is verified. Under these conditions the term $\gamma_i^2 \vartheta_{we}^2/M^2$ in eqn (5.62) is also $O(1/N)$, and may be neglected. The gain formula for wave and undulator fields of any polarization and direction within the limitation of eqn (5.63) then becomes simply

$$g = \frac{|\boldsymbol{A}_{ueff,n}(0) \cdot \hat{\boldsymbol{E}}_w^*|^2}{A_u^2} g_0. \tag{5.64}$$

When the polarization of the undulator is other than circular, amplification is possible, just as is spontaneous radiation, on harmonics of the fundamental resonance wavenumber. Amplification of harmonics may be calculated from eqns (5.62)–(5.64), provided that v_n is read in the place of v in the basic gain formula (5.23), and is governed by the same effective vector potential $\boldsymbol{A}_{ueff,n}$ that determines the intensity of spontaneous emission. In particular, we recall from Section 4.5 that when the polarization is linear and the dimensionless undulator vector potential is large $\boldsymbol{A}_{ueff,n}$ decays with the harmonic number n only as $n^{-1/3}$, so that in principle, amplification of fairly high harmonics is possible.

Notice that for operation on higher harmonics the definition (5.25) of the gain length is correct but (5.26) is not, because k does not equal k_r but rather nk_r.

5.9 Gain of a finite-cross-section free-electron laser

The free-electron laser has until now been assumed to be fed by an infinite-cross-section electron beam of a given current density, and to be amplifying infinite-cross-section plane waves of a given power density. In this section we shall investigate how the gain formula is to be modified in order to take into account the finite cross-section of a real electron beam. Of course, since the available current and power are now finite, plane waves cannot be amplified, if not infinitesimally, by a transversely finite FEL, and a finite-cross-section input wave has to be considered.

In this section we shall outline the general procedure that may be followed practically in order to apply Madey's theorem to a finite-cross-section FEL, and to calculate the gain corresponding to any given input electron- and laser-beam configurations. In the next two sections we shall consider the two limiting cases of very wide and very narrow electron beams.

The gain of an FEL working in the small-signal small-gain regime may be calculated, according to eqns (5.3) and (5.6), as the ratio between the negative work, $-W_{wr}$, done per unit time by the input wave upon the radiator current and the power, P_{in}, flowing in the input wave. In order to obtain from Madey's theorem the work done by the input wave upon the electrons in the undulator, we need only put the appropriate expressions of the undulator and input wave fields in the expression (5.39) of V and evaluate the integrals of eqns (5.48) along the unperturbed motion. It may save some time to first evaluate the auxiliary integral

$$U = \int_0^L \left\{ \Phi_w^* - A_{wz}^* + \frac{1}{\gamma_i} [\boldsymbol{P}_{\perp i} - \text{Re}(\boldsymbol{A}_u e^{ik_u z})] \cdot \boldsymbol{A}_{w\perp}^* \right\} e^{-ikt_0(z)}\, dz, \quad (5.65)$$

defined so that

$$\int_0^L V_0\, dz = mc^2\, \text{Re}(U). \quad (5.66)$$

By exchanging the differentiation and integration signs in eqns (5.48) one easily finds $\mathscr{E}_1 = -mc^2\, \text{Re}(\partial U/\partial t_i)$ and $\boldsymbol{P}'_{\perp 1} = mc^2\, \text{Re}(\partial U/\partial \boldsymbol{x}_i)$. In particular, since t_i appears only inside the expression of t_0, and additively there, $\partial U/\partial t_i = -ikU$, and thus

$$\langle \mathscr{E}_1^2 \rangle = (kmc^2)^2 |U|^2/2, \quad (5.67)$$

Madey's theorem (5.47) may then be recast in terms of U as

$$\langle W_{we}\rangle = \frac{kmc^2}{2}\left[\frac{k}{2}\frac{\partial |U|^2}{\partial \gamma_i} + \mathrm{Im}\left(\frac{\partial U^*}{\partial \boldsymbol{P}_{\perp i}}\cdot\frac{\partial U}{\partial \boldsymbol{x}_i}\right)\right]. \tag{5.68}$$

It must be remarked that the second term (the one deriving from the variation of transverse momentum) of eqn (5.68) is actually negligible under most practical conditions, as may be understood from the following argument. In Section 5.8 we have shown that the transverse terms in the generalized Madey theorem give only a contribution of relative order $1/N$ to the gain (diagonal element of the transition matrix) for all plane waves travelling at an angle not larger than the angle $\Delta\vartheta_{we}$ for which the value of the detuning parameter v (and with it the gain) is sensibly altered. It may be shown that the same is true for the off-diagonal transition matrix coefficients that couple plane waves of different orientations to one another. Since the gain corresponding to an arbitrary input wave may always be calculated by applying the transition matrix to its plane-wave expansion, it may be expected that gain will be maximum for those input-wave configurations the plane-wave spectrum of which is mainly constituted by waves propagating at an angle less than $\Delta\vartheta_{we}$. For such an input wave the contribution of transverse terms to Madey's theorem is $O(1/N)$. Therefore the 1-D form of Madey's theorem may very often be applied to 3-D systems.

Equation (5.68) gives the average work done by the input wave on a single electron of given initial position. The total work per unit time done by the input wave upon the electron beam may be calculated by integrating eqn (5.68) over the cross-section of the beam, of density $n_e = j_b/e$ electrons per unit time and transverse area. Thus:

$$W_{wr} = \int \frac{j_b}{e}\langle W_{we}\rangle\, \mathrm{d}^2\boldsymbol{x}_i = \int \frac{kmc^2 j_b}{2e}\left[\frac{k}{2}\frac{\partial |U|^2}{\partial \gamma_i} + \mathrm{Im}\left(\frac{\partial U}{\partial \boldsymbol{P}_{\perp i}}\cdot\frac{\partial U}{\partial \boldsymbol{x}_i}\right)\right]\mathrm{d}^2\boldsymbol{x}_i. \tag{5.69}$$

With the aid of eqns (5.3) and (5.6) it is then immediately possible to calculate gain, as $g = -W_{wr}/P_{\mathrm{in}}$.

According to the above procedure, the calculation of FEL gain for given input wave and electron beams of finite cross-section may be reduced to the calculation of the two integrals appearing in eqns (5.65) and (5.69). In general, these two integrals must be calculated numerically.

5.10 Wide-beam gain formula and filling factor

A quite common method to incorporate the effects of finite cross-section in the calculation of gain of a conventional laser is to multiply the one-dimensional gain formula by a 'filling factor', roughly defined as the ratio of volumes of the region of overlap between the optical resonator mode and the active medium and of the resonator mode itself.

By analogy one is led to expect that the gain of a finite-cross-section free-electron laser may be written, at least approximately, as

$$g = FF \times g_0, \tag{5.70}$$

where g_0 is the basic one-dimensional FEL gain of eqn (5.23), evaluated for the current density j_{b0} existing at the e-beam centre, and FF is a filling factor expressed by a ratio of volumes occupied by the electron and laser beams. We shall see, however, that a formula such as (5.70) may be adopted for an FEL only under rather heavy restrictions.

For eqn (5.70) to be valid it is necessary that different regions within the active medium contribute independently to the gain. In an FEL, in order to be allowed to separate the contributions to the total gain deriving from different portions of the electron beam cross-section one must assume that:

(a) the amplitude Δx_u of the electron wiggling motion (given by eqn (3.29)) is much smaller than the width, w, of the electromagnetic wave; and

(b) the angular spread of electrons in the input e-beam is sufficiently small for the drift accumulated over the whole undulator length to be negligible compared to w.

When these conditions are verified it is possible to neglect the variation of $\boldsymbol{x}$ with z in the integral of eqn (5.65) and simply replace $\boldsymbol{A}_w[\boldsymbol{x}(z), z]$ by $\boldsymbol{A}_w(\boldsymbol{x}_i, z)$. This is, however, not yet sufficient to take $\boldsymbol{A}_w$ out of the integral. One has to add the further condition that

(c) the amplitude and direction of the laser electromagnetic wave does not vary significantly over the undulator length, so that it is permissible to write, approximately, $\boldsymbol{A}_{w\perp}(\boldsymbol{x}, z) \simeq \tilde{\boldsymbol{A}}_{w\perp}(\boldsymbol{x})\mathrm{e}^{-ikz}$, with $\tilde{\boldsymbol{A}}_{w\perp}$ independent of z.

Under this additional assumption, $\tilde{\boldsymbol{A}}_{w\perp}(\boldsymbol{x}_i)$ may be taken out of the integral. Moreover, as a side-effect of (b) and (c), it is possible to ignore the contributions of Φ_w, A_{wz}, and transverse momentum $\boldsymbol{P}_\perp$ to eqns (5.65) and (5.68), since these contributions are only significant for waves and electrons moving at large angles (see comment below eqn (5.68)), and the presence of such waves and electrons is excluded by conditions (c) and (b), respectively. The quantity U in eqn (5.65) and the time-averaged work $\langle W_{we}\rangle$ of eqn (5.68) then become the same as those corresponding to a plane wave of amplitude $\tilde{\boldsymbol{A}}_{w\perp}(\boldsymbol{x}_i)$ travelling on axis. W_{we} may therefore be obtained from eqn (5.58), valid for plane waves, by letting $\boldsymbol{k}_\perp = 0$. Thus

$$\langle W_{we}\rangle = \frac{e^2L^2}{16mc^2}\,|\hat{\boldsymbol{E}}_w(\boldsymbol{x}_i)\cdot \boldsymbol{A}_{ueff,n}(0)|^2\,\frac{\partial}{\partial\gamma_i}\frac{\sin^2 v_n}{\gamma_i^2 v_n^2} \tag{5.71}$$

(only one term in the sum has been kept).

$\langle W_{we} \rangle$ must now be integrated over the e-beam cross-section, as in eqn (5.69), and then divided by the input power in order to yield the gain. Taking the factors independent of $\boldsymbol{x}_i$ out of the integral one has

$$g = -\frac{e^2 L^2}{16mc^2} \frac{\partial}{\partial \gamma_i} \frac{\sin^2 v_n}{\gamma_i^2 v_n^2}$$

$$\times \left[\int \frac{j_b}{e} |\tilde{\boldsymbol{E}}_w(\boldsymbol{x}_i) \cdot \boldsymbol{A}_{ueff,n}(0)|^2 \, \mathrm{d}^2\boldsymbol{x}_i\right] \Big/ \left[\int \frac{|\tilde{\boldsymbol{E}}_w(\boldsymbol{x}_i)|^2}{2\zeta} \, \mathrm{d}^2\boldsymbol{x}_i\right]. \qquad (5.72)$$

As is seen, eqn (5.72) may indeed be given the form of eqn (5.70), by just defining the filling factor as

$$FF = \left[\int j_b |\tilde{\boldsymbol{E}}_w(\boldsymbol{x}_i) \cdot \boldsymbol{A}_{ueff,n}(0)|^2 \, \mathrm{d}^2\boldsymbol{x}_i\right] \Big/ \left[j_{bo} A_{ueff,n}^2(0) \int \tilde{E}_w^2(\boldsymbol{x}_i) \, \mathrm{d}^2\boldsymbol{x}_i\right]. \qquad (5.73)$$

Equation (5.73) is the expression of the filling factor appropriate to free-electron lasers, and is often used in the design of these devices (where sometimes the concept of 'filling factor' is replaced by the equivalent one of an 'effective area').

However, as remarked at the beginning of this section, the assumptions under which eqn (5.70) may be used are far more restrictive than in connection with a conventional laser. Indeed, for a conventional laser, in which the individual radiators (atoms or molecules) occupy an infinitesimal volume and are not mutually coupled other than by the laser wave itself, the possibility of introducing a filling factor stems from the assumption of small gain alone. On the contrary, in order to extend the concept to the free-electron laser the assumptions (a), (b), and (c) above are essential, because the radiators (the electrons) move across macroscopic regions of space and create correlations among the fields radiated in different parts of the undulator. This may easily be seen from the fact that gain is proportional to L^3 rather than simply to L, as it should be if amplification of the input radiation took place independently in longitudinally separated regions of space, as in a conventional laser (in the small-gain regime).

Conditions (a) and (b) ensure that, at least in the transverse plane, different electrons travel in separate regions, so that a certain amount of separation is achieved, but it is absolutely impossible, as noticed above, to treat the emission as independent in regions aligned along a direction parallel to the beam axis, because these are necessarily traversed by the same electrons. That is why we were obliged to add condition (c), which ensures the longitudinal homogeneity of the laser beam, and also why the filling factor is defined in eqn (5.73) as a ratio of surface integrals over the cross-section rather than as a ratio of volume integrals.

The necessary condition of longitudinal uniformity of the laser beam is not very often verified in practice, and limits noticeably the applicability of the filling-factor concept to free-electron lasers.

5.11 Optimization of gain for a Gaussian beam

The reason why a wide and uniform laser beam is not of practical utility in an FEL is that a thin beam often allows a greater gain. To understand the effect of beam width on gain let us undertake the calculation of gain for a Gaussian input beam.

The Gaussian beam is an analytical solution of the parabolic wave equation which is very often assumed to represent a 'perfect' laser beam. There are two main reasons for this assumption: first, the Gaussian beam is probably the simplest analytical solution of the source-free wave equation representing a radiation beam of finite width and has, in a well defined sense, the smallest width allowed by diffraction; second, it is an acceptable representation of the fundamental oscillation mode of a stable open resonator (optical cavity formed by two mutually facing mirrors of suitable curvature) with low losses. Therefore it is of particular interest to know how a Gaussian beam is amplified by a free-electron laser, and which choice of Gaussian-beam parameters maximizes the gain.

Let us start out by recalling the main properties of the Gaussian beam. The parabolic (or paraxial) wave equation (4.26), valid whenever only waves propagating at small angles with respect to the z-axis are considered, has a solution corresponding to a δ-function current density, $\boldsymbol{J} = \tilde{\boldsymbol{J}}\delta(\boldsymbol{r} - \boldsymbol{r}')$, given by

$$\boldsymbol{A} = \tilde{\boldsymbol{A}}\mathrm{e}^{-\mathrm{i}kz} = \frac{\tilde{\boldsymbol{J}}}{4\pi\mu_0(z - z')} \exp -\mathrm{i}k\left[z - z' + \frac{(\boldsymbol{x} - \boldsymbol{x}')^2}{2(z - z')}\right] \tag{5.74}$$

(the corresponding electric field was given in eqn (4.23)). Equation (5.74) is also a solution of the source-free wave equation everywhere, except at the position of the source, $(\boldsymbol{x}', z')$, where it has a singularity, and remains a valid solution even when this position is replaced by a complex number. In particular, letting $z' = z_0 - \mathrm{i}z_R$ (and $\boldsymbol{x}' = 0$ for the sake of simplicity) gives the expression

$$\boldsymbol{A} = \frac{\boldsymbol{A}_{gb}}{(z - z_0)/z_R + \mathrm{i}} \exp -\mathrm{i}k\left[z + \frac{\boldsymbol{x}^2}{2(z - z_0 + \mathrm{i}z_R)}\right], \tag{5.75}$$

where $\boldsymbol{A}_{gb} = \tilde{\boldsymbol{J}}\mathrm{e}^{kz_R + \mathrm{i}kz_0}/4\pi\mu_0 z_R$ is a constant complex vector. Equation (5.75) has no singularity for any real value of the co-ordinates and therefore represents a source-free electromagnetic field in all space. Taking the modulus of both sides of eqn (5.75) gives

$$|\boldsymbol{A}| = \frac{|\boldsymbol{A}_{gb}|}{[1 + (z - z_0)^2/z_R^2]^{1/2}} \exp - \frac{(k/z_R)\boldsymbol{x}^2}{2[1 + (z - z_0)^2/z_R^2]}. \tag{5.76}$$

Because of the Gaussian dependence on transverse co-ordinates of its amplitude, the field represented by eqn (5.75) is significantly different from zero only in the proximity of the z-axis, and represents a beam of radiation propagating along this axis: hence the name Gaussian beam.

Let us now introduce some standard definitions. The Gaussian amplitude factor $\exp -(k/z_R)\boldsymbol{x}^2/2[1+(z-z_0)^2/z_R^2]$ may be rewritten as $\exp -\boldsymbol{x}^2/2w^2[1+(z-z_0)^2/z_R^2]$, thus defining w such that $w^2 = z_R/k = \lambda z_R/2\pi$, or also as $\exp -[\boldsymbol{x}/(z-z_0)]^2/2\vartheta_a^2[1+z_R^2/(z-z_0)^2]$, thus defining ϑ_a such that $\vartheta_a^2 = 1/kz_R$. The length w, giving the typical transverse scale of variation of the field amplitude at $z = 0$, is named the 'waist' of the Gaussian beam, while the length z_R, giving the typical longitudinal scale of variation, is named the Rayleigh range. The angle ϑ_a, giving the typical angular scale of variation of the field amplitude for $z \to \infty$, is the aperture of the Gaussian beam. These three quantities are related to one another, as $kz_R = \vartheta_a^{-2} = k^2w^2$. Notice that the Gaussian beam has been obtained as a solution of the paraxial wave equation and can only be assumed to represent an electromagnetic field when the propagation angle is small, i.e. when $\vartheta_a \ll 1$. Being $1/kw = w/z_R = \vartheta_a$, this condition implies that $z_R \gg w \gg \lambda$.

The total power transported by a Gaussian beam, that is the input power of a free-electron laser invested by a Gaussian beam, is (power in mks units, A_{gb} dimensionless)

$$P_{\text{in}} = \pi k z_R (mc^2/e)^2 A_{gb}^2/2\zeta. \tag{5.77}$$

We shall now calculate the amplification experienced by a Gaussian beam in an FEL under the assumption that conditions (a) and (b) of the previous section are verified but (c) is not; that is, that transverse electron motion (both wiggling and drift) is negligible on the scale of the Gaussian beam waist, but noticeable variations of the laser beam cross-section may take place along the undulator. This situation is very important because, as will soon be seen, gain is maximum for a value of the Rayleigh range somewhat smaller than the undulator length, and as the Rayleigh range gives the typical longitudinal scale of variation of the Gaussian beam, longitudinal variations are noticeable in this case.

Let us begin by evaluating the quantity U of eqn (5.65). We shall assume $\boldsymbol{P}_{\perp i} = 0$ and neglect the contributions of A_{wz} and Φ_w since they are associated with waves moving at a wide angle with respect to the z-axis and are negligible near the point of maximum gain. It may be inferred from symmetry considerations that maximum gain will occur when the Gaussian beam waist is located at the centre of the undulator; therefore we shall also set $z_0 = L/2$. Inserting eqn (5.75) into eqn (5.65), with $\Phi_w = A_{wz} = 0$ and $\boldsymbol{P}_{\perp i} = 0$, and setting $\boldsymbol{x} = \boldsymbol{x}_i$ because of assumptions (a) and (b), we obtain

$$U = -\int_0^L \frac{A_{gb}^* \mathrm{e}^{-\mathrm{i}k[t_0(z) - z - \boldsymbol{x}_i^2/2(z - L/2 - \mathrm{i}z_R)]}}{(z - L/2)/z_R - i} \cdot \frac{\mathrm{Re}(A_u \mathrm{e}^{\mathrm{i}k_u z})}{\gamma_i} \,\mathrm{d}z, \tag{5.78}$$

where $t_0(z)$ represents the unperturbed motion of the electron in the undulator and is given by eqn (3.24b). Proceeding similarly to what was done in the evaluation of eqn (4.33), we insert the Fourier series (4.35) (with $\vartheta_{we} = 0$) into eqn (5.78) and obtain

$$U = -\frac{1}{2\gamma_i} e^{-ikt_i} \sum_{n=-\infty}^{\infty} A^*_{gb} \cdot A_{ueff,n}(0)$$

$$\times \int_0^L \frac{\exp i[(nk_u - \kappa)z + kx_i^2/(z - L/2 - iz_R)]}{(z - L/2)/z_R - i} \mathrm{d}z, \quad (5.79)$$

where κ is the usual reduced wavenumber given by $\kappa = kM^2/2\gamma_i^2$. The coefficient $A_{ueff,n}$ is the by now familiar effective undulator vector potential defined by eqn (4.36), the expression of which in terms of Bessel functions was given in Section 4.5. As on previous occasions, only one term in the series, the one for which $nk_u \simeq \kappa$, need be retained.

According to the procedure outlined in Section 5.9, gain can now be obtained by inserting U into eqn (5.69), neglecting the contribution of transverse momentum, and dividing the result by the input power P_{in} of eqn (5.77). On so doing one obtains

$$g = \int -\frac{\partial |U|^2}{\partial \gamma_i} \frac{2j_b(\boldsymbol{x}_i)}{w^2 A_{gb}^2 I_A} \mathrm{d}^2\boldsymbol{x}_i, \quad (5.80)$$

where $w^2 = z_R/k$ is the squared waist of the Gaussian beam, and $I_A = 4\pi mc^2/\zeta e \simeq 17.03$ kA.

The integrals of eqns (5.79) and (5.80) were calculated numerically by Luchini, Prisco and Solimeno (1984) for a Gaussian e-beam current distribution $j_b(\boldsymbol{x}_i) = j_{b0} e^{-x_i^2/2\sigma^2}$. The plot of Fig. 5.2 gives the maximum gain of the Gaussian laser beam as a function of the electron beam width (represented by the e-beam Fresnel number, $F_e = k\sigma^2/4L$).

The two limiting cases of a very narrow and a very wide electron beam are worthy of attention:

1. For $F_e \ll 1$, gain is an absolute maximum and the corresponding optimal value of the Rayleigh range is $z_R \simeq 0.25L$. This is a sort of 'magic number' that the designers of FEL experiments built around accelerators characterized by a not very high current (in particular, storage rings) have struggled to achieve. The problem with it derives from the fact that the optical resonator must necessarily be longer than the undulator and, to avoid interference with other components of the accelerating machine, often two or three times longer. As will be seen in greater detail in Chapter 9, an optical resonator having a Rayleigh range much shorter than its length is very sensitive to misalignment, and a compromise must usually be sought. The gain corresponding to the ideal conditions of an infinitely thin electron beam (of current I_b) and $z_R = 0.25L$ is

$$g_{\max} = 0.93 \frac{I_b}{I_A} \frac{k^2 L^2 M^2 A_{ueff,n}^2}{4\gamma_i^5}. \quad (5.81)$$

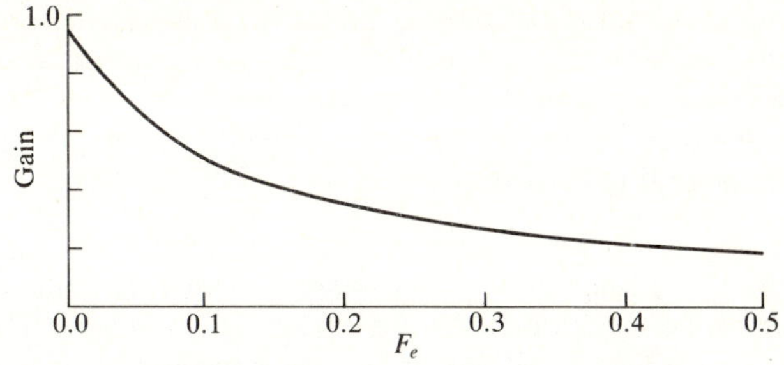

FIG. 5.2. Maximum gain of an FEL for a Gaussian laser beam vs. electron-beam Fresnel number F_e (from Luchini *et al.* 1984).

This is the maximum gain achievable for a given e-beam current. On replacing k by its resonance value, $2n\gamma_i^2 k_u/M^2$, eqn (5.81) may also be rewritten as

$$g_{\max} = 0.93 \frac{I_b}{I_A} \frac{(2\pi nN)^2 A_{ueff,n}^2}{\gamma_i M^2}. \tag{5.82}$$

2. For $F_e \gg 1$ the Rayleigh range z_R is large compared to L, the filling-factor technique is applicable, and the gain may be obtained from the formulae given in the last section. A more accurate result may be obtained by expanding the integrand in eqn (5.79) in a power series in $1/z_R$.

5.12 Effects of momentum and energy spread

The effects of momentum and energy spread of the electrons are easily incorporated into the theory of small-signal, small-gain FEL amplification because of the close connection existing between the gain and spontaneous radiation spectra. In particular, all that needs to be added to the general procedure of Section 5.9 for the calculation of gain is that the current density in eqn (5.69) must be replaced by e times the electron distribution function f in co-ordinate, momentum, and energy space (such that $f\,\mathrm{d}\boldsymbol{x}_i\,\mathrm{d}\boldsymbol{P}_{\perp i}\,\mathrm{d}\gamma_i$ is the number of electrons entering the undulator per unit time with a transverse position in the range $\mathrm{d}\boldsymbol{x}_i$, a transverse momentum in the range $\mathrm{d}\boldsymbol{P}_{\perp i}$, and a γ in the range $\mathrm{d}\gamma_i$), and the integral must be extended over this space rather than over transverse co-ordinates alone.

Since the averaging over transverse co-ordinates, transverse momentum, and energy may be performed indifferently before or after the application of Madey's theorem, and since the same averaging is involved in the calculation of spontaneous radiation, it remains true even in the presence of momentum and energy spread that the spectral gain is approximately proportional to the derivative with respect to wavenumber of the spontaneous radiation spectrum. Therefore all the qualitative conclusions of Section 4.10 concerning spontaneous undulator radiation may be transported to FEL amplification. In particular, the effects of momentum and energy spread may be compounded in a 'detuning-parameter spread', given by eqn (4.76), and the limitations that must be satisfied in order for gain to be negligibly affected are deducible from eqns (4.77)–(4.80):

$$\delta\gamma_i/\gamma_i \ll 1/2nN, \qquad \delta\vartheta_i^2 \ll 2\lambda/L. \tag{5.83}$$

Equations (5.83) must be satisfied together with the condition that the Fresnel number be small, which requires that $\sigma^2 \ll \lambda L$, in order to achieve the best operating conditions and obtain the maximum gain given in eqns (5.81) and (5.82). By combining the angular-spread and cross-section limitations it is possible to obtain the limitation on beam emittance, which is roughly given by the product of the two (a more precise definition is contained in footnote 1 of Chapter 9). In fact, rather than the angular spread or the cross-section separately, the parameter of concern for the free-electron laser designer is the beam emittance, because the electron optics of the accelerating machine that supplies the electron beam can, with relative ease, achieve any combination of angular spread and cross-section compatible with a given emittance. The emittance itself, on the other hand, is an optically invariant parameter, and for a given accelerator can hardly be reduced below certain limits. By multiplying term by term the two inequalities $\sigma \ll (\lambda L)^{1/2}$ and $\delta\vartheta_i \ll (\lambda/L)^{1/2}$ it is easy to obtain that

$$\varepsilon = \sigma\delta\vartheta_i \ll \lambda. \tag{5.84}$$

As is seen, the limitation on beam emittance is given exclusively by the output wavelength that is desired, independently of any of the other undulator and beam parameters. This has some important consequences for the possibility of scaling down the free-electron laser to lower and lower wavelengths. One might observe that λ (more precisely, $\lambda/2\pi$) equals the product $w\vartheta_a$ between the waist and the aperture of an optical Gaussian beam, which could be called the 'emittance' of the Gaussian beam. Equation (5.84) may then be read by saying that the emittance of the electron beam must not exceed the emittance of the laser beam to be amplified.

In order to modify the gain formulae given in the previous sections so as to take into account the effects of momentum and energy spread, one may spare the calculation of any additional integral, just as explained in Section 4.10 with reference to spontaneous radiation, when the resulting detuning distribution is Lorentzian, by giving the complex value $\bar{\nu} + i\delta\nu$ to the detuning parameter, with $\delta\nu$ given by eqn (4.76). The basic gain formula (5.23) then must be written as

$$\begin{aligned} g_0 &= N^3 A_u^2 \gamma_i^{-3} \frac{\pi^2 l_u^2 j_b}{I_A} \operatorname{Re}\left(\frac{\mathrm{d}}{\mathrm{d}\nu} \frac{1 + 2i\nu - e^{2i\nu}}{2\nu^2}\right) \\ &= N^3 A_u^2 \gamma_i^{-3} \frac{\pi^2 l_u^2 j_b}{I_A} \operatorname{Re}\left[\frac{(1 - i\nu)e^{2i\nu} - 1 - i\nu}{\nu^3}\right], \end{aligned} \tag{5.85}$$

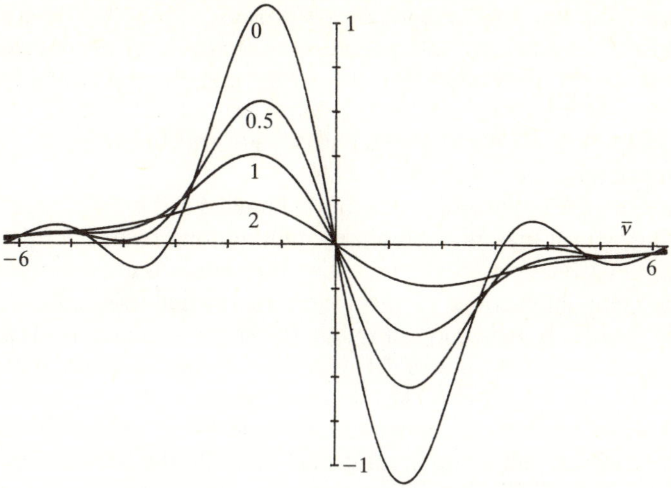

FIG. 5.3. Effect of energy spread upon the FEL gain spectrum. Each curve, representing gain versus detuning parameter $\bar{\nu} = \mathrm{Re}(\nu)$, is labelled with the corresponding value of the detuning-spread parameter $\delta\nu = \mathrm{Im}(\nu)$.

where $(1 + 2i\nu - e^{2i\nu})/\nu^2$ is—we recall—that analytic function g vanishing at infinity in the half-plane $\mathrm{Im}(\nu) > 0$ the real part of which for ν real is $\sin^2\nu/\nu^2$. A plot of the function $\mathrm{Re}\{[(1 - i\nu)e^{2i\nu} - 1 - i\nu]/\nu^3\}$ appearing in eqn (5.85) is shown in Fig. 5.3 for various values of the detuning spread $\delta\nu = \mathrm{Im}(\nu)$.

5.13 Quantum effects

Under the conditions considered in Sections 2.2 and 4.11, that is when the quantum-recoil red shift λ_C/γ is not negligible with respect to the undulator bandwidth λ/N (an operating regime which may be called the 'quantum regime'), quantum effects may be expected to influence the phenomenon of wave amplification by an undulator. Contrarily to what was found in the analysis of spontaneous radiation, however, the spectral distribution of small-signal gain *is* changed by quantum effects, as may be seen from the following description of the quantum version of Madey's theorem.

Several different quantum derivations of Madey's theorem have been given, starting with Madey's (1979) original one, which was fully based on quantum electrodynamics, but which finally considered only a parameter range in which the classical result may be recovered, thus failing to point out the existence of a quantum regime. It was later shown (Gea-Banacloche *et al.* 1984*a,b*; Renieri 1984; Gea-Banacloche and Scully 1985; see also the appendix of Grover and Pantell 1985) that a simplified quantum description of the FEL which still displays all the relevant effects but does not need the formalism of quantum electrodynamics is possible. Since a full description of the quantum approach would require more space than can be reasonably allotted within this section and, on the other hand, the operating conditions under which the quantum regime is entered are out of the reach of present experiments, we shall now only qualitatively illustrate these results, referring the reader to the cited papers for details.

Consider a uniform, one-dimensional, collimated and monoenergetic beam of electrons, which may be represented quantum-mechanically by a perfect plane wave of given momentum, travelling through an undulator. If radiative electromagnetic interactions are neglected, energy and momentum are conserved in such a system and, although the shape of the wave-function is altered inside the undulator, a perfect plane wave with the same momentum is again found after the exit. The electron beam is then in an unbound stationary state in which its momentum and energy are not modified overall, and it is possible, with some precautions, to apply the standard quantum-mechanical perturbative theory of radiation developed for bound states (atomic emission), and calculate in this way both spontaneous and stimulated radiation.

The transition-matrix elements for the spontaneous emission of a photon will be significantly different from zero at those wavenumbers and frequencies for which the corresponding electromagnetic wave matches best the product of the electron wave-functions inside the undulator before and after the emission. The phase-matching conditions associated with this requirement are simply the momentum and energy conservation equations (4.84) that were already used for computing the quantum recoil, suitably modified when the undulator field is strong ($A_u \gtrsim 1$) as to take into account the difference between the electron axial momentum inside and outside the undulator. The resulting spectral distribution of the transition probability is practically coincident, apart from quantum recoil, with the classically calculated spectral distribution of spontaneous emission under all practical conditions, as was discussed in Section 4.11.

For the purpose of calculating stimulated emission it is possible to show, through the introduction of suitable variables, that the standard quantum-mechanical relationship between the spontaneous and stimulated transition probabilities may be applied, namely that the probability of stimulated emission of a photon of frequency ω, accompanied by the transition of an electron from a state with energy $\mathscr{E}$ to one with energy $\mathscr{E} - \hbar\omega$, is n_{ph} times the probability of spontaneous emission of the same photon from an electron with the same energy $\mathscr{E}$, n_{ph} being the number of photons already present in the electromagnetic mode, and conversely that the probability of absorption of a photon of the same frequency with transition of an electron from a state with the same energy $\mathscr{E}$ to one with energy $\mathscr{E} + \hbar\omega$ is n_{ph} times the probability of spontaneous emission of a photon from an electron with energy $\mathscr{E} + \hbar\omega$. Therefore, if $W(\omega, \mathscr{E})$ denotes the probability of spontaneous emission of a photon with frequency ω from an electron with energy $\mathscr{E}$, proportional to the spectral power density of spontaneous radiation S_ω of eqn (4.46), the net number of photons added to a field where n_{ph} photons are already present is

$$n_{ph}[W(\omega, \mathscr{E}) - W(\omega, \mathscr{E} + \hbar\omega)], \tag{5.86}$$

the first term representing stimulated emission and the second absorption. Dividing (5.86) by the number of photons present in the input field, n_{ph}, gives the gain of the undulator as

$$g = W(\omega, \mathscr{E}) - W(\omega, \mathscr{E} + \hbar\omega), \tag{5.87}$$

which is the quantum-mechanical version of Madey's theorem. In fact, if the photon energy $\hbar\omega$ is small compared to the energy acceptance $\mathscr{E}/2N$ of spontaneous emission (which is yet another way of stating the classicality condition (2.15), as may be checked by substituting γmc^2 for $\mathscr{E}$ and $4\pi c\gamma^2/l_u$ for ω) the difference in eqn (5.87) may be replaced by a derivative and one has the classical result

$$g = -(\partial W/\partial \mathscr{E})\hbar\omega, \tag{5.88}$$

where Planck's constant $\hbar$ cancels out as soon as W is expressed in terms of S_ω. In the opposite case, in the quantum regime, the spectral distribution of gain is still antisymmetric, as may be seen from eqn (5.87), but not around the spontaneous-emission frequency. Instead, it has a positive peak at the frequency of spontaneous emission and a negative (absorption) peak at the, higher, frequency where radiation would be emitted by an electron beam of energy $\mathscr{E} + \hbar\omega$. This behaviour is illustrated in Fig. 5.4.

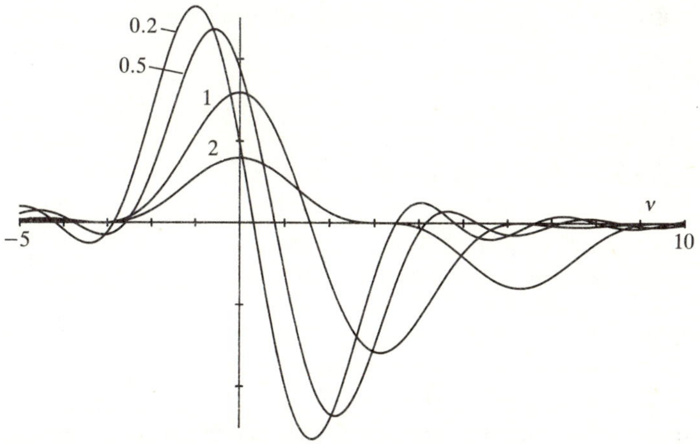

FIG. 5.4. Modification of the gain curve of an FEL in the quantum regime. Each curve is labelled with the corresponding value of the parameter $q = N\lambda_C/\gamma\lambda$ measuring the importance of quantum effects.

5.14 Optical klystron

According to the interpretation of FEL amplification given at the end of Section 5.5, the process may be ideally broken down into three stages: energy modulation of the electron beam by the input wave; transformation of the energy modulation into position modulation; and emission from the resulting space-modulated electron beam. In a purely sinusoidal undulator of the kind we have considered up to now these three processes take place simultaneously, but the above conceptual picture suggests that a possible way to improve FEL gain is to separate the three stages and optimize each. In particular, only the energy modulation and the re-emission processes require a sinusoidal undulator in order to couple the electron beam efficiently to the radiation, while the transformation of energy modulation into position modulation is not particularly enhanced by the undulator. This phenomenon, on the other hand, can be enhanced by a strong uniform magnetic field, just as in a mass spectrometer. The device obtained by interposing a strong uniform magnetic field between two short undulators is named optical klystron, because of the analogy of the above three-stage mechanism with the workings of the microwave klystron, and was first proposed by Vinokurov and Skrinsky (1977) (see also Vinokurov 1977). A typical optical-klystron magnetic-field profile is shown in Fig. 5.5. As may be noticed, the central region of constant magnetic field is preceded and followed by two negative

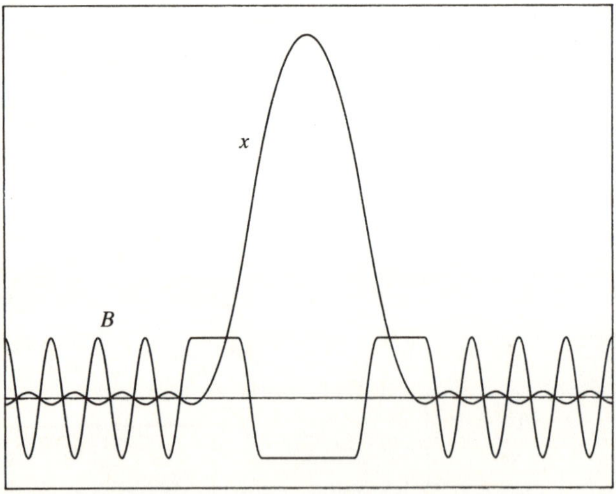

FIG. 5.5. Typical magnetic-field profile of an optical klystron and corresponding electron trajectory $x(z)$. Although their amplitudes have been drawn on the same plot, the field $\boldsymbol{B}$ and the electron transverse displacement vector $\boldsymbol{x}$ are mutually orthogonal in space. The scale of x has been amplified with respect to the abscissa z for representation purposes.

peaks of half the length, which serve to straighten the electron trajectory and make it enter the second undulator on the same line as it entered the first. The unperturbed electron trajectory resulting from this field configuration is also qualitatively shown in Fig. 5.5.

While the above qualitative account explains the reasons for the choice of the optical-klystron field configuration well, the quantitative calculation of its gain is most easily carried out using Madey's theorem. In order to apply Madey's theorem we must first consider the spontaneous radiation of the optical klystron. Let us start by calculating the unperturbed electron motion.

If we assume that the first part of the optical klystron, say $0 < z < z_1$, is a perfect sinusoidal undulator, the motion there is still given by eqns (3.24). Let us call t_1, $\boldsymbol{x}_1$ the time and position of an electron's arrival at $z = z_1$. The central part of the undulator, say $z_1 < z < z_2$, is where a flat field configuration exists. As its purpose is to disperse the electrons with different energies, this part is called the 'dispersive section'. If we assume that the vector potential is a function of z only, even if not sinusoidal, we can still obtain $\boldsymbol{x}(z)$ and $t(z)$ explicitly from eqns (3.22) as

$$\boldsymbol{x} = \boldsymbol{x}_1 + \int_{z_1}^{z} \frac{\boldsymbol{P}_\perp - \boldsymbol{a}_\perp}{\gamma}\,\mathrm{d}z, \quad t = t_1 + \int_{z_1}^{z} \left[1 + \frac{1 + (\boldsymbol{P}_\perp - \boldsymbol{a}_\perp)^2}{2\gamma^2}\right]\mathrm{d}z. \quad (5.89\text{a,b})$$

We shall require that the angle at which the electron leaves the first undulator, and the angle and position at which the electron arrives at the entrance of the second, be the same as if the electron moved in free space between z_1 and z_2; that is, $(\mathrm{d}\boldsymbol{x}/\mathrm{d}z)_{z_2} = (\mathrm{d}\boldsymbol{x}/dz)_{z_1} = \boldsymbol{P}_\perp/\gamma$, $\boldsymbol{x}_2 = \boldsymbol{x}_1 + (\boldsymbol{P}_\perp/\gamma)(z_2 - z_1)$. According to eqn (5.89a), we must then have

$$\boldsymbol{a}(z_2) = \boldsymbol{a}(z_1) = 0, \qquad \int_{z_1}^{z_2} \boldsymbol{a}\,\mathrm{d}z = 0. \qquad (5.90\text{a,b})$$

(This is the reason why the field in the central region cannot just be uniform but must have a central plateau and two lateral peaks of the opposite sign, as in Fig. 5.5.)

For the time t_2 of arrival at the entrance of the second undulator, eqn (5.89b) gives

$$t_2 = t_1 + [1 + (1 + \boldsymbol{P}_\perp^2 + \langle \boldsymbol{a}_\perp^2 \rangle)/2\gamma^2](z_2 - z_1), \qquad (5.91)$$

where $\langle \boldsymbol{a}_\perp^2 \rangle$ is the mean square value of the central-section vector potential, i.e.

$$\langle \boldsymbol{a}_\perp^2 \rangle = \frac{1}{z_2 - z_1} \int_{z_1}^{z_2} \boldsymbol{a}_\perp^2\,\mathrm{d}z.$$

In the second undulator section the electron's motion will be given again by eqns (3.24), but with $\boldsymbol{x}_2$ and t_2 as initial conditions.

To calculate the far-field spontaneous radiation from an optical klystron eqn (4.31) is still valid, and we need only insert the appropriate expression of the electron motion. The radiation field so obtained will be the sum of three contributions, coming from each of the three sections. The contribution of the two undulators in particular may be expressed by the already written eqn (4.38). If, for simplicity, we assume the two undulators to be similar and neglect the contribution of the central region, where radiation is non-resonant and much less effective, the optical-klystron radiation field may be written as

$$\boldsymbol{E}_{rff} = \frac{-ikL_1\zeta e}{2\gamma}\frac{\mathrm{e}^{-ikr}}{4\pi r}\{\mathrm{e}^{-\mathrm{i}(kt_i - \boldsymbol{k}_\perp\cdot\boldsymbol{x}_i)} + \mathrm{e}^{-\mathrm{i}[k(t_2 - z_2) - \boldsymbol{k}_\perp\cdot\boldsymbol{x}_2]}\}$$

$$\times \sum_{n=-\infty}^{\infty} \boldsymbol{A}_{ueff,n}(\vartheta_{we})\mathrm{e}^{-\mathrm{i}\nu_n}\frac{\sin\nu_n}{\nu_n}; \tag{5.92}$$

that is, as the field radiated by a single undulator section multiplied by the sum of the two phase factors determined by the time and position of arrival of the electron at each undulator.

Assuming $\boldsymbol{P}_\perp = 0$ (so that $\boldsymbol{x}_2 = \boldsymbol{x}_i$) and using eqn (5.91), one finds that

$$\mathrm{e}^{-\mathrm{i}(kt_i - \boldsymbol{k}_\perp\cdot\boldsymbol{x}_i)} + \mathrm{e}^{-\mathrm{i}[k(t_2 - z_2) - \boldsymbol{k}_\perp\cdot\boldsymbol{x}_2]} = 2\mathrm{e}^{-\mathrm{i}[k(t_i + t_2)/2 - \boldsymbol{k}_\perp\cdot\boldsymbol{x}_i]}\cos\nu_d, \tag{5.93}$$

with

$$\nu_d = k[(1 + \langle \boldsymbol{a}_1^2\rangle)L_1 + (1 + \langle \boldsymbol{a}_d^2\rangle)L_d]/2\gamma^2, \tag{5.94}$$

where $\langle \boldsymbol{a}_1^2\rangle$ is the mean square value of the vector potential in the first undulator section, of length L_1, and $\langle \boldsymbol{a}_d^2\rangle$ is the mean square value of the vector potential in the central dispersive section, of length L_d. Thus the radiated power is proportional to

$$|\boldsymbol{E}_{rff}|^2 = \left(\frac{kL_1\zeta e}{8\pi\gamma r}\right)^2 4\cos^2\nu_d \sum_{n=0}^{\infty} A^2_{ueff,n}(\vartheta_{we})\frac{\sin^2\nu_n}{\nu_n^2}. \tag{5.95}$$

From eqn (5.95) one can observe that the optical-klystron spontaneous radiation spectrum contains two oscillating factors: the $(\sin^2\nu_n)/\nu_n^2$ factor characteristic of radiation from a single undulator and the new $\cos^2\nu_d$ factor. Since, as will be seen, ν_d varies with frequency much more rapidly than ν_n, the resulting spectrum has around each harmonic a shape like that shown in Fig. 5.6.

We can now see that the optical klystron has a higher gain than a simple undulator acting as an amplifier if we remember that, according to Madey's theorem, gain is proportional to the derivative of the spontaneous-radiation spectrum: even if the maximum amplitude of spontaneous radiation is not in itself greatly enhanced with respect to that emitted by each of the two

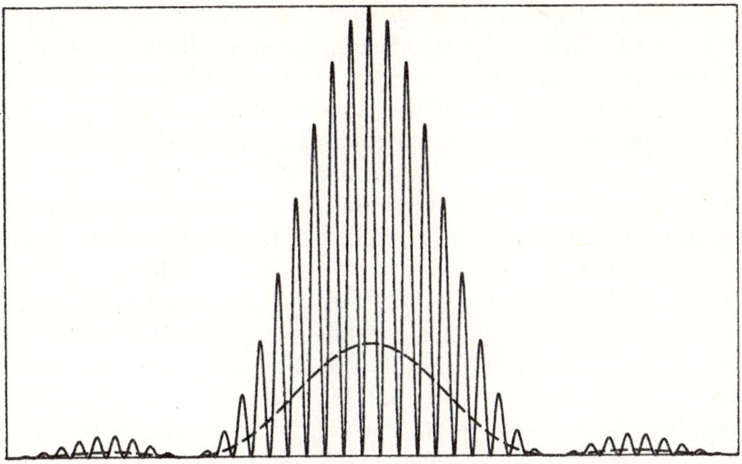

FIG. 5.6. Typical spontaneous-radiation spectrum of an optical klystron (solid line) superposed to the spontaneous-radiation spectrum of each of the two component undulator sections (dashed line).

undulator sections, as is seen in Fig. 5.6, its oscillating behaviour gives rise to a much larger derivative.

Let us now calculate the gain of an optical klystron. According to Madey's theorem, in order to calculate gain we must take the derivative of the mean square first-order energy perturbation, which is proportional to $|E_{rff}|^2$ of eqn (5.95), with respect to γ_i. If we assume that $1 + \langle a_d^2 \rangle \gg 1 + \langle a_1^2 \rangle$, so that the most rapidly varying factor in eqn (5.95) is $\cos^2 v_d$, it is possible to consider only this factor and neglect the derivatives of all the others. Thus one has

$$g = -\frac{\zeta e j_b L_1^2}{8mc^2}\frac{\partial}{\partial \gamma_i}\left[4\cos^2 v_d \sum_{n=0}^{\infty} A_{ueff,n}^2(0)\frac{\sin^2 v_n}{\gamma_i^2 v_n^2}\right]$$

$$\simeq \frac{\zeta e j_b L_1^2 k[(1+\langle a_1^2\rangle)L_1 + (1+\langle a_d^2\rangle)L_d]}{16mc^2\gamma_i^5}$$

$$\times A_{ueff,n}^2(0)\frac{\sin^2 v_n}{v_n^2}\frac{\partial}{\partial v_d}(4\cos^2 v_d), \tag{5.96}$$

where, in addition, only the resonant term in the sum has been kept. Equation (5.96) may be put in a form similar to the basic gain formula (5.23) of a uniform undulator upon remembering that the constant $4\pi mc^2/\zeta e$ has been given the name Alfvén current, I_A, and using the definition (5.94) of v_d. This form is

$$g = \frac{v_d}{2\pi}\frac{L_1^2}{l_u^2}A_{ueff,n}^2\gamma_i^{-3}\frac{\pi^2 l_u^2 j_b}{I_A}\frac{\sin^2 v_n}{v_n^2}\frac{\partial}{\partial v_d}(4\cos^2 v_d). \tag{5.97}$$

Notice that, in comparison with eqn (5.23), the factor N^3 has been replaced by $(\nu_d/2\pi)(L_1^2/l_u^2)$, whereas all other factors are either the same or of order unity. Therefore the ratio of the gain of an optical klystron to that of an ordinary uniform undulator of the same length is of the order of $\nu_d L_1^2/2\pi l_u^2 N^3$.

In order to estimate this ratio, we need a relation between the r.m.s. vector potential in the dispersive and in the undulator sections of the optical klystron. Let us consider two limiting cases. If the magnetic-field distribution in the dispersive section is given by exactly one period of a sinusoid having a different period but the same maximum amplitude as the sinusoid which describes the field in the undulator sections, as in Fig. 5.7(c), it may be easily calculated that the ratio $\langle \boldsymbol{a}_d^2 \rangle / \langle \boldsymbol{a}_1^2 \rangle$ equals the square of the ratio of the respective periods, i.e. $(L_d/l_u)^2$. If, in the opposite limit, the magnetic field in the dispersive section follows a square wave like that of Fig. 5.7(b), the ratio $\langle \boldsymbol{a}_d^2 \rangle / \langle \boldsymbol{a}_1^2 \rangle$ may be calculated to be $(\pi^2/6)(L_d/l_u)^2$, or approximately 1.5 times as much as before. $\langle \boldsymbol{a}_1^2 \rangle$, on the other hand, is simply $A_u^2/2$, A_u being, as usual, the amplitude of the undulator vector potential.

From eqn (5.96) and the above considerations it may be seen that the ratio of the gain of an optical klystron to that of an ordinary undulator is of the order of $L_1^2 L_d^3 / l_u^2 L_{\text{tot}}^3$.

We may go a step further and optimize this ratio by adjusting the magnetic-field distribution in the optical klystron under the constraint of an imposed maximum magnetic field. In order to do so one only needs to maximize the product $L_1^2 L_d^3$ under the constraint $2L_1 + L_d = L_{\text{tot}}$. It may be easily calculated that the maximum is attained for $L_1 = L_{\text{tot}}/5$ and $L_d = 3L_{\text{tot}}/5$, and is $(81/3125)L_{\text{tot}}^5$. Therefore, the gain of an optimized optical klystron increases as the fifth power of the ratio L_{tot}/l_u, which for a uniform

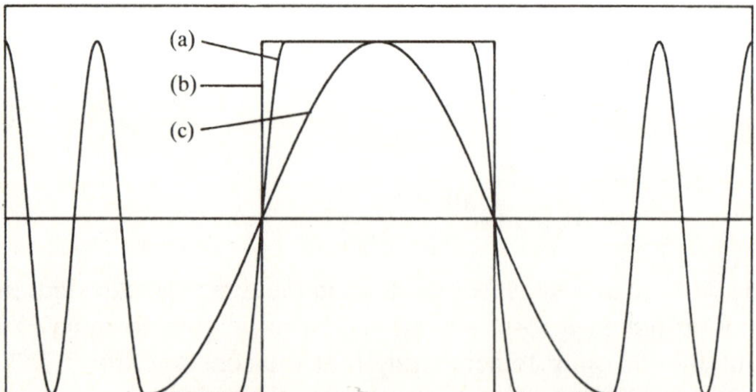

FIG. 5.7. A possible practical magnetic-field profile in the dispersive section of an optical klystron (a) shown together with the two limiting cases of a square (b) and sinusoidal (c) profile.

undulator would represent the number of periods N, rather than as N^3. Taking into account the small factor 81/3125, we may conclude by saying that the ratio of the gain of an optimized optical klystron to that of a comparable uniform undulator is approximately $(N/6)^2$.

The above gain increase may be quite important, but is not without a penalty. In order to exploit this gain the electron beam energy spread must be such that the phase v_d does not deviate from its optimum value by a quantity comparable to or larger than π. This requires a relative energy spread $\delta\gamma_i/\gamma_i \ll \pi\gamma_i^2/k\langle \boldsymbol{a}_d^2\rangle L_d$, that is approximately $0.4N^2$ times smaller (on the assumption that $\langle \boldsymbol{a}_1^2\rangle \gg 1$ so that $\langle \boldsymbol{a}_1^2\rangle \simeq M^2$) than that tolerable by a uniform undulator of equal length, and $N^{4/3}$ times smaller than that tolerable by a uniform undulator that is long enough to give the same gain.

Since energy spread is always a severe practical limiting factor of FEL performance, its smaller energy-spread tolerance makes the optical klystron much less useful than the foregoing gain estimate might make it appear, for if the gains of an optical klystron and a conventional undulator are compared on the basis of equal energy-spread acceptance rather than of equal length the conventional undulator wins out. It is therefore clear that the typical application range of optical klystrons is where external factors in the design constrain the length of the undulator to values smaller than those which would allow a complete exploitation of the existing energy spread [as of eqn (5.83), there is a definite relation between energy-spread acceptance and number of periods, i.e., length, of an undulator]. Under these circumstances the choice of an optical-klystron configuration optimized for the given length *and* energy spread affords a larger gain. This situation typically occurred in FEL experiments designed around already existing storage rings, where energy spread is comparatively low but space available in the straight sections of the ring is limited (see, e.g., Section 11.4).

Notes

1. The concept of smallness here is, of course, a relative one. The power levels below which the FEL operates linearly, which will be determined in Chapter 7, are not at all 'small' by usual standards.
2. Since A_u and A_w are circularly polarized and $A_u \cdot A_w = 0$, it follows that $\mathrm{Re}(A_u e^{ik_u z}) \cdot \mathrm{Re}(A_w e^{ik(t-z)}) = \mathrm{Re}\{A_u^* \cdot A_w e^{i[k(t-z)-k_u z]}\}/2$.
3. It is preferable to let e denote the unsigned value of the electron charge, so that j_b and I_A are defined as positive quantities: e being squared in eqn (5.19), the actual sign of the charge of the radiating particles is irrelevant.
4. The integral in eqn (5.54) must be extended from $-\infty$ to ∞ and not just from 0 to L, because when the directions of motion of electron and wave are not exactly parallel a perturbation to the electron energy exists, even outside the undulator. This perturbation gives an oscillatory contribution to the integral which prevents it from converging in the ordinary sense. This difficulty is similar to that found in connection with the far-field radiation integral (4.19), and may be solved by replacing the infinitely wide plane wave by one falling smoothly to zero at infinity, and by letting the range of this wave become infinite only after having calculated the

integral. We shall, however, bypass the problem by re-using the final results of the previous chapter directly.

References

Gea-Banacloche, J. and Scully, M. O. (1985). Quantum theory of the free-electron laser in the laboratory frame. *FEL*84, 100–107.

Gea-Banacloche, J., Moore, G. T., and Scully, M. O. (1984*a*). Prospects for an X-ray free-electron laser. *FEL*83, 393–401.

Gea-Banacloche, J., Moore, G. T., and Scully, M. O. (1984*b*). In *Free electron generation of extreme ultraviolet coherent radiation* (ed. J. M. J. Madey and C. Pellegrini). AIP Conf. Proc. no. 118, p. 161.

Grover, L. K. and Pantell, R. H. (1985). Simplified analysis of free-electron lasers using Madey's theorem. *IEEE J. Quantum Electron.* **21**, 944–51.

Krinsky, S. (1985). Action-angle variables and fluctuation–dissipation relations for a driven quantum oscillator. *Phys. Rev. A* **31**, 1267–72.

Krinsky, S., Wang, J. M., and Luchini, P. (1982). Madey's gain-spread theorem for the free-electron laser and the theory of stochastic processes. *J. Appl. Phys.* **53**, 5453–8.

Luchini, P. and Solimeno, S. (1985). Gain and mode-coupling in a three-dimensional free-electron laser: a generalization of Madey's theorem. *IEEE J. Quantum Electron.* **21**, 952–65.

Luchini, P., Prisco, G., and Solimeno, S. (1984). Optimization of the gain of a free-electron laser for a gaussian beam. *FEL*83, 283–8.

Madey, J. M. J. (1979). Relationship between mean radiated energy, mean squared radiated energy and spontaneous power spectrum in a power series expansion of the equation of motion in a free-electron laser. *Nuovo Cimento* **50B**, 64–88.

Renieri, A. (1984). Free-electron generation of extreme ultraviolet coherent radiation: theoretical aspects. In *Free electron generation of extreme ultraviolet coherent radiation* (ed. J. M. J. Madey and C. Pellegrini). AIP Conf. Proc. no. 118, pp. 1–11.

Vinokurov, N. A. (1977). *Proc. 10th Int. Conf. High Energy Charged Particle Accelerators*, Serpukhov, Vol. 2, p. 454.

Vinokurov, N. A. and Skrinsky, A. N. (1977). Preprint INP 77–59, Novosibirsk.

6

LARGE GAIN

When the radiated field intensity in an FEL is comparable to the input field, that is when gain is not small compared to unity, it is no longer possible to study the electron motion independently of the radiation process. In this chapter we shall address the situation where the gain is not small but the field itself *is* small, so as to permit the use of linearized equations of motion. The characteristic feature of this operating regime is that the motion of each electron does not take place in an externally imposed electromagnetic field but rather in the field generated by all the other electrons. It is then preferable to abandon the single-particle description of the previous chapters, where the motion of every single electron could be studied independently of the others, and pass to a collective approach in which the position and velocity of the electrons is described by a suitable distribution function in phase space and the linearized Boltzmann equation for this distribution function and the Maxwell equations for the electromagnetic field must be solved self-consistently.

The analysis of large-gain FELs bears a resemblance to the theory of travelling wave tubes, as was seen in Chapter 1, and many properties of these devices will be found again.

6.1 The Boltzmann equation

To study the collective behaviour of a beam of electrons shone through an undulator it is necessary to introduce the concept of a distribution function, as is usually done in the kinetic theory of gases and plasmas (an electron beam is in fact a sort of non-neutralized, fast-moving plasma). The distribution function $f(\boldsymbol{r}, \boldsymbol{P}, t)$ is a statistical quantity, analogous to a probability density, such that $f\, d^3\boldsymbol{r}\, d^3\boldsymbol{P}$ is the average number of electrons which can be found in the element $d^3\boldsymbol{r}\, d^3\boldsymbol{P}$ of phase space (the six-dimensional space of co-ordinates and momenta) at time t. The evolution in time of the distribution function is governed by the Boltzmann equation, the derivation of which can be found in textbooks (e.g., Chapman and Cowling 1970; Landau and Lifschitz 1981), expressing a balance between the electrons entering and exiting a given element of phase space owing to free motion, external interactions, and internal collisions.

For the purposes of the present analysis it is useful to adopt a modified definition of the distribution function and, correspondingly, a different form of the Boltzmann equation. Since the electrons in a beam, contrarily to those in a high-temperature plasma, have a prevalent motion in a single direction and are kept together mostly by inertia rather than by a container or confining field, a formulation of the Boltzmann equation adopting the axial co-ordinate z rather than time as the independent variable is preferable. Such a formulation will now be derived, capitalizing on the generalized Hamilton equations of Section 3.1.

Let us introduce a new six-dimensional phase space formed by the variables t, $\boldsymbol{x}$, $\mathscr{E}$, $\boldsymbol{P}_\perp$, and let $\boldsymbol{c}$ denote a vector in this space. Just as in standard phase space, we define a distribution function $f(\boldsymbol{c})$ as a density of particles, such that $f\,\mathrm{d}t\,\mathrm{d}\mathscr{E}\,\mathrm{d}^2\boldsymbol{x}\,\mathrm{d}^2\boldsymbol{P}_\perp$ represents the average number of electrons which pass at an axial position z in a time $\mathrm{d}t$, with an energy lying in the range $\mathrm{d}\mathscr{E}$, a transverse position contained in the element $\mathrm{d}^2\boldsymbol{x}$, and a transverse momentum contained in the element $\mathrm{d}^2\boldsymbol{P}_\perp$. An evolution equation for this distribution function can be written, just like the ordinary Boltzmann equation, by imposing the conservation of particles in phase space; that is, if one thinks of the particles as a six-dimensional fluid, by writing the particle balance equation

$$\partial f/\partial z + \mathrm{div}(f\dot{\boldsymbol{c}}) = 0. \tag{6.1}$$

(z is the independent variable, which plays the role of time in our phase space.) $\dot{\boldsymbol{c}}$ is the six-dimensional 'velocity' of particles, that is the vector formed by the derivatives with respect to z of t, $\boldsymbol{x}$, $\mathscr{E}$, $\boldsymbol{P}_\perp$. These derivatives are given by the Hamilton equations in the generalized form (3.4) and therefore eqn (6.1) can be written as

$$\frac{\partial f}{\partial z} + \frac{\partial}{\partial t}\left(f\frac{\partial \mathscr{K}}{\partial \mathscr{E}}\right) - \frac{\partial}{\partial \mathscr{E}}\left(f\frac{\partial \mathscr{K}}{\partial t}\right) - \frac{\partial}{\partial \boldsymbol{x}}\left(f\frac{\partial \mathscr{K}}{\partial \boldsymbol{P}_\perp}\right) + \frac{\partial}{\partial \boldsymbol{P}_\perp}\left(f\frac{\partial \mathscr{K}}{\partial \boldsymbol{x}}\right) = 0. \tag{6.2}$$

As $\partial^2\mathscr{K}/\partial t\partial\mathscr{E} - \partial^2\mathscr{K}/\partial\mathscr{E}\partial t - \partial^2\mathscr{K}/\partial\boldsymbol{x}\partial\boldsymbol{P}_\perp + \partial^2\mathscr{K}/\partial\boldsymbol{P}_\perp\partial\boldsymbol{x} = 0$, because of the equality of mixed derivatives,[1] eqn (6.2) may also be written as

$$\frac{\partial f}{\partial z} + \frac{\partial \mathscr{K}}{\partial \mathscr{E}}\frac{\partial f}{\partial t} - \frac{\partial \mathscr{K}}{\partial t}\frac{\partial f}{\partial \mathscr{E}} - \frac{\partial \mathscr{K}}{\partial \boldsymbol{P}_\perp}\frac{\partial f}{\partial \boldsymbol{x}} + \frac{\partial \mathscr{K}}{\partial \boldsymbol{x}}\frac{\partial f}{\partial \boldsymbol{P}_\perp} = 0. \tag{6.3}$$

Equation (6.3) is the modified Boltzmann equation that we want. When the motion of the particles is not completely deterministic and random collisions are allowed for, the Boltzmann collision integral must appear on the r.h.s. of eqn (6.3). However, we shall not take collisions into account.

6.2 Particle–wave interaction

In order to study the interaction between the electron beam and a co-propagating electromagnetic wave we need to consider, together with eqn (6.3), an equation describing the evolution of the electromagnetic field. This is the wave equation (4.2), where the current density of the electron beam must be inserted.

The current density of the electron beam is given in terms of the distribution function by the integral over $\boldsymbol{P}_\perp$ and $\mathscr{E}$ of the number of particles per unit volume of given momentum and energy that are found inside a volume element $\mathrm{d}^2\boldsymbol{x}\,\mathrm{d}z$ at a time t, i.e. $f\,\mathrm{d}t/\mathrm{d}z$, times the electron charge e times the electron velocity $\mathrm{d}\boldsymbol{r}/\mathrm{d}t$. Therefore eqn (4.2) may be rewritten as

$$\Delta_2 \boldsymbol{a} - \frac{1}{c^2}\frac{\partial^2 \boldsymbol{a}}{\partial t^2} = -\mu_0 \boldsymbol{j} = -\mu_0 e \int \frac{\mathrm{d}\boldsymbol{r}}{\mathrm{d}z} f\,\mathrm{d}^2\boldsymbol{P}_\perp\,\mathrm{d}\mathscr{E}. \tag{6.4}$$

We shall find it convenient, just as in the previous chapters, to work with dimensionless quantities, defined so that e, m, and c disappear from our calculations. After introducing the dimensionless vector potential, i.e. the mks one divided by mc/e, together with the dimensionless momentum $\boldsymbol{P}/mc$ and energy $\gamma = \mathscr{E}/mc^2$, we are led to define a new distribution function, f, as $\mu_0 e^2 mc^2 f$; rewriting the wave equation in terms of the dimensionless vector potential and this new distribution function, and expressing $\mathrm{d}\boldsymbol{r}/\mathrm{d}z$ through the Hamilton equations (3.4a), which give $\mathrm{d}z/\mathrm{d}z = 1$ and $\mathrm{d}\boldsymbol{x}/\mathrm{d}z = -\partial\mathscr{H}/\partial\boldsymbol{P}_\perp$, we obtain

$$\Delta_2 \boldsymbol{a}_\perp - \frac{\partial^2 \boldsymbol{a}_\perp}{\partial t^2} = \int \frac{\partial \mathscr{H}}{\partial \boldsymbol{P}_\perp} f\,\mathrm{d}^2\boldsymbol{P}_\perp\,\mathrm{d}\gamma, \qquad \Delta_2 a_z - \frac{\partial^2 a_z}{\partial t^2} = -\int f\,\mathrm{d}^2\boldsymbol{P}_\perp\,\mathrm{d}\gamma. \tag{6.5, 6.6}$$

Notice that the axial component of the e-beam current density in mks units, which we shall want to refer to later, may be expressed in terms of the new, rescaled, distribution function (which, incidentally, has the dimensions of a length raised to -2) as

$$j_z = \frac{mc}{\mu_0 e}\int f\,\mathrm{d}^2\boldsymbol{P}_\perp\,\mathrm{d}\gamma = \frac{I_A}{4\pi}\int f\,\mathrm{d}^2\boldsymbol{P}_\perp\,\mathrm{d}\gamma, \tag{6.7}$$

where I_A is the Alfvén current introduced in Section 5.5, approximately equal to 17.03 kA.

As was said at the beginning, in this chapter we do not consider the radiated field small in comparison to the input field, but we do consider small the effects of the combined field upon the electron motion. This means that we can write the Hamiltonian $\mathscr{H}$ again as the sum of an unperturbed Hamiltonian $\mathscr{H}_0$ containing only the undulator field and a small perturbation $\mathscr{H}_1$ due to the, now unknown, wave field. Correspondingly, we seek the distribution function as the sum of an unperturbed one, f_0, and a small correction, f_1. A straightforward perturbative expansion of eqns (6.3), (6.5), and (6.6) gives the four equations (in dimensionless form):

$$\frac{\partial f_0}{\partial z} + \frac{\partial \mathscr{H}_0}{\partial \gamma}\frac{\partial f_0}{\partial t} - \frac{\partial \mathscr{H}_0}{\partial t}\frac{\partial f_0}{\partial \gamma} - \frac{\partial \mathscr{H}_0}{\partial \boldsymbol{P}_\perp}\frac{\partial f_0}{\partial \boldsymbol{x}} + \frac{\partial \mathscr{H}_0}{\partial \boldsymbol{x}}\frac{\partial f_0}{\partial \boldsymbol{P}_\perp} = 0, \tag{6.8}$$

$$\frac{\partial f_1}{\partial z}+\frac{\partial \mathscr{K}_0}{\partial \gamma}\frac{\partial f_1}{\partial t}-\frac{\partial \mathscr{K}_0}{\partial t}\frac{\partial f_1}{\partial \gamma}-\frac{\partial \mathscr{K}_0}{\partial \boldsymbol{P}_\perp}\frac{\partial f_1}{\partial \boldsymbol{x}}+\frac{\partial \mathscr{K}_0}{\partial \boldsymbol{x}}\frac{\partial f_1}{\partial \boldsymbol{P}_\perp}$$

$$=-\frac{\partial \mathscr{K}_1}{\partial \gamma}\frac{\partial f_0}{\partial t}+\frac{\partial \mathscr{K}_1}{\partial t}\frac{\partial f_0}{\partial \gamma}+\frac{\partial \mathscr{K}_1}{\partial \boldsymbol{P}_\perp}\frac{\partial f_0}{\partial \boldsymbol{x}}-\frac{\partial \mathscr{K}_1}{\partial \boldsymbol{x}}\frac{\partial f_0}{\partial \boldsymbol{P}_\perp}, \tag{6.9}$$

$$\Delta_2 \boldsymbol{a}_{1\perp}-\frac{\partial^2 \boldsymbol{a}_{1\perp}}{\partial t^2}=\int\left(\frac{\partial \mathscr{K}_1}{\partial \boldsymbol{P}_\perp} f_0+\frac{\partial \mathscr{K}_0}{\partial \boldsymbol{P}_\perp} f_1\right) \mathrm{d}^2\boldsymbol{P}_\perp\, \mathrm{d}\gamma, \tag{6.10}$$

$$\Delta_2 a_{1z}-\frac{\partial^2 a_{1z}}{\partial t^2}=-\int f_1\, \mathrm{d}^2\boldsymbol{P}_\perp\, \mathrm{d}\gamma. \tag{6.11}$$

In the ultrarelativistic approximation the dimensionless Hamiltonian is given by eqn (3.15). Writing the electromagnetic potentials as the sum of unperturbed potentials $\varphi_0 = 0$ and $\boldsymbol{a}_0$ due to the undulator and perturbations φ_1 and $\boldsymbol{a}_1$ due to the propagating electromagnetic wave, and perturbatively expanding eqn (3.15) with respect to $\boldsymbol{a}_1$, we obtain $\mathscr{K}_0 = \gamma + a_{0z} - [1 + (\boldsymbol{P}_\perp - \boldsymbol{a}_0)^2]/2\gamma$ and $\mathscr{K}_1 = a_{1z} - \varphi_1 + \boldsymbol{a}_{1\perp}\cdot(\boldsymbol{P}_\perp - \boldsymbol{a}_0)/\gamma$. Here $\boldsymbol{a}_0$ is to be considered known, while $\boldsymbol{a}_1$ represents the solution of eqns (6.8)-(6.11), and φ_1 can be obtained from $\boldsymbol{a}_1$ by means of the gauge condition (4.3). The system of equations (6.8)-(6.11) is thus closed and the mathematical statement of the problem is complete. The solution of this system will be the object of the rest of this chapter.

6.3 The one-dimensional, collimated, monoenergetic beam model

As has already appeared from the previous chapters, the simplest picture of an FEL which is still meaningful is obtained when considering a one-dimensional undulator fed by a one-dimensional, transversely infinite, beam of electrons, all with zero transverse momentum and the same energy. In this chapter too, we shall begin with such a picture.

Let us, therefore, consider a situation in which the undulator field and all the unknowns, namely f_0, f_1 and $\boldsymbol{a}_1$, do not depend on $\boldsymbol{x}$. As the unperturbed Hamiltonian $\mathscr{K}_0$ does not depend on either $\boldsymbol{x}$ or t, eqn (6.8) is identically satisfied by any unperturbed distribution function $f_0(\gamma, \boldsymbol{P}_\perp)$ independent of $\boldsymbol{x}$, t, and z, which means that at zero order the initial energy and momentum distribution of the electron beam is maintained throughout the undulator. We shall assume f_0 to be proportional to $\delta(\boldsymbol{P}_\perp)\delta(\gamma - \bar{\gamma})$, so as to represent a perfectly collimated and monoenergetic beam.

Under the above conditions the terms containing derivatives with respect to $\boldsymbol{x}$ and the term $(\partial\mathscr{K}_0/\partial t)(\partial f_1/\partial\gamma)$ disappear from eqn (6.9). Since the r.h.s. of eqn (6.9) contains the derivative $\partial f_0/\partial\gamma$, that is the derivative of a δ function, an expression of f_1 may be sought in the form of the sum of two terms proportional, respectively, to $\delta(\boldsymbol{P}_\perp)\delta(\gamma - \bar{\gamma})$ and $\delta(\boldsymbol{P}_\perp)\delta'(\gamma - \bar{\gamma})$, δ' denoting the derivative of the δ function. Substituting $f_0 = (4\pi j_b/I_A)\delta(\boldsymbol{P}_\perp)\delta(\gamma - \bar{\gamma})$ (where j_b is the e-beam current density, in accordance with eqn (6.7)) and $f_1 = c_0(t, z)\delta(\boldsymbol{P}_\perp)\delta(\gamma - \bar{\gamma}) + c_1(t, z)\delta(\boldsymbol{P}_\perp)\delta'(\gamma - \bar{\gamma})$ into eqn (6.9) and collecting coefficients of $\delta(\gamma - \bar{\gamma})$ and $\delta'(\gamma - \bar{\gamma})$, one obtains the two equations[2]

$$\frac{\partial c_0}{\partial z} + \left(1 + \frac{1 + a_0^2}{2\bar{\gamma}^2}\right)\frac{\partial c_0}{\partial t} + \frac{1 + a_0^2}{\bar{\gamma}^3}\frac{\partial c_1}{\partial t} = -\frac{4\pi j_b}{I_A}\frac{\partial}{\partial t}\frac{\boldsymbol{a}_0 \cdot \boldsymbol{a}_1}{\bar{\gamma}^2} \tag{6.12}$$

and

$$\frac{\partial c_1}{\partial z} + \left(1 + \frac{1 + a_0^2}{2\bar{\gamma}^2}\right)\frac{\partial c_1}{\partial t} = \frac{4\pi j_b}{I_A}\frac{\partial}{\partial t}\left(a_{1z} - \varphi_1 - \frac{\boldsymbol{a}_0 \cdot \boldsymbol{a}_{1\perp}}{\bar{\gamma}}\right). \tag{6.13}$$

Equations (6.10) and (6.11), after the same substitution, become

$$\frac{\partial^2 \boldsymbol{a}_{1\perp}}{\partial z^2} - \frac{\partial \boldsymbol{a}_{1\perp}}{\partial t^2} = \frac{4\pi j_b}{I_A}\frac{\boldsymbol{a}_{1\perp}}{\bar{\gamma}} + c_0\frac{\boldsymbol{a}_0}{\bar{\gamma}} - c_1\frac{\boldsymbol{a}_0}{\bar{\gamma}^2} \tag{6.14}$$

and

$$\frac{\partial^2 a_{1z}}{\partial z^2} - \frac{\partial^2 a_{1z}}{\partial t^2} = -c_0. \tag{6.15}$$

We are thus left with the linear system of eqns (6.11)–(6.15) to be solved for c_0, c_1, and $\boldsymbol{a}_1$, all functions of t and z only. Since these equations are linear and have time-independent coefficients, they admit time-harmonic complex solutions in which all the unknowns depend on time as e^{ikt} (just as in the previous chapters the variable t, which represents time multiplied by c, is multiplied by the wavenumber $k = \omega/c$ rather than by the angular frequency ω). Writing $\boldsymbol{a}_1 = \mathrm{Re}(A_1 e^{ikt})$, $c_0 = \mathrm{Re}(C_0 e^{ikt})$, and $c_1 = \mathrm{Re}(C_1 e^{ikt})$, and obtaining φ_1 from the gauge condition (4.3), which gives (in dimensionless form) $\varphi_1 = \mathrm{Re}[-(1/ik)(\mathrm{d}A_{1z}/\mathrm{d}z)e^{ikt}]$, transforms eqns (6.12)–(6.15) into

$$\frac{\mathrm{d}C_0}{\mathrm{d}z} + \left(1 + \frac{1 + a_0^2}{2\bar{\gamma}^2}\right)ikC_0 + \frac{1 + a_0^2}{\bar{\gamma}^3}ikC_1 = -\frac{4\pi j_b}{I_A}ik\frac{\boldsymbol{a}_0 \cdot \boldsymbol{A}_{1\perp}}{\bar{\gamma}^2} \tag{6.16}$$

$$\frac{\mathrm{d}C_1}{\mathrm{d}z} + \left(1 + \frac{1 + a_0^2}{2\bar{\gamma}^2}\right)ikC_1 = \frac{4\pi j_b}{I_A}\left(ikA_{1z} + \frac{\mathrm{d}A_{1z}}{\mathrm{d}z} - ik\frac{\boldsymbol{a}_0 \cdot \boldsymbol{A}_{1\perp}}{\bar{\gamma}}\right), \tag{6.17}$$

$$\frac{\mathrm{d}^2 \boldsymbol{A}_{1\perp}}{\mathrm{d}z^2} + k^2 \boldsymbol{A}_{1\perp} = \frac{4\pi j_b}{I_A}\frac{\boldsymbol{A}_{1\perp}}{\bar{\gamma}} + C_0\frac{\boldsymbol{a}_0}{\bar{\gamma}} - C_1\frac{\boldsymbol{a}_0}{\bar{\gamma}^2}, \tag{6.18}$$

$$\frac{\mathrm{d}^2 A_{1z}}{\mathrm{d}z^2} + k^2 A_{1z} = -C_0. \tag{6.19}$$

6.4 Basic large-gain FEL description

We are now going to solve eqns (6.16)–(6.19) in an approximate way which will enable us to display the main features of the large-gain FEL regime. The results will be refined and generalized in the next sections. As was also true in the previous chapters, the simplest case is that of a circularly polarized sinusoidal undulator field represented by the vector potential $\boldsymbol{a}_0 = \mathrm{Re}(\boldsymbol{A}_u e^{ik_u z})$, with the complex vector $\boldsymbol{A}_u$ such that $A_u^2 = 0$, and we shall consider this case first.

Equations (6.16)–(6.19) represent a typical wave-coupling problem, $4\pi j_b/I_A$ being the coupling parameter. In fact, for $4\pi j_b/I_A \to 0$ eqns (6.16) and (6.17) decouple from eqns (6.18) and (6.19) and, in the case of circular polarization, when a_0^2 is constant, their solutions take the form of travelling waves with propagation constant $(1 + M^2/2\bar{\gamma}^2)k$, M^2 representing, as usual, $1 + a_0^2 = 1 + A_u^2/2$. (More precisely, the system of eqns (6.16) and (6.17) has two degenerate solutions, having the same propagation constant: two independent solutions are, e.g., $C_0 = e^{-i(1+M^2/2\bar{\gamma}^2)kz}$, $C_1 = 0$, and $C_0 = (M^2/\bar{\gamma}^3)kze^{-i(1+M^2/2\bar{\gamma}^2)kz}$, $C_1 = -e^{-i(1+M^2/2\bar{\gamma}^2)kz}$.) On the other hand, eqns (6.18) and (6.19) in the limit for $4\pi j_b/I_A \to 0$, when C_0 and C_1 themselves become negligible, have travelling wave solutions proportional to $e^{\pm ikz}$.

When the coupling parameter $4\pi j_b/I_A$ is non-zero but small, the system of eqns (6.16)–(6.19) may be expected to have four independent travelling-wave solutions with propagation constants not far from those of the uncoupled equations. In general, these four propagation constants are all different (the degeneracy disappears) and complex; some of them may acquire an imaginary part of the appropriate sign for the corresponding waves to be amplified as they proceed along the undulator. One has then obtained an amplifying device; that is, the FEL.

It may be easily imagined that the interaction between the particle and wave equations, and therefore the amplification of the FEL, will be largest when the equations 'resonate'; that is, when the r.h.s. of each equation is a travelling wave with a propagation constant close to that of the free solution of the same equation. Let us see when this resonance may occur.

If the expression $A_w e^{-i\beta z}$, with $\beta \simeq k$, is assumed for $\boldsymbol{A}_1$, the product $\boldsymbol{a}_0 \cdot \boldsymbol{A}_1$ appearing in the r.h.s. of eqns (6.16) and (6.17) may be written as $[\boldsymbol{A}_u \cdot \boldsymbol{A}_w e^{-i(\beta - k_u)z} + \boldsymbol{A}_u^* \cdot \boldsymbol{A}_w e^{-i(\beta + k_u)z}]/2$ and, as such, contains two terms with wavenumber, respectively, $\beta - k_u$ and $\beta + k_u$: the first of these may never be resonant; the second falls close to the free wavenumber of eqns (6.16) and (6.17), $k(1 + M^2/2\bar{\gamma}^2)$, when $\kappa = kM^2/2\bar{\gamma}^2 \simeq k_u$. Not surprisingly, this condition, which may also be written as $k \simeq 2\bar{\gamma}^2 k_u/M^2 = k_r$, is the same undulator resonance condition that we are already familiar with.

It may be easily verified that when $\kappa \simeq k_u$ a resonance appears in eqn (6.18) as well. In particular, the wavenumber $\beta + k_u$ of C_0 and C_1 combines in the r.h.s. with the wavenumber k_u of $\boldsymbol{a}_0$ to produce the sum and difference $\beta + 2k_u$

and β; of these the first is non-resonant, while the second resonantly enhances the wave with wavenumber β that was assumed to exist in the first place. Equation (6.19), on the other hand, has only a term with wavenumber $\beta + k_u$ on its r.h.s. and is never resonant.

Under near-resonance conditions, eqns (6.16)–(6.19) may be simplified by neglecting the non-resonant terms. In particular, we may neglect A_{1z} altogether, because eqn (6.19) is non-resonant, and in the product $\boldsymbol{a}_0 \cdot \boldsymbol{A}_{1\perp}$ we may keep only the term $\boldsymbol{A}_{1\perp} \cdot \boldsymbol{A}_u^* \mathrm{e}^{-ik_u z}/2$. Equation (6.17) then becomes

$$\mathrm{d}C_1/\mathrm{d}z + (1 + M^2/2\bar{\gamma}^2)ikC_1 = -ik(4\pi j_b/I_A)\boldsymbol{A}_{1\perp} \cdot \boldsymbol{A}_u^* \mathrm{e}^{-ik_u z}/2\bar{\gamma}; \quad (6.20)$$

i.e., for $\boldsymbol{A}_{1\perp} = \boldsymbol{A}_w \mathrm{e}^{-i\beta z}$ and $C_1 = C_{w1}\mathrm{e}^{-i(\beta + k_u)z}$,

$$C_{w1} = (4\pi j_b/I_A)k\boldsymbol{A}_u^* \cdot \boldsymbol{A}_w/2\bar{\gamma}(\beta - k - \mu), \quad (6.21)$$

where μ is the beat wavenumber, introduced in the previous chapter, given by

$$\mu = kM^2/2\bar{\gamma}^2 - k_u = \kappa - k_u. \quad (6.22)$$

Both $\beta - k$ and μ are small near resonance, so that the factor $1/(\beta - k - \mu)$ is large. Next, combining eqn (6.20) with eqn (6.16) gives

$$\frac{\mathrm{d}}{\mathrm{d}z}(C_0 - C_1/\bar{\gamma}) + ik(1 + M^2/2\bar{\gamma}^2)(C_0 - C_1/\bar{\gamma}) = -\frac{M^2}{\bar{\gamma}^3} ikC_1; \quad (6.23)$$

i.e., for $C_0 = C_{w0}\mathrm{e}^{-i(\beta + k_u)z}$,

$$C_{w0} - C_{w1}/\bar{\gamma} = kM^2 C_{w1}/\bar{\gamma}^3(\beta - k - \mu). \quad (6.24)$$

In eqn (6.18), since $\boldsymbol{A}_{1\perp} \sim \mathrm{e}^{-i\beta z}$ and $\beta \simeq k$, one may approximate $\mathrm{d}^2\boldsymbol{A}_{1\perp}/\mathrm{d}z^2 + k^2\boldsymbol{A}_{1\perp}$ by $-2ik(\mathrm{d}\boldsymbol{A}_{1\perp}/\mathrm{d}z + ik\boldsymbol{A}_{1\perp})$; moreover, the term $(4\pi j_b/I_A)\boldsymbol{A}_{1\perp}/\bar{\gamma}$ may be neglected with respect to C_{w0} and C_{w1}, which contain the large factor $(\beta - k - \mu)^{-1}$, and in the expression of $\boldsymbol{a}_0$, $\boldsymbol{a}_0 = (\boldsymbol{A}_u \mathrm{e}^{ik_u z} + \boldsymbol{A}_u^* \mathrm{e}^{-ik_u z})/2$, only the first term need be retained because the second is non-resonant. Equation (6.18) then becomes

$$-2ik(\mathrm{d}\boldsymbol{A}_{1\perp}/\mathrm{d}z + ik\boldsymbol{A}_{1\perp}) = (C_0 - C_1/\bar{\gamma})\boldsymbol{A}_u \mathrm{e}^{ik_u z}/2\bar{\gamma}, \quad (6.25)$$

i.e.

$$-2k(\beta - k)\boldsymbol{A}_w = (C_{w0} - C_{w1}/\bar{\gamma})\boldsymbol{A}_u/2\bar{\gamma}. \quad (6.26)$$

Inserting eqn (6.21) into eqn (6.24) and the result into eqn (6.26) one obtains $-2k(\beta - k)\boldsymbol{A}_w = (4\pi j_b/I_A)k^2M^2\boldsymbol{A}_u\boldsymbol{A}_u^* \cdot \boldsymbol{A}_w/4\bar{\gamma}^5(\beta - k - \mu)^2$, which implies that $\boldsymbol{A}_w$ is proportional to (has the same polarization as) $\boldsymbol{A}_u$ and

$$(\beta - k)(\beta - k - \mu)^2 = -\pi M^2 A_u^2 k j_b/2\bar{\gamma}^5 I_A. \quad (6.27)$$

Equation (6.27) is the dispersion equation, giving implicitly the propagation constant β of travelling waves in the FEL system as a function of the free-space wavenumber k. On remembering the definition of the gain length (5.25),

$$L_g = (2\bar{\gamma}^5 I_A/\pi M^2 A_u^2 k j_b)^{1/3}, \tag{6.28}$$

where $\bar{\gamma}$ has been written instead of γ_i, and introducing the dimensionless parameters $p = (\beta - k)L_g$ and $\nu = \mu L_g$, eqn (6.27) may be rewritten as

$$p(p - \nu)^2 + 1 = 0. \tag{6.29}$$

Equation (6.29) is similar to the dispersion equation arising in the theory of travelling wave tubes, as explained in Chapter 1. For $\nu = 0$ its roots are the three cubic roots of -1, $p = -1$, $0.5 + i0.866$, and $0.5 - i0.866$. The three waves corresponding to these roots are one of constant amplitude, one amplified, and one damped.

The three waves are simultaneously excited at the entrance of the undulator, in a manner which will be analysed later in Section 6.10, and then propagate independently inside. The wave with a positive imaginary part of β increases its amplitude exponentially and, in a long enough undulator, overrides the other two components. The gain of a long undulator may therefore be written as the amplification factor of this single wave, i.e., since the imaginary part of β is given by the imaginary part of p divided by L_g, as

$$G = K e^{2\,\mathrm{Im}(p)L/L_g}, \tag{6.30}$$

where L is as usual the undulator length and K is an input coupling parameter, independent of L, the value of which will be determined in Section 6.10 to be $|(p - \nu)/(3p - \nu)|^2$. G is the total power gain $P_{\text{out}}/P_{\text{in}}$ (i.e., the square of the amplitude gain, whence the factor of 2 in the exponent) and

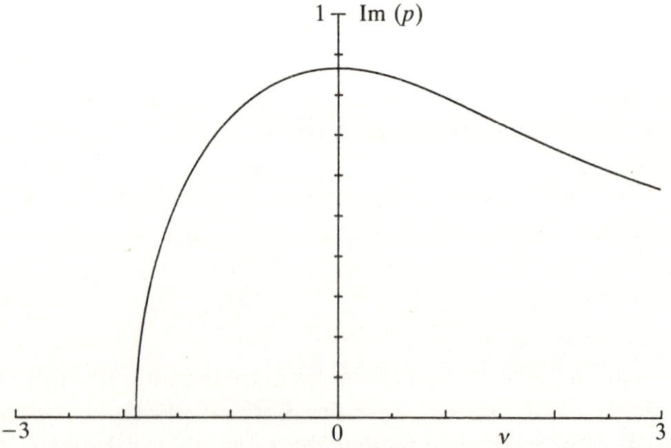

FIG. 6.1. Plot of amplification rate Im(p) vs. detuning parameter ν for the amplified mode of a large-gain FEL.

equals the incremental gain g calculated in Chapter 5 plus 1. The quantity $\mathrm{Im}(p)$ is plotted versus ν for the amplified wave in Fig. 6.1. Notice that, contrarily to what happens in the small-gain regime, maximum amplification takes place for $\nu = 0$.

Equation (6.30) replaces eqn (5.27) in the large-gain regime and, contrarily to eqn (5.27), is only valid when $G \gg 1$, i.e. when $L \gg L_g$. Because of the exponential dependence of gain upon L, the large-gain regime is often also referred to in the literature as the exponential-gain regime.

The maximum gain, corresponding to the optimal detuning $\nu = 0$, is given by

$$G_{\max} = \tfrac{1}{9}e^{1.732L/L_g}. \tag{6.31}$$

It should be remarked that the definition of the detuning parameter appropriate to the large-gain regime, $\nu = \mu L_g$, differs from the one appropriate to the small-gain regime, which may be written as $\nu = \mu L/2$, by the substitution of L_g for $L/2$. Correspondingly, the relative bandwidth over which gain is present, which in the small-gain case was $\delta\lambda/\lambda \simeq 1/N = l_u/L$ becomes, in the large-gain regime,

$$\delta\lambda/\lambda \simeq l_u/2L_g. \tag{6.32}$$

We shall often find in the following that, as a rule of thumb, many results of the small-gain regime may be qualitatively adapted to the large-gain one simply by replacing L by $2L_g$.

6.5 The complete dispersion equation of a one-dimensional circularly polarized FEL

The description given in the previous section is, under most circumstances, adequate to represent the large-gain behaviour of a one-dimensional circularly polarized FEL. However, it is also possible to give an exact solution of eqns (6.9)–(6.11) in the case of a one-dimensional circularly polarized FEL with a collimated but not necessarily monoenergetic beam.

Let us start again from eqns (6.8)–(6.11). As was remarked at the beginning of Section 6.3, for a one-dimensional system eqn (6.8) is identically satisfied by a distribution function $f_0(\boldsymbol{P}_\perp, \gamma)$ independent of t, $\boldsymbol{x}$, and z. Here we shall assume $f_0 = F_0(\gamma)\delta(\boldsymbol{P}_\perp)$, with $F_0(\gamma)$ an unspecified function. Equations (6.9)–(6.11), which are linear, have time-independent coefficients and admit time-harmonic solutions. We therefore set $f_1 = \mathrm{Re}[F_1(\gamma, z)\delta(\boldsymbol{P}_\perp)e^{ikt}]$ and $\boldsymbol{a}_1 = \mathrm{Re}[\boldsymbol{A}_1(z)e^{ikt}]$. For the scalar potential the gauge condition (4.3) gives (in dimensionless form) $\varphi_1 = \mathrm{Re}[-(1/ik)(\mathrm{d}A_{1z}/\mathrm{d}z)e^{ikt}]$. The undulator vector potential is, as usual, $\boldsymbol{a}_0 = \mathrm{Re}(\boldsymbol{A}_u e^{ik_u z})$. With these expressions eqns (6.9)–(6.11) become

$$\frac{\partial F_1}{\partial z} + \left(1 + \frac{M^2 + \mathrm{Re}(A_u^2 e^{2ik_u z})/2}{2\gamma^2}\right) ikF_1$$

$$= \left[ikA_{1z} + \frac{dA_{1z}}{dz} - \frac{ik}{\gamma}\boldsymbol{A}_{1\perp}\cdot\mathrm{Re}(\boldsymbol{A}_u e^{ik_u z})\right]\frac{dF_0}{d\gamma}, \tag{6.33}$$

$$\frac{d^2\boldsymbol{A}_{1\perp}}{dz^2} + k^2\boldsymbol{A}_{1\perp} = \int [F_0\boldsymbol{A}_{1\perp} + F_1\,\mathrm{Re}(\boldsymbol{A}_u e^{ik_u z})]/\gamma\, d\gamma, \tag{6.34}$$

$$\frac{d^2A_{1z}}{dz^2} + k^2A_{1z} = -\int F_1\, d\gamma. \tag{6.35}$$

Equations (6.33)–(6.35) are linear differential equations with periodic coefficients. The solutions of equations of this kind are governed by Floquet's theory (as explained, e.g., by Stoker 1950), according to which there exists in general a set of linearly independent solutions which, except for degenerate cases, are given by the product of an exponential factor $e^{-i\beta z}$ and a periodic function with the same period as the coefficients. These solutions are the equivalent of the exponential modes that characterize the case of constant coefficients, and represent travelling waves. However, contrarily to the case of constant coefficients, there is no general method enabling us to find these solutions.

Nevertheless, when the undulator field is circularly polarized it may be verified that solutions of the form $\boldsymbol{A}_{1\perp} = \boldsymbol{A}_w e^{-i\beta z} + \boldsymbol{A}'_w e^{-i(\beta+2k_u)z}$, $F_1 = F_w(\gamma)e^{-i(\beta+k_u)z}$, $A_{1z} = A_{wz}e^{-i(\beta+k_u)z}$, where $\boldsymbol{A}_w$, $\boldsymbol{A}'_w$, $F_w(\gamma)$, and A_{wz} are coefficients independent of z, satisfy eqns (6.33)–(6.35) exactly.[3] Substituting the above expressions into eqns (6.33)–(6.35), where $\mathrm{Re}(\boldsymbol{A}_u e^{ik_u z})$ may be written as $(\boldsymbol{A}_u e^{ik_u z} + \boldsymbol{A}_u^* e^{-ik_u z})/2$, and choosing circularly polarized constant vectors $\boldsymbol{A}_w$ and $\boldsymbol{A}'_w$ with such orientations that $\boldsymbol{A}_w$ is proportional to $\boldsymbol{A}_u$ and $\boldsymbol{A}'_w$ to $\boldsymbol{A}_u^*$, so that $\boldsymbol{A}_w\cdot\boldsymbol{A}_u - \boldsymbol{A}'_w\cdot\boldsymbol{A}_u^* = \boldsymbol{A}_u^2 = 0$ (as explained in the comment to eqn (3.21)), one obtains the following set of linear equations in the coefficients $\boldsymbol{A}_w$, $\boldsymbol{A}'_w$, $F_w(\gamma)$, and A_{wz}:

$$(k - \beta + \mu)F_w = \left[(k - \beta - k_u)A_{wz} - k\frac{\boldsymbol{A}_u^*\cdot\boldsymbol{A}_w + \boldsymbol{A}_u\cdot\boldsymbol{A}'_w}{2\gamma}\right]\frac{dF_0}{d\gamma}, \tag{6.36}$$

$$(k^2 - \beta^2)\boldsymbol{A}_w = \left(\int_0^\infty \frac{F_0}{\gamma}\, d\gamma\right)\boldsymbol{A}_w + \left(\int_0^\infty \frac{F_w}{\gamma}\, d\gamma\right)\frac{\boldsymbol{A}_u}{2}, \tag{6.37}$$

$$[k^2 - (\beta + 2k_u)^2]\boldsymbol{A}'_w = \left(\int_0^\infty \frac{F_0}{\gamma}\, d\gamma\right)\boldsymbol{A}'_w + \left(\int_0^\infty \frac{F_w}{\gamma}\, d\gamma\right)\frac{\boldsymbol{A}_u^*}{2}, \tag{6.38}$$

$$[k^2 - (\beta + k_u)^2]A_{wz} = -\int_0^\infty F_w \, d\gamma, \qquad (6.39)$$

where, as before, $\mu = kM^2/2\gamma^2 - k_u$.

The compatibility condition of this linear homogeneous system of equations, i.e. the dispersion equation, may be found by successive substitutions. The result is

$$\{(\beta^2 - k^2 + k_p^2)^{-1} + [(\beta + 2k_u)^2 - k^2 + k_p^2]^{-1}\}^{-1}$$
$$= -(kA_u^2/4)[I_2 + I_1^2/(k + \beta + k_u - I_0)], \qquad (6.40)$$

where $k_p^2 = \int_0^\infty F_0/\gamma \, d\gamma$ ($\omega_p = ck_p$ is the plasma frequency, below which electromagnetic waves cannot propagate in the beam) and, for $r = 0, 1, 2$,

$$I_r = \int_0^\infty \frac{dF_0/d\gamma}{\gamma^r(\beta - k - \mu)} \, d\gamma. \qquad (6.41)$$

Equation (6.40) is the exact dispersion equation of travelling waves in a one-dimensional sinusoidal undulator fed by a uniform, collimated electron beam with a general energy distribution $F_0(\gamma)$. In the particular case of a monoenergetic beam, when $F_0 = (4\pi j_b/I_A)\delta(\gamma - \bar{\gamma})$, the expression of k_p^2 is $4\pi j_b/\bar{\gamma} I_A$, and that of I_r is

$$I_r = \frac{4\pi j_b}{I_A} \frac{2k_u + r(\beta - k) + (2 - r)\mu}{\bar{\gamma}^{r+1}(\beta - k - \mu)^2}. \qquad (6.42)$$

6.6 Limits of validity of the basic large-gain FEL description

Equation (6.40) puts us in a position to analyse the limits of validity of the approximations performed in Section 6.4. The assumptions that those approximations were based upon were that the e-beam current j_b be small enough for the propagation constant β to stay near to that of the uncoupled equations, and that the beat wavenumber μ be small enough that a resonance might arise between the electromagnetic wave and the linearized Boltzmann equation. We can now make these two assumptions quantitative by requiring that both $|\beta - k|$ and $|\mu|$ be small compared to k_u. In particular, the first assumption may be written, since $|\beta - k|$ is of the order of $1/L_g$, as

$$L_g \gg l_u; \qquad (6.43)$$

i.e., on using the expression (5.26) of the gain length, as

$$j_b \ll \bar{\gamma}^3 I_A / A_u^2 l_u^2. \qquad (6.44)$$

Let us estimate the relative order of magnitude of the different terms in the dispersion equation (6.40).

The term k_p^2 may be written, for a monoenergetic beam, as $4\pi j_b/\bar{\gamma} I_A$ and, according to the expression (5.26) of L_g, equals $2\bar{\gamma}^2 l_u/\pi A_u^2 L_g^3$. It is therefore $O(M^2 l_u^2/A_u^2 L_g^2)$ times smaller than $\beta^2 - k^2$, which may be approximated as $2k(\beta - k)$ and is of the order of $4\pi/\lambda L_g \simeq 8\pi\bar{\gamma}^2/M^2 l_u L_g$.

The term $[(\beta + 2k_u)^2 - k^2 + k_p^2]^{-1}$ is approximately $1/4kk_u = l_u/8\pi k$, when $|\beta - k| \ll k_u$, and is therefore of the order of l_u/L_g times smaller than $|2k(\beta - k)|^{-1} \simeq L_g/2k$. If both this and the previous term are neglected, the l.h.s. of eqn (6.40) reduces to $2k(\beta - k)$.

With regard to I_0, I_1, and I_2, the numerator of eqn (6.42) may be approximated simply as $2k_u$. Moreover, for $|\beta - k - \mu| \simeq 1/L_g$, I_0 turns out to be of the order of $8\pi j_b L_g^2 k_u/\bar{\gamma} I_A \simeq 8\bar{\gamma}^2/A_u^2 L_g$ and is $O(M^2 l_u/A_u^2 L_g)$ times smaller than $k + \beta + k_u \simeq 2k \simeq 8\pi\bar{\gamma}^2/M^2 l_u$. By a similar argument it can be seen that $I_1^2/2k$ is $O(M^2 l_u/A_u^2 L_g)$ times smaller than I_2. The r.h.s. of eqn (6.40) may therefore be approximated by simply

$$-kA_u^2 I_2/4 \simeq -\pi j_b k^2 M^2 A_u^2/I_A \bar{\gamma}^5(\beta - k - \mu)^2$$

(for $k_u \simeq kM^2/2\bar{\gamma}^2$).

After the above approximations eqn (6.40) reduces to eqn (6.27). From the above discussion it can be seen that, provided $A_u^2/M^2 = O(1)$ (which is true when $A_u^2 \gtrsim 1$, as it is in most free-electron lasers), the approximate dispersion equation (6.27) is valid to within an error $O(l_u/L_g)$, as expected. This result may be compared with the error of the approximate small-gain-regime gain formula (5.23), which is $O(1/N) = O(l_u/L)$. Again, the gain length is found to play the same role in large-gain FELs as the undulator length does in small-gain ones.

Even if there may be practical difficulties in achieving such a current, one may wonder what would happen were the current so strong that condition (6.43) were violated. We shall not describe this situation, which may be called the 'strong coupling' regime, in any detail, because little research has been performed on it and, in addition, the phenomenology is so radically altered that we feel it must be considered to be outside the scope of undulator theory. Suffice it to say that in the strong-coupling regime the undulator emission is no longer characterized by well defined harmonic lines, as may be inferred from eqn (6.32), which shows that line width becomes comparable with line separation. In fact, the sinusoidal undulating motion of the electrons plays so minor a role that such a device probably should not be called an undulator in the first place. This is so because when $L_g \lesssim l_u$ the changes of the electromagnetic field that occur over a length comparable with the undulator period are considerable: in figurative language one could say that the field does not have the time to realize that the undulator is a periodic structure.

Notice that eqn (6.40) still holds in the strong-coupling regime, but no term is negligible in this case.

6.7 Space-charge effects (Raman regime)

Two types of neglected terms were discovered in the error analysis just performed, one that is proportional to a power of l_u/L_g and another that is proportional to a power of l_u/L_g multiplied by M^2/A_u^2. The distinction between these two types is unnecessary when $A_u^2 \gtrsim 1$ because then $M^2/A_u^2 = O(1)$. Although in a magnetic undulator of the kind used in today's experiments the condition of $A_u^2 \gtrsim 1$ is quite easy to achieve, and there is all the interest in making A_u high because the gain of the FEL increases with it, a considerable amount of research has been performed on the situation in which A_u^2 is small and the terms of the second type indicated above are not negligible, mainly because of the theoretical liaisons that exist with the theory of plasma waves. Such a situation may arise, for instance, when an electromagnetic wave rather than a static magnetic field is used as the undulator or, in microwave applications, when the greater spacing that must be allowed between the undulator poles at the larger wavelengths lowers the achievable field intensity. The modifications of the dispersion equation (6.27) that take place in this case have been given the name of space-charge effects, because in a reference frame moving with the average electron velocity (the E-frame) they may be ascribed to the electrostatic field generated by the e-beam space charge.

An approximate dispersion equation including space-charge effects may be obtained from the exact eqn (6.40) by neglecting terms of order l_u/L_g but not terms of order $M^2 l_u/A_u^2 L_g$ or $M^2 l_u^2/A_u^2 L_g^2$.

Since we do not want to neglect k_p^2 in the l.h.s. of eqn (6.40), let us define a renormalized k as $k' = (k^2 - k_p^2)^{1/2}$ (this is the wavenumber of transverse electromagnetic waves propagating inside the electron beam in the absence of the undulator). Thus, as both β and k' are close to k, the term $\beta^2 - k^2 + k_p^2 = \beta^2 - k'^2 = (\beta + k')(\beta - k')$ may be approximated as $2k(\beta - k')$. Provided that $|\beta - k'| \ll k_u$, the term $[(\beta + 2k_u)^2 - k^2 + k_p^2]^{-1}$ may still be approximated as $\frac{1}{4}kk_u$, and is much smaller than $|2k(\beta - k')|^{-1}$. The l.h.s. of eqn (6.40) therefore reduces to $2k(\beta - k')$.

In the r.h.s. of eqn (6.40) neither I_0 nor I_1 may be neglected, but in the expression (6.41) of I_r the factor $1/\gamma^r$ may be evaluated at an average γ, $\bar{\gamma}$, and moved out of the integral. Moreover, the sum $k + \beta + k_u$ may be approximated as simply $2k$. Equation (6.40) then becomes

$$2k(\beta - k') = -(kA_u^2/4\bar{\gamma}^2)2kI_0/(2k - I_0). \tag{6.45}$$

In the particular case of a monoenergetic beam eqn (6.45) evaluates to

$$2k(\beta - k') = \frac{-\pi k^2 A_u^2 j_b}{\bar{\gamma}^5 I_A[(\beta - k - \mu)^2 - M^2 k_p^2/2\bar{\gamma}^2]}, \tag{6.46}$$

which may be written in terms of the dimensionless variables $p = (\beta - k)L_g$ and $\nu = \mu L_g$ as

$$[(p - \nu)^2 - a^2](p + b) + 1 = 0, \tag{6.47}$$

where

$$a^2 = M^2 k_p^2 L_g^2/2\bar{\gamma}^2 = M^2 l_u/\pi A_u^2 L_g$$

and

$$b = (k - k')L_g \simeq k_p^2 L_g/2k = M^2 l_u^2/A_u^2 L_g^2.$$

Equation (6.47) may be thought of as eqn (6.29) corrected for the effects of space charge, and is uniformly valid with respect to A_u. It is again a cubic equation in p which gives three roots corresponding to three independent travelling waves and reduces to eqn (6.27) when $M^2 l_u/A_u^2 L_g \ll 1$ as expected.

In the opposite case, when $M^2 l_u/A_u^2 L_g \gg 1$, the effects of space charge are predominant. This situation is usually called the 'Raman regime' of the FEL, because the role played in it by space charge is similar to the one it has in the Raman effect. (When it is to be opposed to 'Raman', the regime in which $M^2 l_u/A_u^2 L_g \ll 1$ is called 'Compton'.) In the Raman regime a and possibly b are large parameters, and it becomes possible to give an approximate solution of eqn (6.47). In particular, an interesting situation (resonance) occurs when the detuning parameter v is such that one of the two factors $(p - v - a)$ and $(p - v + a)$ of $[(p - v)^2 - a^2]$ comes close to $(p + b)$, i.e. when either $v \simeq -a - b$ or $v \simeq a - b$. In the first case, $v \simeq -a - b$, we expect p to be close to $v + a$, and therefore we may replace the non-resonant factor, $p - v + a$, by $(v + a) - v + a = 2a$. Equation (6.47) then becomes a quadratic equation, the solution of which is

$$p \simeq (v + a - b)/2 \pm [(v + a + b)^2/4 - 1/2a]^{1/2}. \tag{6.48}$$

Since $v + a + b$ is by assumption a small number, eqn (6.48) gives two complex solutions, corresponding to a growing and a decaying wave. In the second case, when $v \simeq a - b$, the solution may be obtained simply by substituting $-a$ for a in eqn (6.48). In this case the sum under the square root sign is always positive and p is always real. Therefore the phenomenon of FEL amplification takes place only when $v \simeq -a - b$. In particular, the amplification factor, $\mathrm{Im}(p)$, of the growing wave is maximum when $v + a + b = 0$, and is in this case $(1/2a)^{1/2}$.

As the dimensionless variable p is not of order unity in the Raman regime, the gain length L_g of eqn (6.28) is not a particularly significant quantity. By combining the expressions of L_g and a we can define a Raman gain length, as

$$L_{g(\mathrm{Raman})} = (2a)^{1/2} L_g = (8\bar{\gamma}^3 I_A/\pi A_u^4 k_u^2 j_b)^{1/4}. \tag{6.49}$$

As a must be a large number for eqn (6.48) to be valid, $L_{g(\mathrm{Raman})}$ is greater than L_g. In terms of $L_{g(\mathrm{Raman})}$ the FEL power gain at optimal detuning is simply

$$G_{\max(\mathrm{Raman})} = K e^{2L/L_{g(\mathrm{Raman})}}. \tag{6.50}$$

K being an input coupling parameter independent of the undulator length L. The range of validity of eqn (6.50) is determined by the two conditions that

$a \gg 1$, so that eqn (6.48) is valid, and $L_{g(\text{Raman})} \gg l_u$, so that the approximations leading to the simplified dispersion equation (6.47) are allowed. In terms of input-beam current density these two conditions may be written as

$$A_u^4 \bar{\gamma}^3 I_A / l_u^2 \ll j_b \ll \bar{\gamma}^3 I_A / A_u^4 l_u^2. \qquad (6.51)$$

When the left limit of eqn (6.51) is exceeded the FEL falls back into the Compton regime, in which space charge is negligible; when the right limit is exceeded it goes over into the strong-coupling regime where only the exact dispersion equation (6.40) is valid. Equation (6.47) is uniformly valid across the Compton and Raman regimes.

As it is immediately apparent, eqn (6.51) only makes sense when $A_u \ll 1$. In fact, as was said before, only in this case there exists a Raman regime; when $A_u \gtrsim 1$ space charge is negligible for all currents up to the strong-coupling regime. Notice also that, since we do not want A_u to be too small because amplification would then be insignificant, the current density required to satisfy the left condition of eqn (6.51) becomes inpractically high as soon as $\bar{\gamma}$ is even moderately large. For this reason only FELs working with relatively low-energy electron beams may fall in the Raman regime.

6.8 The plasma-theoretical picture of the Raman FEL

In the last section we arrived at the definition of space-charge effects and the Raman regime, starting from the exact dispersion equation (6.40). It is also possible, however, to depict the working of an FEL in the Raman regime as a mode-coupling phenomenon, in the same way as we did in Section 6.4, with the only difference being that the small coupling parameter is now A_u rather than $4\pi j_b / I_A$. This is consistent with the fact that space charge is only significant for small A_u. Let us sketch this alternative picture, which is also found in the literature (e.g., ch. 4 of Marshall 1985).

When $A_u \to 0$, that is when the undulator disappears, the problem described by eqns (6.9)–(6.11) becomes that of the propagation of electromagnetic waves in the presence of the electron beam alone, that is of an electron gas moving at relativistic speed along the z-axis. In this case the one-dimensional equations (6.16)–(6.19) split into two sets. One, formed by eqn (6.18) alone, describes transverse electromagnetic waves which, as is well known, have propagation constant $\beta = \pm(k^2 - k_p^2)^{1/2}$. The other, formed by eqns (6.16), (6.17), and (6.19), describes longitudinal electrostatic, or space-charge, waves with propagation constant $\beta' \simeq k(1 + M^2/2\bar{\gamma}^2) \pm k_p/\sqrt{2\bar{\gamma}}$. This propagation constant may be obtained directly from eqns (6.16), (6.17), and (6.19), or by relativistically transforming into the laboratory frame the dispersion equation of electrostatic waves in an electron gas at rest, which is $\omega(\beta) = \omega_p$ for any β.

A non-zero undulator field introduces oscillating terms with wavenumber $\beta' \pm k_u$ into the r.h.s. of eqn (6.18) and terms with wavenumber $\beta \pm k_u$ into the r.h.s. of eqns (6.16) and (6.17), thus creating the possibility of a resonance between electromagnetic and electrostatic waves with $\beta - \beta' = \pm k_u$, much in the same way as the electron current was seen in Section 6.4 to cause a resonance between vacuum electromagnetic waves with $\beta = k$ and electron density waves with propagation constant $k(1 + M^2/2\bar{\gamma}^2)$ when $\kappa = kM^2/2\bar{\gamma}^2 \simeq k_u$. The difference $\beta - \beta'$ is $(k^2 - k_p^2)^{1/2} - k - \kappa \pm k_p/\sqrt{2\bar{\gamma}}$, and equals $-k_u$ when $\mu = \kappa - k_u = \pm k_p/\sqrt{2\bar{\gamma}} - [k - (k^2 - k_p^2)^{1/2}]$. This is the same resonance condition that was written before in dimensionless form as $\nu = \pm a - b$. Exploiting this resonance one can, as was done in Section 6.4, neglect the non-resonant terms in eqns (6.16)–(6.19) and obtain eqns (6.48)–(6.50) again. Since the Raman regime is characterized by the resonance between one electromagnetic and one electrostatic wave, it is also sometimes called the 'two-wave' regime. In contrast, the Compton regime, which is characterized by the interaction between an electromagnetic wave and two electron density waves, is called the 'three-wave' regime.

The plasma-theoretical approach to the FEL has elicited research on resonances between more elaborate types of plasma waves in the presence of an undulator. Let us quote, for one, the resonance between electron cyclotron waves in a uniform axial magnetic field and electromagnetic waves. The possibility of combining this interaction, which by itself already leads to wave amplification, with the FEL mechanism so as to obtain a larger overall gain has been studied by several authors (e.g., Bernstein and Friedland 1981; McMullin and Bekefi 1982; Freund *et al.* 1982; Grossman, Marshall, and Schlesinger 1983).

6.9 Extension to any polarization

Up to this point we have always explicitly or implicitly assumed that we are working with a circularly polarized undulator and a circularly polarized electromagnetic wave. All the previous results except eqn (6.40) are, however, of wider validity, and may be extended to undulator and wave fields of any polarization. Moreover, just as in the small-gain regime, we shall find that when the polarization is other than circular operation of the FEL near harmonics of the fundamental resonance wavenumber k_r becomes possible.

Let us start out again from eqns (6.33)–(6.35), which were written without specifying the polarization of $\boldsymbol{A}_u$. As explained in Section 6.5, these are linear differential equations with periodic coefficients and, according to Floquet's theory, admit travelling-wave solutions in which each unknown is given by the product of an exponential factor $e^{-i\beta z}$ and a periodic function with period l_u. (Notice that β is defined only up to an integral multiple of k_u. For $f(z)e^{i\beta z}$, where $f(z)$ is any periodic function, may always be written as $f'(z)e^{i(\beta + mk_u)z}$, with $f'(z) = f(z)e^{-imk_u z}$ being another periodic function.) In the case of general polarization a closed-form solution such as the one given in Section 6.5 does not exist, but under condition (6.43) it is possible to obtain an approximate solution based on the resonant character and weak coupling of the electromagnetic and Boltzmann equations.

Let us therefore assume that eqn (6.43) holds, i.e. $l_u/L_g \ll 1$. Let us also assume, for the sake of simplicity, that $A_u^2 \gg l_u/L_g$, so that space-charge effects are negligible. Then eqn (6.35) and the term proportional to $ikA_{1z} + dA_{1z}/dz$ in eqn (6.33) may be discarded. It is not difficult to show, by an argument similar to the following, that the results of Sections 6.7 and 6.8 concerning space-charge effects are also valid for general polarization.

In order to give an approximate solution of eqns (6.33) and (6.34), we begin by formally integrating eqn (6.33) as if the r.h.s. of this equation were known. According to the general formula for solving linear first-order differential equations,

$$F_1 = F_{1\text{free}}(\gamma, z) \int_{z_0}^{z} \frac{-ik}{\gamma} \frac{dF_0}{d\gamma} A_{1\perp} \cdot \mathrm{Re}(A_u e^{ik_u z}) F_{1\text{free}}^{-1} \, dz, \tag{6.52}$$

where $F_{1\text{free}}(\gamma, z)$, the free solution of the l.h.s. of eqn (6.33), is given by

$$F_{1\text{free}} = \exp\left[-ik\left(1 + \frac{M^2}{2\gamma^2}\right)z - ik\,\mathrm{Re}\left(\frac{A_u^2}{8i\gamma^2 k_u} e^{2ik_u z}\right)\right] \tag{6.53}$$

and z_0 is arbitrary. Now, according to Floquet theory (Stoker 1950), A_1 must be expressed by a periodic function of z with spatial frequency k_u multiplied by the exponential factor $e^{-i\beta z}$; on the other hand, according to eqn (6.53), $F_{1\text{free}}^{-1}$ is given by a periodic function of z with spatial frequency k_u multiplied by the exponential factor $\exp[-i(1 + M^2/2\gamma^2)kz]$. All together, the integrand in eqn (6.52) is the product of a periodic function with spatial frequency k_u and the exponential factor $\exp[i(k + kM^2/2\gamma^2 - \beta)z]$, and admits the Fourier series expansion:

$$A_{1\perp} \cdot \mathrm{Re}(A_u e^{ik_u z}) F_{1\text{free}}^{-1} = \sum_{n=-\infty}^{\infty} C_n e^{i(k-\beta+\mu_n)z} \tag{6.54}$$

where $\mu_n = kM^2/2\gamma^2 - nk_u$ and

$$C_n = \frac{1}{l_u} \int_0^{l_u} A_{1\perp} \cdot \mathrm{Re}(A_u e^{ik_u z}) \exp\left[i(\beta + nk_u)z + ik\,\mathrm{Re}\left(\frac{A_u^2}{8i\gamma^2 k_u} e^{2ik_u z}\right)\right] dz. \tag{6.55}$$

The assumption of near resonance implies that $|k - \beta|$ is small compared to k_u and, for a particular n, $|\mu_n|$ is small compared to k_u, so that $|k - \beta + \mu_u|$ is also small compared to k_u. When this happens, the nth term of the Fourier series (6.54), which appears under the sign of integral in eqn (6.52), will be multiplied after integration by a factor of $1/i(k - \beta + \mu_n)$, which is at least $k_u/|k - \beta + \mu_n|$ times larger than the similar factor multiplying any other term. We are therefore allowed to reduce the series to this single low-frequency term. Just as in the small-gain case, we shall refer to operation in a condition where $|\mu_n| \ll k_u$ for a particular value of n as operation on the nth harmonic.

Having reduced the series of eqn (6.41) to the nth harmonic only, we are left with an expression of F_1 that is proportional to the free solution $F_{1\mathrm{free}}$ multiplied by the low-frequency exponential factor $\mathrm{e}^{\mathrm{i}(k-\beta+\mu_n)z}$, i.e.

$$F_1 \simeq -\frac{k}{\gamma(k-\beta+\mu_n)}\frac{\mathrm{d}F_0}{\mathrm{d}\gamma}C_n \exp\left[-\mathrm{i}(\beta+nk_u)z - \mathrm{i}k\,\mathrm{Re}\left(\frac{A_u^2}{8\mathrm{i}\gamma^2 k_u}\mathrm{e}^{2\mathrm{i}k_u z}\right)\right]. \tag{6.56}$$

Notice that, in compliance with Floquet theory, F_1 turns out to be given by a periodic function of z with spatial frequency k_u multiplied by the exponential factor $\mathrm{e}^{-\mathrm{i}\beta z}$.

A similar approximation may be carried out in order to obtain $\boldsymbol{A}_{1\perp}$ from eqn (6.34). Since this is a constant-coefficient differential equation, it is possible to apply Fourier analysis directly. On remembering that both $\boldsymbol{A}_{1\perp}$ and the r.h.s. of eqn (6.34) must be expressed by periodic functions of z with spatial frequency k_u multiplied by $\mathrm{e}^{-\mathrm{i}\beta z}$, the formal solution of eqn (6.34) may be written in the form of a Fourier series as

$$\boldsymbol{A}_{1\perp} = \sum_{m=-\infty}^{\infty} \boldsymbol{C}'_m \mathrm{e}^{-\mathrm{i}(\beta+mk_u)z}/[k^2-(\beta+mk_u)^2] \tag{6.57}$$

where

$$\boldsymbol{C}'_m = \frac{1}{l_u}\int_0^{l_u}\left(\int_0^\infty \frac{F_1}{\gamma}\,\mathrm{d}\gamma\right)\mathrm{Re}(\boldsymbol{A}_u \mathrm{e}^{\mathrm{i}k_u z})\mathrm{e}^{\mathrm{i}(\beta+mk_u)z}\,\mathrm{d}z. \tag{6.58}$$

(The term $\boldsymbol{A}_{1\perp}\int_0^\infty F_0/\gamma\,\mathrm{d}\gamma = k_p^2\boldsymbol{A}_{1\perp}$ of eqn (6.34) has been omitted because this term is of the order of magnitude of space-charge effects. It may be included, if necessary, by substituting $k' = (k^2-k_p^2)^{1/2}$ in the place of k as in Section 6.7).

With reference to eqn (6.57) the hypothesis of near resonance is that in one term of the Fourier series the denominator $k^2-(\beta+mk_u)^2$ should be much smaller than in the others. When this happens we can neglect all but this term (which can always be made to correspond to $m=0$ by properly defining β). Since when $|k-\beta|$ is small $k^2-\beta^2 \simeq 2k(k-\beta)$, we may assume for $\boldsymbol{A}_{1\perp}$ the approximate expression

$$\boldsymbol{A}_{1\perp} \simeq \boldsymbol{C}'_0\mathrm{e}^{-\mathrm{i}\beta z}/2k(k-\beta). \tag{6.59}$$

Notice that the r.h.s. of eqn (6.59), like that of eqn (6.56), may be thought to be given by the free solution of the relevant equation (which is just a complex exponential function in this case) multiplied by a low-frequency exponential factor.

In order to determine the coefficients C_n and C_0' together with the dispersion equation for β, all that remains to be done is to insert the expression (6.59) of $A_{1\perp}$ into eqn (6.55) and the expression (6.56) of F_1 into eqn (6.58).

Equation (6.58), for $m = 0$, gives

$$C_0' = -C_n \int_0^\infty A_{ueff,n} \frac{k\,\mathrm{d}F_0/\mathrm{d}\gamma}{2\gamma^2(k-\beta+\mu_n)}\,\mathrm{d}\gamma, \tag{6.60}$$

where

$$A_{ueff,n} = \frac{1}{l_u}\int_0^{l_u} \Big\{(A_u \mathrm{e}^{\mathrm{i}k_u z} + A_u^* \mathrm{e}^{-\mathrm{i}k_u z}) \times \exp\Big[-\mathrm{i}nk_u z - \mathrm{i}k\,\mathrm{Re}\Big(\frac{A_u^2}{8\mathrm{i}\gamma^2 k_u}\,\mathrm{e}^{2\mathrm{i}k_u z}\Big)\Big]\Big\}\,\mathrm{d}z. \tag{6.61}$$

Equation (6.55) gives

$$C_n = C_0' \cdot A_{ueff,n}^*/4k(k-\beta). \tag{6.62}$$

with $A_{ueff,n}$ expressed again by eqn (6.61). The condition for the compatibility of eqns (6.60) and (6.62), i.e. the dispersion equation, is[4]

$$4k(k-\beta) = \int_0^\infty A_{ueff,n}^2 \frac{k\,\mathrm{d}F_0/\mathrm{d}\gamma}{2\gamma^2(k-\beta+\mu_n)}\,\mathrm{d}\gamma \tag{6.63}$$

which, in the monoenergetic case, differs from eqn (6.27) only by the substitution of $A_{ueff,n}$ for A_u.

The vector $A_{ueff,n}$ defined by eqn (6.61) is the same effective undulator vector potential (4.36) which appeared in the theory of spontaneous emission and small gain, and its properties have already been described in Section 4.5. Therefore it is true in the large-gain regime also that the behaviour of an undulator having any polarization and working on any harmonic is similar to that of a circularly polarized one having a vector potential A_u equal to $A_{ueff,n}$ and a detuning parameter v equal to v_n. This statement is true whenever the detuning parameter v is close to its resonance value in both the Compton and Raman regimes, although we have proved it only in the first. It fails in the strong-coupling regime, because when $|\beta - k| \gtrsim k_u$ reducing the Fourier series of eqns (6.54) and (6.57) to just one term is not allowed.

Among the consequences of the above equivalence is that the considerations of Sections 5.8 and 4.5 on harmonic amplification in the small-gain regime and relative harmonic intensity extrapolate directly to the large-gain regime. Notice, however, that harmonic amplification is negligible in the

Raman regime, because the Raman regime is limited to $A_u \ll 1$ and harmonic amplification requires $A_u \gtrsim 1$.

6.10 Transition between the small-gain and large-gain regimes

After studying the properties of travelling waves supported by the FEL structure, we are now going to deal with how these waves are excited. In order to do so, we shall have to consider the initial conditions associated with the electromagnetic wave and Boltzmann equations. For the sake of simplicity we shall only consider the case of a monoenergetic and collimated electron beam in which, according to Section 6.4, three different waves can propagate in the FEL. The procedure to be used in the presence of energy and momentum spread is analogous but may involve a continuum of waves.

The boundary conditions that must be associated with the basic equations of the large-gain FEL, eqns (6.8)–(6.11), are that the zero-order distribution function f_0 should coincide at the undulator entrance, $z = 0$, with the distribution function of the oncoming electron beam; that the first-order distribution function f_1 should be zero at the same position; and that the electromagnetic vector potential should match any input waves that are imposed on to the system. For the one-dimensional, collimated, monoenergetic beam model of Section 6.3, formed by eqns (6.16)–(6.19), the corresponding conditions are that C_0 and C_1 be zero at $z = 0$ and that for $z < 0$ and $z > L$, where $\boldsymbol{a}_0 = 0$ and the electromagnetic vector potential may be written as the sum of two terms proportional, respectively, to e^{-ikz} and to e^{ikz}, the coefficient of the term representing an incoming wave, i.e. e^{-ikz} for $z < 0$ and e^{ikz} for $z > L$, be imposed.

As was shown in Section 6.4 for the case of circular polarization, near resonance the system of equations (6.16)–(6.19) may be approximated by eqns (6.20), (6.23), and (6.25). It has been shown in Section 6.9 that the same is true for any polarization, provided that $A_{ueff,n}$ is substituted for A_u. Since eqn (6.25) is of lower order than eqn (6.18), one less boundary condition can be imposed. The choice of the condition to be eliminated follows from the hypothesis that underlies eqn (6.25), i.e. that the field configuration be close to a forward-propagating wave. Equation (6.25) is unable to represent a backward-propagating electromagnetic wave, and therefore it is impossible to impose the condition that such a wave be present for $z > L$.

Summing up, the problem we have is to solve the system of eqns (6.20), (6.23), and (6.25), with the boundary conditions $C_0 = C_1 = 0$ and $A_1 = A_{wi}$ at $z = 0$, A_{wi} being the vector potential of the input electromagnetic wave to be amplified by the FEL. The solution of this problem may be written as a linear superposition of the three travelling waves that were determined to exist, for a given k, in Section 6.4:

$$A_1 = A_{w1}\mathrm{e}^{-i\beta_1 z} + A_{w2}\mathrm{e}^{-i\beta_2 z} + A_{w3}\mathrm{e}^{-i\beta_3 z}, \tag{6.64}$$

where β_1, β_2, and β_3 are the three roots of eqn (6.27) and A_{w1}, A_{w2}, and A_{w3} are vector coefficients with the same direction as A_u ($A_{ueff,n}$ when the polarization is not circular), and amplitudes to be determined with the aid of the initial conditions. C_0 and C_1 have similar expressions, with coefficients related to A_{w1}, A_{w2}, and A_{w3} by eqns (6.21) and (6.24).

In order to be able to satisfy the initial condition on A_1 with eqn (6.64) we must assume that the input field has the same direction and polarization as A_u ($A_{ueff,n}$). (Actually, the equations have one more mode which has $\beta = k$, does not interact with the electron beam in any way, and matches a possible component of the input field in the direction orthogonal to that of A_u. We need not be concerned with this component.) The initial condition on A_1 may then be imposed as a condition on amplitudes, and simply gives

$$A_{w1} + A_{w2} + A_{w3} = A_{wi}. \tag{6.65}$$

The condition $C_0(0) = 0$ implies, according to eqn (6.21), that

$$(4\pi j_b/2\gamma I_A)A_u^* \cdot [A_{w1}/(\beta_1 - k - \mu) + A_{w2}/(\beta_2 - k - \mu) + A_{w3}/(\beta_3 - k - \mu)] = 0,$$

where we have put into evidence the factor independent of β. In terms of the dimensionless quantities $p = (\beta - k)L_g$ and $\nu = \mu L_g$ this equation may be written as

$$A_{w1}/(p_1 - \nu) + A_{w2}/(p_2 - \nu) + A_{w3}/(p_3 - \nu) = 0. \tag{6.66}$$

Analogously, by using eqn (6.24), the condition $C_0 - C_1/\gamma = 0$, which is equivalent to $C_1 = 0$, may be written as

$$A_{w1}/(p_1 - \nu)^2 + A_{w2}/(p_2 - \nu)^2 + A_{w3}/(p_3 - \nu)^2 = 0. \tag{6.67}$$

Some algebraic manipulation[5] gives the solution of eqns (6.65)–(6.67) as

$$A_{wr} = \frac{p_r - \nu}{3p_r - \nu} A_{wi} \qquad (r = 1, 2, 3). \tag{6.68}$$

Putting eqn (6.68) back into eqn (6.64) then gives the solution of our problem.

The above solution enables us to write a formula for the gain of an FEL which is uniformly valid for any length. (Recall that eqn (6.28) is only valid for $L \gg L_g$, and eqn (5.27) is only valid for $L \ll L_g$.) The power gain G, defined as the ratio between the values of $|A_1|^2$ at the exit ($z = L$) and entrance ($z = 0$) of the undulator, turns out to be

$$G = \left| \sum_{r=1}^{3} \frac{p_r - \nu}{3p_r - \nu} e^{-ip_r L/L_g} \right|^2. \tag{6.69}$$

Equation (6.69) is plotted in Fig. 6.2. For $L/L_g \gg 1$, when the exponential factor corresponding to the root with a positive imaginary part exceeds the others by far, eqn (6.69) attains the limit form of eqn (6.28), with $K =$

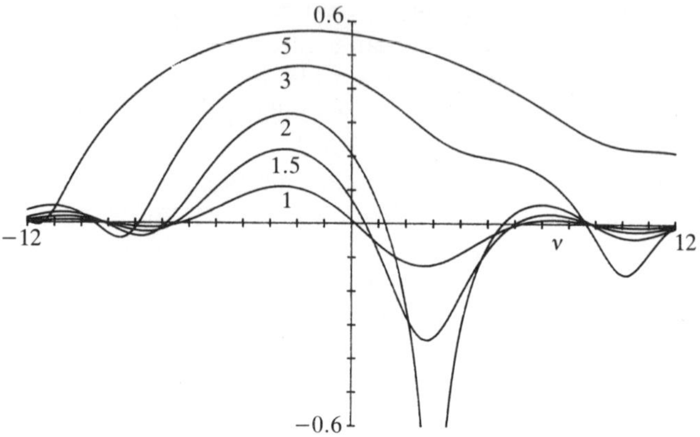

FIG. 6.2. Transition between the small-gain and large-gain FEL regimes. Plot of $(L_g/L)\log G$ vs. $(L_g/L)\nu$ for various values of L/L_g. Each curve is labelled with the corresponding value of L/L_g. It may be observed that for $L/L_g = 1$ the small-gain curve represented in Fig. 5.1 and for $L/L_g = 5$ the large-gain one of Fig. 6.1 are approached.

$|(p - \nu)/(3p - \nu)|^2$, while for $L/L_g \ll 1$ we expect it to approach the small-gain result of eqn (5.27).

Since eqn (5.27) was obtained in a completely different framework, it is instructive to verify how the small-gain limit is attained. To this end we need to consider a situation in which the ratio L/L_g is small, but at least one of the p_r is so large that the exponent $p_r L/L_g$ is still of order unity. Such a situation is realized when the detuning parameter ν itself is large, of the order of L_g/L. (This is consistent with what was observed in Section 6.4, that the ratio between the detuning parameters appropriate to the small-gain and large-gain regimes is $L/2L_g$.) When ν is large the roots of the dispersion equation (6.29) may be approximated by perturbation analysis, which shows that one root approaches zero, or in a higher approximation $-1/\nu^2$, and two approach ν or, more precisely, $\nu \pm i\nu^{-1/2} + 1/2\nu^2$. Equation (6.69) may then be approximated as

$$G \simeq \left| \left(1 - \frac{2}{\nu^3}\right) e^{iL/\nu^2 L_g} + \left(\frac{i}{2\nu^{3/2}} + \frac{1}{\nu^3}\right) e^{(-i\nu + \nu^{-1/2})L/L_g} + \left(-\frac{i}{2\nu^{3/2}} + \frac{1}{\nu^3}\right) e^{(-i\nu - \nu^{-1/2})L/L_g} \right|^2, \tag{6.70}$$

i.e. upon letting $e^{iL/\nu^2 L_g} \simeq 1 + iL/\nu^2 L_g$ and $e^{\pm \nu^{-1/2} L/L_g} \simeq 1 \pm \nu^{-1/2} L/L_g$,

$$\begin{aligned} G &\simeq \left| 1 - \frac{2}{\nu^3} + \frac{iL}{\nu^2 L_g} + \left(\frac{2}{\nu^3} + \frac{iL}{\nu^2 L_g}\right) e^{-i\nu L/L_g} \right|^2 \\ &\simeq 1 + [4\cos(\nu L/L_g) - 4 + 2(\nu L/L_g)\sin(\nu L/L_g)]/\nu^3. \end{aligned} \tag{6.71}$$

On replacing vL/L_g by $2v$, according to the definition of v adopted in the previous chapter, eqn (5.27) is obtained again.

6.11 Three-dimensional large-gain theory and optical guiding

Let us now remove the assumption of a transversely infinite electron beam and study the three-dimensional solutions of eqns (6.8)–(6.11). On doing so the three plane-wave modes that compose the 1-D solution will be replaced by three infinite sets of non-planar modes. Since the electron beam now has finite width, these modes, if they exist, must be in some sense confined to the proximity of the electron beam itself. We express this concept by saying that the modes are 'guided.' Guiding may be very noticeable in practice, in the sense that the modes may be very different from, and possibly much thinner than, a free-space Gaussian mode. We shall retain the assumption of an ideal sinusoidal undulator field.

The three-dimensional large-gain model of the free-electron laser is formed by eqns (6.8)–(6.11). The general solution of eqn (6.8) for the ultrarelativistic Hamiltonian $\mathscr{H}_0 = \gamma - [1 + (\boldsymbol{P}_\perp - \boldsymbol{a}_{0\perp})^2]/2\gamma$, with $\boldsymbol{a}_{0\perp} = \mathrm{Re}(\boldsymbol{A}_u \mathrm{e}^{\mathrm{i}k_u z})$, may be found by the method of characteristics (the characteristic equations of the Boltzmann equation are simply the Hamilton equations of motion), and is

$$f_0(z, \boldsymbol{x}, t, \boldsymbol{P}_\perp, \gamma) = F\left\{\boldsymbol{x} - \int_0^z \frac{\boldsymbol{P}_\perp - \boldsymbol{a}_0}{\gamma}\,\mathrm{d}z,\ t - \int_0^z \left[1 + \frac{1 + (\boldsymbol{P}_\perp - \boldsymbol{a}_0)^2}{2\gamma^2}\right]\mathrm{d}z,\ \boldsymbol{P}_\perp, \gamma\right\}, \tag{6.72}$$

where F is an arbitrary function of its four arguments which coincides with the initial distribution function $f_0(0, \boldsymbol{x}, t, \boldsymbol{P}_\perp, \gamma)$. Guided modes will be obtained when f_0 is a periodic function of z. In order for f_0 to be so, F must be independent of t and zero for non-zero values of $\boldsymbol{P}_\perp$, i.e. proportional to $\delta(\boldsymbol{P}_\perp)$.[6] Therefore we restrict our attention to f_0 of the form

$$f_0 = F_0\left(\boldsymbol{x} + \int_0^z \frac{\boldsymbol{a}_0}{\gamma}\,\mathrm{d}z, \gamma\right)\delta(\boldsymbol{P}_\perp), \tag{6.73}$$

F_0 denoting an arbitrary function of two arguments, coinciding with the position and energy distribution of electrons at the beginning of the undulator. In the previous sections we chose F_0 to be independent of $\boldsymbol{x}$. In the particular case of a monoenergetic input electron beam

$$f_0 = \frac{4\pi}{I_A} j_b\left(\boldsymbol{x} + \int_0^z \frac{\boldsymbol{a}_0}{\bar{\gamma}}\,\mathrm{d}z\right)\delta(\gamma - \bar{\gamma})\delta(\boldsymbol{P}_\perp), \tag{6.74}$$

where $j_b(\boldsymbol{x})$ is the current density in the cross-section of the input beam.

f_0 and $\mathscr{H}_0$ being periodic functions of z and independent of time, the coefficients of the linear system of equations (6.9)–(6.11) are all independent of time and periodic in z with period k_u. Just as in the one-dimensional case, Floquet's theory then tells us that solutions exist where a generic variable, say $\boldsymbol{a}_1$, takes on the form $\boldsymbol{a}_1(z, \boldsymbol{x}, t) = \mathrm{Re}[\mathrm{e}^{\mathrm{i}(kt-\beta z)}\boldsymbol{A}_1(z, \boldsymbol{x})]$, with $\boldsymbol{A}_1$ a periodic function of its first argument z. Solutions of this form are non-planar travelling waves and repeat themselves, apart for a phase and amplitude change, every period. They will be called by definition 'modes', characterized by a wavenumber k and a propagation constant β.

Modes of this kind share many properties with the previously studied one-dimensional modes, in particular that for any given wavenumber k only certain, generally complex, eigenvalues of the propagation constant β are allowed (notice that here too the propagation constant is only defined up to an integral multiple of k_u). Each mode is characterized by an exponential amplification or attenuation with increasing z, at a rate determined by the imaginary part of β. The mode with the highest positive imaginary part of β is particularly important because it is amplified the most, and in a sufficiently long undulator will outgrow all the others. Far enough from the entrance the laser field will practically coincide with this mode, and since the transverse section of a mode is periodic in z it will appear to be confined in a sausage-like shape. It will become a 'guided' laser beam.

As has been remarked already in the one-dimensional case, there is no general method of calculating the solution of a system of differential equations with periodic coefficients. This is *a fortiori* true for eqns (6.9)–(6.11), which are partial differential equations. Although it is not impossible to reduce the problem to equations with coefficients independent of z by exploiting the resonance conditions and neglecting non-resonant terms, in a similar manner to what was done in Section 6.8, the resulting equations are still quite complicated. An alternative solution, which is adopted in the great majority of the research papers published to date, is to restrict one's attention to relatively wide electron beams and replace the 3-D zero and first-order Boltzmann equations (6.8) and (6.9) by their 1-D counterparts, retaining the full 3-D form of the wave equation (6.10) only.

In order to find the conditions under which the above approximation is legitimate, let us start by recognizing that there are three different transverse scales of length in the problem. One scale, fixed by the initial conditions, is the width σ_e of the input electron beam. A second scale, characteristic of the Boltzmann equation, is the radius of the unperturbed electron orbit $\Delta x_u = A_u/\gamma k_u$ (3.29). A third scale, characteristic of diffraction phenomena associated with the electromagnetic wave equation, may be written as $w = (L_g/2k)^{1/2}$, on assuming that, as in the 1-D case, the typical longitudinal scale is the gain length L_g. In terms of these three scales the previous 1-D analysis may be expected to be valid when σ_e is much larger than both Δx_u and w (this expectation will be confirmed later). When σ_e becomes comparable with w

diffraction effects can no longer be neglected, but if both σ_e and w are much larger than Δx_u it is conceivable that the Boltzmann equation may still be treated as one-dimensional.

For the zero-order Boltzmann equation (6.8) the validity of the above considerations may be immediately deduced from the exact solution (6.73): if the width of the electron beam is much larger than Δx_u, i.e. the maximum of $\int_0^z \boldsymbol{a}_0/\gamma\, \mathrm{d}z$, this term may be neglected and the zero-order distribution function may be treated as independent of z. For eqn (6.9) a term-by-term analysis, which we omit for the sake of brevity, confirms the validity of the same criterion.

Thus, under the assumptions that $\Delta x_u \ll \sigma_e$ and $\Delta x_u \ll w$, the problem of the 3-D evolution of an electromagnetic wave in a large-gain FEL may be reduced to the single partial differential equation (6.10), which we may rewrite for a time-harmonic field $\boldsymbol{a}_1 = \mathrm{Re}(\boldsymbol{A}_1 \mathrm{e}^{ikt})$ as

$$\Delta_{2\perp}\boldsymbol{A}_{1\perp} + \frac{\partial^2 \boldsymbol{A}_{1\perp}}{\partial z^2} + k^2 \boldsymbol{A}_{1\perp} = \int_0^\infty -\frac{\boldsymbol{a}_0}{\gamma} F_1\, \mathrm{d}\gamma, \tag{6.75}$$

where $\Delta_{2\perp}$ denotes the Laplacian operator in the transverse plane and F_1 is related to $\boldsymbol{A}_1$ by the solution of the 1-D Boltzmann equation (6.33) (without space-charge terms):

$$\frac{\partial F_1}{\partial z} + \mathrm{i}k\left(1 + \frac{M^2 + \mathrm{Re}(\boldsymbol{A}_u^2 \mathrm{e}^{2ik_u z})/2}{2\gamma^2}\right)F_1 = -\frac{ik}{\gamma}\boldsymbol{A}_{1\perp}\cdot \mathrm{Re}(\boldsymbol{A}_u \mathrm{e}^{ik_u z})\frac{\partial F_0}{\partial \gamma}. \tag{6.76}$$

In the case of circular polarization, when $\boldsymbol{A}_u^2 = 0$ and $\boldsymbol{A}_u \cdot \boldsymbol{A}_{1\perp} = 0$, eqns (6.75) and (6.76) have a solution of the form $\boldsymbol{A}_{1\perp} = \boldsymbol{A}_w(\boldsymbol{x})\mathrm{e}^{-i\beta z}$, $F_1 = F_w \mathrm{e}^{-i(\beta + k_u)z}$, where $F_w = k(\partial F_0/\partial\gamma)\boldsymbol{A}_u^* \cdot \boldsymbol{A}_w / 2\gamma(\beta - k - \mu)$ and $\boldsymbol{A}_w$ satisfies the equation

$$\Delta_{2\perp}\boldsymbol{A}_w + (k^2 - \beta^2)\boldsymbol{A}_w = -k\boldsymbol{A}_u \boldsymbol{A}_u^* \cdot \boldsymbol{A}_w \int_0^\infty \frac{\partial F_0/\partial\gamma}{4\gamma^2(\beta - k - \mu)}\,\mathrm{d}\gamma \tag{6.77}$$

which, after an integration by parts,[7] may be rewritten as

$$\Delta_{2\perp}\boldsymbol{A}_w + (k^2 - \beta^2)\boldsymbol{A}_w = k\boldsymbol{A}_u \boldsymbol{A}_u^* \cdot \boldsymbol{A}_w \int_0^\infty \frac{kM^2 F_0}{4\gamma^5(\beta - k - \mu)^2}\,\mathrm{d}\gamma. \tag{6.78}$$

In the case of general polarization, eqns (6.75) and (6.76) must be treated as equations with periodic coefficients in z, and their exact solution has the form of a Fourier series such as eqn (6.57), but with coefficients $\boldsymbol{C}'_m$ that are functions of $\boldsymbol{x}$. However, just as in Section 6.9, under weak-coupling and near-resonance conditions, i.e. when $1/L_g \ll k_u$ and $|\mu_n| = |\kappa - nk_u| \ll k_u$, it is possible to approximate the series by a single, resonant, term and reduce the problem, even in the case of operation on higher harmonics, to an equation similar to eqn (6.78) but with $\boldsymbol{A}_{ueff,n}$ in the place of $\boldsymbol{A}_u$.

Equation (6.78) is a vector equation. By projecting it onto the direction of $\boldsymbol{A}_u$ and the direction orthogonal to it, i.e. by multiplying it once with $\boldsymbol{A}_u$ and then with $\boldsymbol{A}_u^*$, it can be seen that the orthogonal component does not interact with the FEL at all (because $\boldsymbol{A}_u^* \cdot \boldsymbol{A}_w = 0$), whereas for the parallel component the product $\boldsymbol{A}_u \boldsymbol{A}_u^* \cdot \boldsymbol{A}_w$ in eqn (6.78) may also be written as $A_u^2 \boldsymbol{A}_w$. As the orthogonal component is uninteresting, in the following we shall assume $\boldsymbol{A}_w$ to have only the component parallel to $\boldsymbol{A}_u$, and we shall write eqn (6.78) as a scalar equation, denoting by A_w the complex amplitude of this component (not the modulus $|\boldsymbol{A}_w|$, which is real).

In particular, for a monoenergetic beam, when F_0 is given by $[4\pi j_b(\boldsymbol{x})/I_A]\delta(\gamma - \bar{\gamma})$, eqn (6.78) becomes

$$\Delta_{2\perp} A_w + (k^2 - \beta^2) A_w = \frac{4\pi j_b(\boldsymbol{x})}{I_A} \frac{k^2 M^2 A_u^2}{4\bar{\gamma}^5 (\beta - k - \mu)^2} A_w. \tag{6.79}$$

In eqns (6.78) and (6.79) the difference $k^2 - \beta^2$ may be approximated, as usual, by $2k(k - \beta)$. By introducing the gain length, calculated with respect to the on-axis current as $L_g = [2\bar{\gamma}^5 I_A / \pi M^2 A_u^2 k j_b(0)]^{1/3}$, and the dimensionless quantities $p = (\beta - k)L_g$ and $v = \mu L_g$, we may rewrite eqn (6.79) in dimensionless form as

$$\Delta_{2\perp} A_w - [p + J(\boldsymbol{x})/(p - v)^2] A_w = 0, \tag{6.80}$$

where J is the normalized current-density profile $j_b(\boldsymbol{x})/j_b(0)$, and transverse co-ordinates have been non-dimensionalized with respect to the diffraction characteristic width $w = (L_g/2k)^{1/2}$.

6.12 Solutions of the mode equation

Equation (6.80) is a partial differential equation containing a parameter, p, which must be solved with radiation boundary conditions specifying that A_w behaves at infinity as an outgoing wave, i.e. as $e^{-i\alpha|\boldsymbol{x}|}$ for some α with a positive real part. Solutions exist only for discrete eigenvalues of the parameter p, and these solutions are the FEL propagation modes.

Equation (6.80) may be solved analytically only for very special forms of the current density profile $J(\boldsymbol{x})$. These include the unphysical, although not uninformative, parabolic profile $J = 1 - |\boldsymbol{x}|^2/\sigma^2$, for which eqn (6.80) reduces to the parabolic-cylinder equation, and the stepwise profile $J = 1$ for $|\boldsymbol{x}| < \sigma$ and 0 for $|\boldsymbol{x}| > \sigma$, which was studied by Moore (1984). For other profiles, e.g. for the Gaussian profile $J = e^{-|\boldsymbol{x}|^2/\sigma^2}$ that is typical of the electron beam emitted by many accelerating machines, only a numerical or an approximate analytical solution is possible.

Among the approximate methods that have been proposed to solve eqn (6.80) one of the most flexible and general is based on an extension of the Rayleigh–Ritz variational principle. Let us examine this method, which was first applied on a somewhat heuristic basis by Xie and Deacon (1986), and later justified and extended by Luchini and Motz (1988), in some detail.

In order to construct the variational principle, let us multiply eqn (6.80) by A_w and integrate over the whole $\boldsymbol{x}$ plane. By Green's formula, $A_w \Delta_2 A_w = \nabla \cdot (A_w \nabla A_w) - (\nabla A_w)^2$. Since, at least for the growing modes, which have A_w vanishing at infinity,[8] the integral of the divergence term is zero, one has

$$\int (\nabla_\perp A_w)^2 \, \mathrm{d}^2\boldsymbol{x} + p \int A_w^2 \, \mathrm{d}^2\boldsymbol{x} + (p - \nu)^{-2} \int J A_w^2 \, \mathrm{d}^2\boldsymbol{x} = 0. \tag{6.81}$$

Equation (6.81) may be read as an algebraic equation for p and can be used to determine p if A_w is given. But, in addition, eqn (6.81) can be used to determine A_w itself, because it can be shown that the l.h.s. of eqn (6.81) is an extremum with respect to variations of A_w (i.e. its variational first derivative is zero) just when A_w is the solution of eqn (6.80). In fact, upon giving an infinitesimal variation $\delta A_w(\boldsymbol{x})$ to $A_w(\boldsymbol{x})$ the l.h.s. of eqn (6.81) undergoes a variation equal to

$$\int 2\nabla_\perp A_w \cdot \nabla_\perp \delta A_w \, \mathrm{d}^2\boldsymbol{x} + \int \left[p + \frac{J(\boldsymbol{x})}{(p - \nu)^2} \right] 2 A_w \delta A_w \, \mathrm{d}^2\boldsymbol{x}. \tag{6.82}$$

By Green's formula, where again the contribution of the divergence term is zero, eqn (6.82) may be transformed into

$$\int -2\Delta_{2\perp} A_w \delta A_w \, \mathrm{d}^2\boldsymbol{x} + \int \left[p + \frac{J(\boldsymbol{x})}{(p - \nu)^2} \right] 2 A_w \delta A_w \, \mathrm{d}^2\boldsymbol{x}, \tag{6.83}$$

and this last expression is identically zero for any $\delta A_w(\boldsymbol{x})$ when and only when A_w satisfies eqn (6.80).

Just like the original Rayleigh–Ritz principle (for which see, e.g., Argence and Kahan 1964), from which it differs in that eqn (6.80) is not linear with respect to the eigenvalue parameter p and is not self-adjoint, this variational principle may be used to determine approximate solutions within a pre-assigned family of functions. To this aim it is sufficient to insert a trial expression of A_w containing a number of free parameters into eqn (6.81) and then extremize its l.h.s. with respect to the free parameters. The more free parameters there are, the better is the approximation that may be attained.

As an illustration of the above technique, and also for its intrinsic interest, let us determine a simple approximation for the case of a circular Gaussian current-density distribution represented by $J = \mathrm{e}^{-|\boldsymbol{x}|^2/\sigma^2}$. As the trial expression of A_w we shall also assume a Gaussian, given by $A_w = (2\pi w^2)^{-1} \mathrm{e}^{-|\boldsymbol{x}|^2/2w^2}$, leaving w^2 as a complex free parameter. Substituting these expressions of J and A_w into eqn (6.81) gives

$$1/w^2 + p + \sigma^2/(p - \nu)^2(w^2 + \sigma^2) = 0. \tag{6.84}$$

Differentiating the l.h.s. of eqn (6.84) with respect to w^2 and equating the result to zero (i.e. imposing the extremum condition) gives the additional equation:

$$1/w^4 + \sigma^2/(p-\nu)^2(w^2+\sigma^2)^2 = 0. \qquad (6.85)$$

Equations (6.84) and (6.85) together determine both w^2 and p. In particular, eqn (6.85) yields

$$1 + \sigma^2/w^2 = \pm \mathrm{i}\sigma/(p-\nu), \qquad (6.86)$$

where the $\pm$ sign must be chosen so that $\mathrm{Im}(w^2) < 0$ (namely, the $+$ sign if $\mathrm{Re}(p-\nu) > 0$, and the $-$ sign if $\mathrm{Re}(p-\nu) < 0$) in order to comply with the radiation condition. Inserting eqn (6.86) into eqn (6.84) gives, after a few calculations, the cubic equation

$$p(p-\nu)^2 + [1 \pm \mathrm{i}(p-\nu)/\sigma]^2 = 0, \qquad (6.87)$$

where the $\pm$ sign is the same as in eqn (6.86).

Equation (6.87) is the approximate dispersion equation of the modes (eigensolutions) of eqn (6.80). It gives three eigenvalues of p, of which one may be checked to have a positive imaginary part (and thus to correspond to a growing mode) and two a negative one (decaying modes). For $\sigma \to \infty$, when the electron beam becomes infinitely wide, eqn (6.87) may be readily seen to approach the 1-D dispersion equation (6.29), so that the eigenvalues of the three modes tend to the three 1-D eigenvalues.

The precision of the approximation leading to eqn (6.87) may be improved by choosing a trial expression of A_w with more than one free parameter, and repeating the same procedure.[9] Such a calculation would also reveal the existence of more than three modes. In fact, as pointed out by Luchini and Solimeno (1986), who calculated an approximation for the eigenvalues of the whole set of the modes by means of a WKB technique, eqn (6.80) has an infinite number of modes, which may be subdivided into three sets according to which root of the 1-D dispersion equation their eigenvalues tend to for $\sigma \to \infty$. One infinite set of modes turns out to be characterized by a positive and two by a negative imaginary part of p.

A comparison of the roots of eqn (6.87) with the eigenvalues yielded by a numerical solution of the mode equation (6.80) (Luchini 1987) shows that the approximation obtained for the two decaying modes is poor. This is not completely unexpected, not so much because of the difficulties involved in proving the validity of the variational principle (the integrals that appear in eqn (6.81) are not defined in the ordinary sense, and must be reinterpreted in a generalized manner to be meaningful), but because the decaying modes explode for $|\boldsymbol{x}| \to \infty$ with a law very different from the assumed Gaussian one; fortunately, however, the decaying modes are relatively uninteresting. The propagation constant of the fundamental growing mode, the one with the largest amplification rate and which, in a long undulator, prevails over

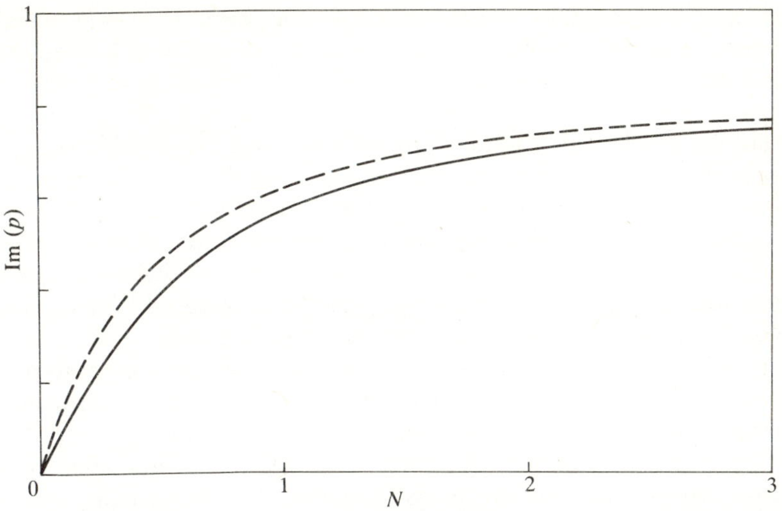

FIG. 6.3. Amplification rate of the fundamental guided mode for a Gaussian transverse distribution of the electron beam. Solid line, approximate dispersion equation (6.87); dashed line, numerical solution. The parameter N that appears as abscissa is related to the dimensionless e-beam radius σ as $N = (2/\pi)\sigma$ (from Luchini 1987).

the others, turns out to be approximated quite well by the corresponding root of eqn (6.87), as may be seen in Fig. 6.3. This fastest-growing mode determines, once the initial transient is exhausted, the mode shape and amplification factor of the amplified laser beam.

We may now obtain a complete picture of 3-D effects in a large-gain FEL by distinguishing different ranges of electron-beam width.

When the dimensionless e-beam width σ (i.e. $\sigma/(L_g/2k)^{1/2}$ in dimensional terms) is much greater than 1, eqn (6.87) tends to become eqn (6.29) and 3-D effects are small. The amplification rate $\mathrm{Im}(p)$ of a large-gain free-electron laser, which when put into eqn (6.30) gives the gain, is approximated well by a root of the one-dimensional dispersion equation (6.29); moreover, several higher modes have approximately the same growth rate as the fundamental one and therefore the laser beam shape is highly dependent on input conditions.

When $\sigma \simeq 1$ the amplification rate of the fundamental mode must be calculated from a 3-D theory, and is approximated fairly well by a dispersion equation such as eqn (6.87), derived from the variational principle. In this case higher-order modes are significantly less amplified than the fundamental one and disappear after a few gain lengths: the laser beam appears to be guided.

When $\Delta x_u/(L_g/2k)^{1/2} \ll \sigma \ll 1$ guiding continues to occur but, as may be seen in Fig. 6.3, the amplification rate of the fundamental mode is severely reduced (for equal on-axis current density) with respect to its 1-D value. It is, however, somewhat unfair to compare the two cases of $\sigma \gg 1$ and $\sigma \ll 1$ on the basis of equal current density, since the total current is very different. It was shown by Moore (1984), by determining the asymptotic solution of eqn (6.80) for $\sigma \to 0$, that if total current is kept constant rather than current density (which is also not quite fair because it is difficult to reduce the cross-section while maintaining the same current) the amplification rate increases instead, although very slowly, with reducing σ. This slow (logarithmic) increase in the limit of $\sigma \to 0$ is not revealed by eqn (6.87), which gives a finite limit, because A_w actually develops a logarithmic singularity which is badly represented by the Gaussian assumed as a trial function.

Finally, when $\sigma \lesssim \Delta x_u/(L_g/2k)^{1/2}$, eqn (6.80) itself fails because it is no longer allowed to treat the Boltzmann equation as one-dimensional. This range of parameters is still the subject of current research.

Let us end this section by observing that, since for constant on-axis current density the amplification rate tends to a constant for $\sigma \gg 1$ and goes to zero for $\sigma \ll 1$ and, for constant total current, it goes to zero for $\sigma \gg 1$ and diverges logarithmically for $\sigma \ll 1$, for the current vs. cross-section curve of an actual accelerating machine, which falls somewhere between constant current and constant current density, there will be an optimal point of maximum amplification rate at some σ of order unity. We may say, therefore, that under the best operating conditions the e-beam cross-section radius in a large-gain FEL must be of the same order of magnitude as the diffraction-limited width of the laser beam $(L_g/2k)^{1/2}$.

6.13 Effects of energy spread and mode cut-off

When the electron beam is not monoenergetic the FEL mode equation (6.80) must be replaced by the more general eqn (6.78). However, just as in the spontaneous-radiation and small-gain cases (Sections 4.10 and 5.12), if the energy distribution of the electrons is described by a Lorentzian curve we may re-utilize the results already obtained for the monoenergetic beam by giving a complex value to the detuning parameter.

To see why this is so, first observe that when F_0 is a Lorentzian function of energy, $F_0 = [4\pi j_b(\mathbf{x})/I_A]\delta\gamma/\pi[\delta\gamma^2 + (\gamma - \bar{\gamma})^2]$, the integral of eqn (6.78) is one of the form (4.83). Since for $\text{Im}(\beta) > 0$, i.e. for the growing modes, the pole of $(\beta - k - \mu)^{-2}$ falls in the half-plane of $\text{Im}(\mu) > 0$ which corresponds to $\text{Im}(\gamma) < 0$, the function that multiplies F_0 in eqn (6.78) has no singularities in the upper half of the complex γ-plane. Therefore, the conditions for the validity of eqn (4.84) are fulfilled, and the integral of eqn (6.78) equals its monoenergetic-beam value calculated for the complex γ given by $\bar{\gamma} + i\delta\gamma$.[10]

By linearizing the expression $\nu = \mu L_g = (M^2k/2\gamma^2 - k_u)L_g$ with respect to $\gamma - \bar{\gamma}$, the complex detuning parameter corresponding to $\gamma = \bar{\gamma} + i\delta\gamma$ may be approximately written as

$$\nu = \bar{\nu} - i\delta\nu \simeq \left(\frac{M^2k}{2\bar{\gamma}^2} - k_u\right)L_g - \frac{M^2kL_g}{\bar{\gamma}^3}\,\delta\gamma. \tag{6.88}$$

Thus the mode equation in the presence of a Lorentzian energy spread is simply eqn (6.80) with the complex value of ν given by eqn (6.88), and all the results of the monoenergetic-beam calculations may be immediately extended by giving this value to ν.

Figure 6.4 shows a plot of amplification rate Im(p) of the fundamental growing mode versus e-beam cross-section radius σ, as obtained from the approximate dispersion equation (6.87) for various values of Im(ν) and the optimal value (the one giving rise to maximum amplification rate) of Re(ν). The optimal value of Re(ν) itself is shown in Fig. 6.5.

As might be expected, given the dimensionless character of the parameter ν, in order for the presence of energy spread not to lead to a noticeable gain reduction $-$Im(ν) must be much smaller than unity. From eqn (6.88) it may be seen that a value of $\delta\nu$ much smaller than unity corresponds to a relative energy spread $\delta\gamma/\bar{\gamma}$ such that

$$\delta\gamma/\bar{\gamma} \ll \bar{\gamma}^2/M^2kL_g \simeq l_u/4\pi L_g. \tag{6.89}$$

This is the limitation on energy spread that must be respected for a large-gain FEL to work at best. Once again we find a qualitative similarity with the

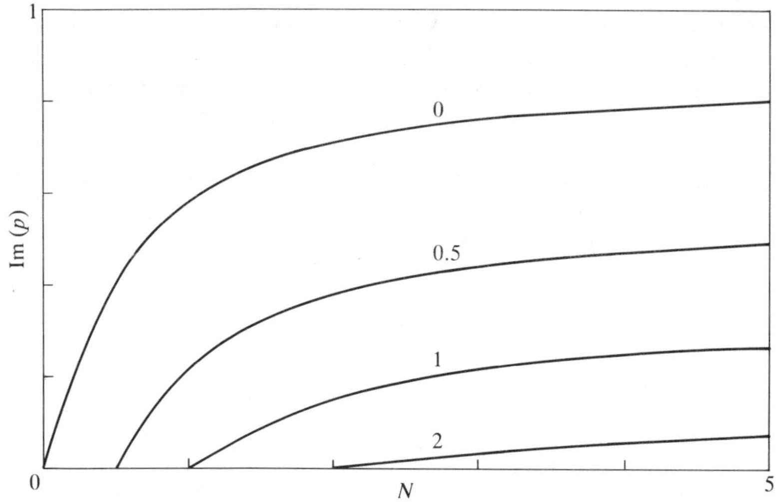

FIG. 6.4. Amplification rate of the fundamental guided mode in the presence of energy spread. Each curve is labelled with the corresponding value of the detuning-spread parameter $\delta\nu = -$Im(ν) (from Luchini 1987).

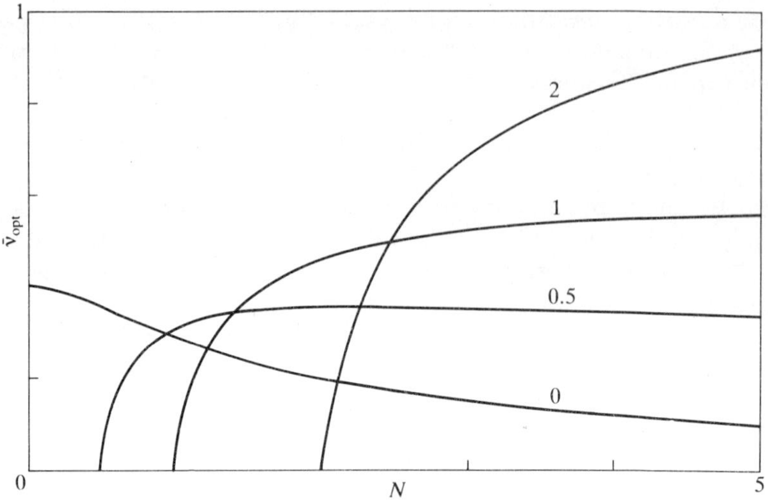

FIG. 6.5. Optical value of the detuning parameter $\bar{\nu} = \mathrm{Re}(\nu)$ giving rise to the amplification rate of Fig. 6.4 (from Luchini 1987).

small-gain result (5.83), provided that twice the gain length L_g is substituted for the undulator length L.

It is also remarkable that the curves corresponding to $\mathrm{Im}(\nu) \neq 0$ in Fig. 6.4 cross the $\mathrm{Im}(p) = 0$ axis for a non-zero value of σ, contrarily to what happens for $\mathrm{Im}(\nu) = 0$. This implies that below a well-defined value of σ (or, equivalently, above a well-defined value of $-\mathrm{Im}(\nu)$), which may be called the cut-off value, the FEL structure does not support any more growing modes and ceases to be an amplifier. The threshold condition for amplifier operation yielded by the approximate dispersion equation (6.87) is very simple:

$$\sigma > \delta\nu. \tag{6.90}$$

Unless eqn (6.90) is satisfied, all FEL modes are cut off. In addition, even when eqn (6.90) is satisfied only a finite number of modes have a positive amplification factor (Luchini 1987). This is in contrast with the, somewhat peculiar, case of the monoenergetic beam, where an infinite number of modes are amplified no matter how small σ is.

6.14 Backward wave and backward-wave instability

In the one-dimensional modal analysis of the FEL we derived the dispersion equation (6.27) under the assumption that $\beta \simeq k$ and $\mu \simeq 0$, which was shown to correspond to the ordinary undulator resonance conditions. As will be remembered, this resonance arises from the existence of a common root of the dispersion equations of the uncoupled wave and electron-beam transport equations, $\beta^2 - k^2 = 0$ and $\beta - k - \mu = 0$. There is, however, a second common solution, in correspondence of the point where $\beta = -k$ and $\mu = -2k$ (Liever *et al.* 1981). This second solution gives rise to the backward-wave resonance, so called because it results from the coupling of a backward-propagating (i.e. negative-β) electromagnetic wave with the electron beam.

Just as in travelling-wave tubes, where this phenomenon is well known, when gain is very high an unwanted backward wave may enter into competition with the forward wave and disturb the operation of the FEL. We shall now examine the properties of the backward wave with reference to the basic case of a 1-D circular-polarization FEL which was considered in Section 6.4, assuming also, for the sake of simplicity, that space-charge effects are negligible. The results we shall obtain may be extended by the techniques shown in the previous sections.

First, let us calculate the actual backward-wave resonance wavelength. The condition $\mu = -2k$ together with the definition (6.22) of μ gives $k = k_u/(2 + M^2/2\gamma^2) \simeq k_u/2$. Therefore, the backward-wave resonance wavelength is always (for $\gamma \gg 1$) close to twice the undulator period and approximately independent of γ. It is thus much larger than the forward-wave resonance wavelength $M^2 l_u/2\gamma^2$, which scales as $1/\gamma^2$ and makes the FEL useful as an optical-wavelength amplifier. This situation is very different from that found in connection with travelling-wave tubes, where the wavelengths of the forward and backward waves fall in the same spectral region and the backward wave has useful applications. On the contrary, the backward wave is definitely a parasite in an FEL.

An approximate dispersion equation for the backward wave may be obtained by approximating the exact eqn (6.40) under the new near-resonance conditions $\beta \simeq -k$ and $k \simeq k_u/2$. We shall put $\beta = -k_u/2 - \delta$, $k = k_u/2 + \varepsilon$, where $|\delta|$ and $|\varepsilon|$ are both assumed to be small compared with k_u. By leaving out space-charge terms, the l.h.s. of eqn (6.40) may be approximated as $\beta^2 - k^2 \simeq k_u(\delta - \varepsilon)$ and the r.h.s. as $-(kA_u^2/4)I_2$. Taking the value of I_2 from eqn (6.42) one may rewrite eqn (6.40) approximately as

$$k_u(\delta - \varepsilon) = (\pi A_u^2 k_u j_b/\bar{\gamma}^3 I_A)(\varepsilon + \delta)/(\delta + \varepsilon + M^2 k_u/4\bar{\gamma}^2)^2. \qquad (6.91)$$

Since we are looking for conditions under which the amplification of the backward wave is strong, so as to interfere with forward-wave FEL operation, we shall assume that $|\delta + \varepsilon| \gg M^2 k_u/4\bar{\gamma}^2$. (This assumption may be verified *a posteriori*. Since, as will be seen, $|\delta + \varepsilon|$ is of the order of $1/L$, the necessary condition is that $M^2/\bar{\gamma}^2 \ll l_u/L = 1/N$. This is, for instance, true in an optical-wavelength FEL having both $\bar{\gamma}$ and N of the order of hundreds.) Equation (6.91) then reduces to

$$\delta^2 = \varepsilon^2 + \pi A_u^2 j_b/\bar{\gamma}^3 I_A, \qquad (6.92)$$

which is the sought-for approximate backward-wave dispersion equation.

Both roots δ of eqn (6.92) are real for real ε, so that neither mode is amplified or damped. One should not conclude, however, that the FEL does not amplify the backward wave at all, because the effects of boundary conditions still have to be taken into account. In this connection an important difference arises with respect to the forward wave. The boundary conditions associated with the exact Boltzmann and wave equations (6.9)–(6.11) read that the distribution-function perturbation, f_1, should be zero at the entrance of the undulator, and the electromagnetic field should match any input waves that are imposed on to the system. As discussed in Section 6.10, in a one-dimensional model two input waves ought to be considered, one coming from the left and one from the right of the undulator; but in the approximation that replaces $\beta^2 - k^2$ by $2k(\beta - k)$ the possibility of accounting for a wave coming from the right is lost and a single condition must be imposed on the vector potential at the left end of the undulator. Therefore, the boundary conditions appropriate to the forward wave, which were considered in Section 6.10, are initial conditions, all on the same side of the undulator. On the contrary, in the approximation adopted here of considering $\beta^2 - k^2 \simeq -2k(\beta + k)$, the forward wave is lost, and a single electromagnetic boundary condition must be imposed at the right end of the undulator. In this case we have a true boundary-value problem, with part of the conditions imposed on one side of the undulator and part on the other.

The presence of boundary conditions at both ends implies that the system incorporates a feedback mechanism, and is capable of amplification even if the individual modes have real propagation constants. Even more important, such a system with feedback may undergo self-oscillation, generating an output even in the absence of any input. The occurrence of self-oscillation at the backward-wave resonance frequency is the phenomenon referred to as backward-wave instability. In travelling-wave tubes the backward-wave instability is put to good use for making oscillators (the so-called backward-wave oscillator, BWO), but in FELs it must be carefully avoided.

In order to determine the threshold for the onset of the backward-wave instability and be able to avoid it, we must add homogeneous boundary conditions to the system of differential equations and try to satisfy them with a linear combination of propagation modes for suitable, possibly complex, eigenvalues of frequency. If an eigenvalue with a negative imaginary part is found, this corresponds to a mode growing in time and to the start-up of self-oscillation.

Let us therefore consider an electromagnetic vector potential expressed by a linear combination of the two modes with frequency $c(k_u/2 + \varepsilon)$ and propagation constants $-k_u/2 \pm \delta$, i.e.

$$A = A^+ e^{i(k_u/2 + \delta)z} + A^- e^{i(k_u/2 - \delta)z}. \tag{6.93}$$

The electromagnetic boundary condition stating that no wave is present at the right end of the undulator ($z = L$) is:

$$A^+ e^{i(k_u/2 + \delta)L} + A^- e^{i(k_u/2 - \delta)L} = 0. \tag{6.94}$$

The perturbation of the electron distribution function is proportional, for each mode, to $A^{\pm}/(\beta - k - \mu) \simeq -A^{\pm}/(\varepsilon \pm \delta)$. Therefore, the condition asserting that this perturbation is zero at the entrance of the undulator ($z = 0$) may be written as

$$A^{+}/(\varepsilon + \delta) + A^{-}/(\varepsilon - \delta) = 0. \tag{6.95}$$

Upon equating to zero the determinant of the linear system of equations (6.94) and (6.95), so as to make possible a non-zero solution, one easily obtains the eigenvalue equation

$$e^{2i\delta L} = (\varepsilon - \delta)/(\varepsilon + \delta), \tag{6.96}$$

which must be solved together with the dispersion equation (6.92) to determine both δ and ε. With varying $\pi A_u^2 j_b/\bar{\gamma}^3 I_A$ the roots of this system of equations move in the complex plane; a particular interest is attached to points where ε becomes real, because those represent the boundary between a stable behaviour (characterized by $\mathrm{Im}(\varepsilon) > 0$) and an unstable one ($\mathrm{Im}(\varepsilon) < 0$). If ε is real δ is too, as may be seen from eqn (6.92), and then the only possibility to satisfy eqn (6.96) is with $\varepsilon = 0$ and $\delta L = \pi/2 + n\pi$, n being an arbitrary integer. Equation (6.92) then requires that

$$\pi A_u^2 L^2 j_b/\bar{\gamma}^3 I_A = (\pi/2 + n\pi)^2. \tag{6.97}$$

Every time that the l.h.s. of eqn (6.97) increases beyond one of the values assumed by the r.h.s. for integral n, one backward-wave oscillation mode goes from stable to unstable. All modes are stable only if

$$A_u^2 L^2 j_b/\bar{\gamma}^3 I_A < \pi/4. \tag{6.98}$$

Equation (6.98) is thus the threshold condition for the onset (or rather the non-onset) of the backward-wave instability.

It is interesting to restate eqn (6.98) in terms of the gain length L_g. Eliminating j_b between eqn (6.98) and eqn (5.26) gives

$$L/L_g < (\pi^3 L_g/2l_u)^{1/2}. \tag{6.99}$$

As long as the limitation of eqn (6.99) is not exceeded the backward-wave instability does not take place. Notice that eqn (6.99) gives an explicit limitation on forward-wave gain, since gain is given by the exponential of L/L_g; the larger is the number of undulator periods contained in one gain length, the higher is the gain that can be achieved without exciting the backward-wave instability.

Equation (6.99) also shows once again the somewhat pathological character of the strong-coupling regime, in which L_g is comparable with the undulator period: if $L_g \simeq l_u$ the backward-wave instability takes place even in a one-period-long undulator.

Notes

1. The above identity in vector notation becomes $\mathrm{div}(\dot{c}) = 0$, and implies that the evolution in phase space of the particles is similar to the behaviour of an incompressible fluid; that is, that the phase space volume occupied by a given set of particles remains constant with z (the independent variable). This is Liouville's theorem, which therefore remains valid even with our non-standard definition of phase space.
2. The following properties of the δ and δ' functions have been used (where $f(x)$ is any function): $f(x)\delta(x-x_0) = f(x_0)\delta(x-x_0)$, $f(x)\delta'(x-x_0) = f(x_0)\delta'(x-x_0) - f'(x_0)\delta(x-x_0)$.
3. The existence of solutions of this kind, and the general simplification of undulator theory that arises in the case of circular polarization, may be ascribed to a particular symmetry property of circularly polarized fields. In fact, a circularly polarized field is left unchanged by an infinitesimal helicoidal rigid motion of suitable pitch. For other than circular polarization there is only invariance with respect to a finite translation of an integral multiple of the period, but not to any infinitesimal rigid motion.
4. Just as eqn (6.27), eqn (6.63) is accurate to within an error $O(l_u/L_g)$.
5. The straightforward solution of the system (6.65)–(6.67) gives, for instance, for A_{w1}, $A_{w1} = A_{wi}x_2x_3/(x_1 - x_2)(x_1 - x_3)$, where we denote $1/(p_r - v)$ by x_r. (Similar expressions for the other components may be obtained by rotation of indices.) In order to put the solution into the form of eqn (6.68) one may use the algebraic properties of the dispersion equation (6.29), which in terms of the variable $x = 1/(p - v)$ becomes $x^3 + vx + 1 = 0$. The general relations between the roots and the coefficients of a polynomial equation tell us that for this equation $x_1 + x_2 + x_3 = 0$ and $x_1x_2x_3 = -1$. Eliminating x_2 and x_3 from the expression of A_{w1} with the help of these two equations, and restoring the variable p, gives eqn (6.68).
6. This requirement stems quite naturally from the consideration that an electron beam with a momentum distribution of non-zero width spreads out in its motion, and therefore cannot sustain modes of constant width in an infinitely long FEL. In practice, in a finite FEL, the allowable momentum spread is determined by the undulator length.
7. In differentiating $1/\gamma^2(\beta - k - \mu)$ with respect to γ we, just as in the 1-D case, neglect the derivative of $1/\gamma^2$ with respect to that of $1/(\beta - k - \mu)$. Inclusion of the former would only add a correction of the order of l_u/L_g; that is, of the same order as other neglected terms.
8. Far enough from the axis, where the current $J(\boldsymbol{x})$ fades to zero, eqn (6.80) reduces to the free wave equation $\Delta_{2\perp}A_w - pA_w = 0$, the solutions of which behave asymptotically for $|\boldsymbol{x}| \to \infty$ as $e^{\pm p^{1/2}|\boldsymbol{x}|}/|\boldsymbol{x}|^{1/2}$. The radiation condition requires that the square root of p with a negative imaginary part be selected. From a simple analysis of the signs that occur in the four quadrants of the complex plane it may be seen that this square root has a negative real part when p has a positive imaginary part and vice versa, independently of the sign of $\mathrm{Re}(p)$. Therefore, A_w vanishes exponentially when $|\boldsymbol{x}| \to \infty$ for the growing modes (those with $\mathrm{Im}(p) > 0$) and explodes exponentially for the decaying ones (those with $\mathrm{Im}(p) < 0$).
9. It is also easy to extend the calculation to the case, frequently encountered in practice, of a beam with an elliptical cross-section. All it takes is to assume the expression $e^{-x_1^2/\sigma_1^2 - x_2^2/\sigma_2^2}$ for J and the expression $e^{-x_1^2/w_1^2 - x_2^2/w_2^2}$ for A_w, w_1 and w_2 being two independent free parameters, and differentiate the result of putting these expressions into eqn (6.81) once with respect to w_1 and once with respect to w_2. Of course, the resulting system of three algebraic equations in w_1, w_2, and p is more complicated than eqns (6.84) and (6.85), and does not reduce to a single cubic equation, but it may still be easily inverted numerically.
10. It is not immediately possible to obtain a similar result for the decaying modes because of the peculiar character of these non-integrable modes. In order to do so one has to interpret the modes (both growing and decaying) as the residues corresponding to the poles of the transfer function of the system of eqns (6.75) and (6.76), obtained through a Laplace transform with respect to z. It then turns out that the integral in eqn (6.78) must run, in the complex γ-plane,

always above the pole of $(\beta - k - \mu)^{-2}$. Therefore the use of $\gamma = \bar{\gamma} + i\delta\gamma$ in the monoenergetic equations is correct for both the growing and decaying modes, even if the latter have $\mathrm{Im}(\beta) < 0$.

References

Argence, E. and Kahan, Th. (1964). *Théorie des guides et cavités électromagnétiques.* Dunod, Paris.

Bernstein, I. B. and Friedland, L. (1981). Theory of the free-electron laser in combined helical pump and axial guide fields. *Phys. Rev. A* **23**, 816–23.

Chapman, S. and Cowling, T. D. (1970). *The mathematical theory of non-uniform gases.* Cambridge University Press, Cambridge.

Freund, H. P., Sprangle, P., Dillenburg, D., Da Jornada, E. H., Schneider, R. S., and Liberman, B. (1982). Collective effects on the operation of free-electron lasers with an axial guide field. *Phys. Rev. A* **26**, 2004.

Grossman, A., Marshall, T. C., and Schlesinger, S. P. (1983). A new millimeter wave free electron laser using a relativistic beam with spiraling electrons. *Phys. Fluids* **26**, 337.

Landau, L. D. and Lifschitz, E. M. (1981). *Physical kinetics.* Pergamon Press, Oxford.

Liever, P. C., Lin, A. T., and Dawson, J. M. (1981). Theory of an absolute instability of a finite-length free-electron laser. *Phys. Rev. A* **23**, 1251–5.

Luchini, P. (1987). More on optical guiding in an FEL. *FEL*86, 150–53.

Luchini, P. and Motz, H. (1988). Effects of bending on free-electron laser performance. *FEL*87, 334–9.

Luchini, P. and Solimeno, S. (1986). Optical guiding in an FEL. *FEL*85, 413–7.

Marshall, T. C. (1985). *Free electron lasers.* Macmillan, New York.

McMullin, W. A. and Bekefi, G. (1982). Stimulated emission from relativistic electrons passing through a spatially periodic longitudinal magnetic field. *Phys. Rev. A* **25**, 1826.

Moore, G. T. (1984). High-gain small-signal modes of the free-electron laser. *Optics Communs.* **52**, 46–51.

Moore, G. T. (1985). The high-gain regime of the free electron laser. *Nucl. Instr. Meth. Phys. Res.* **A239**, 19–28.

Stoker, J. J. (1950). *Nonlinear vibrations.* Interscience, New York, section VI.2.

Xie, M. and Deacon, D. A. G. (1986). Theoretical study of FEL active guiding in the small signal regime. *FEL*85, 426–31.

7

SATURATION

With an ever increasing optical field, a point is eventually reached where the linear approximation of the previous chapters fails. The free-electron laser then enters the large-field, non-linear operating regime.

In this chapter we analyse this non-linear regime assuming, just as in Chapter 5, a small-gain condition, which allows us to neglect the radiated field with respect to the input field as far as electron motion is concerned. The next chapter is devoted to the large-signal, large-gain regime, where variations of the radiation field may not be neglected and tapered undulators become essential.

7.1 Calculation of gain and energy extraction in the non-linear small-gain regime

As explained in Section 5.3, under the assumption of small gain it is possible to neglect the field radiated by the electrons in the undulator when calculating the motion of the electrons themselves, and identify the field acting upon them with the input field. The power, P_{inc}, added to the input wave may then be obtained from eqn (5.6) as the opposite of the work done by the input field upon the electrons moving in it. This work in turn equals the difference between the final and initial energy of the electrons going through the undulator:

$$P_{\text{inc}} = -W_{wr} = -m_e c^2 n_e \langle \gamma_f - \gamma_i \rangle, \tag{7.1}$$

where n_e, the number of electrons entering the undulator per unit time, equals the e-beam current divided by the charge of the electron, and the angular brackets denote averaging over the initial electron distribution.

Equation (7.1) may look like a trivial statement of energy conservation, but there is more to it, and it should be noted that its practical applicability is generally limited to the small-gain regime. In fact, whereas it is true that the total electromagnetic power radiated always coincides with the energy lost by the electrons per unit time, in a general non-linear setting this power is radiated in an uncontrolled fashion, and possibly not in the same direction, nor even at the same frequency, which the input comes in. In other words, although the total electromagnetic radiation emitted certainly abides by energy conservation, energy conservation alone does not enable us to distinguish 'usable' from 'lost' radiation. On the contrary, the reciprocity theorem, as shown in Section 5.6, tells us that under the assumption of small gain P_{inc} coincides at first order with the 'useful' power radiated into the same electromagnetic mode that the input comes in.

On the other hand, linearity of the equations of electron motion is not required for the validity of eqn (1.1), which can be used freely to calculate power radiated by the FEL in the non-linear small-gain regime if only a non-perturbative way is found to solve the equations of electron motion and determine $\gamma(z)$.

Once the radiation power increase P_{inc} is calculated, gain may be immediately obtained from its definition (5.3):

$$g = G - 1 = P_{\text{inc}}/P_{\text{in}}. \tag{7.2}$$

In particular, in the 1-D case of perfectly collimated, transversely infinite laser and electron beams, the input power, per unit area, is given in terms of the dimensionless vector potential A_w by $P_{\text{in}} = k^2 m_e^2 c^4 A_w^2/2\zeta e^2$; combining this with eqns (7.2) and (7.1), where $n_e = j_b/e$, gives

$$g = \frac{8\pi j_b}{k^2 I_A} \frac{\langle \gamma_i - \gamma_f \rangle}{A_w^2},$$

I_A being the Alfvén current which was introduced in Section 5.5.

At saturation, however, gain ceases to be an interesting parameter. More significant is the yield or energy extraction, η; that is, the fraction of the electron-beam energy which is coupled out into the radiation beam. Since in the small-gain regime the energy coupled into the radiation beam coincides at first order with the total energy lost by the electrons, the energy extraction is simply

$$\eta = \langle \gamma_i - \gamma_f \rangle / (\langle \gamma_i \rangle - 1) = \langle \gamma(0) - \gamma(L) \rangle / (\langle \gamma(0) \rangle - 1). \tag{7.3}$$

Of course, as may be seen by comparison of the previous two equations, once η has been calculated gain is simply given by

$$g = \frac{8\pi j_b}{k^2 I_A} \frac{\langle \gamma_i \rangle - 1}{A_w^2} \eta. \tag{7.4}$$

Therefore we shall mostly mention energy extraction in the following sections.

7.2 A general description of the non-linear interaction between electrons and waves

Many well known devices based on the interaction between electrons and electromagnetic waves, including travelling-wave tubes (see, e.g., Webber 1957), most particle accelerators (see, e.g., Courant and Snyder 1958; Sands

1970; Lawson 1980, 1988), and the undulator, may in principle be reduced to the conceptual model of a running electron acted upon by a longitudinal force (which may be of electric and/or magnetic origin) having a wave-like dependence on space and time, i.e.

$$f_z = F \sin(\omega t - \beta z). \tag{7.5}$$

When the phase velocity ω/β of this wave is less than the speed of light, a moving reference frame exists where the wave pattern is steady. In this (primed) frame the force (7.5) becomes $f'_{z'} = -F' \sin \beta' z'$, and may be derived from a sinusoidal potential given by

$$U = -(F'/\beta')\cos \beta' z', \tag{7.6}$$

sometimes called the ponderomotive potential or quasi-potential. In the moving frame the electron moves like a ball rolling over a sinusoidal rail, with an initial velocity bound right or left according to whether its initial velocity in the laboratory is higher or lower than the phase velocity of the wave. When the electron and wave velocities are exactly the same, the initial electron velocity in the moving frame is zero and if, in addition, the initial position z'_i is at a potential minimum the electron will sit still there, moving with uniform motion in the laboratory. The quantity $\xi = \beta' z'$ is named the 'phase' of the electron in particle-accelerator theory, and the value ξ_s at which the potential of eqn (7.6) has a minimum is the 'synchronous' phase.

In general, the electrons of a beam the time structure of which is characterized by a scale that is large compared to the period of the wave will have a rather well defined initial velocity but all possible initial phases. Provided that the initial velocity is close enough to that of the wave, a part of the electrons will not have sufficient energy to escape from one of the troughs of the potential pattern and will start oscillating back and forth in it, performing what are called 'synchrotron oscillations' in particle-accelerator theory, because they played an important role for the first time in the theory of the synchrotron. (This term has become usual in FEL theory as well.) Electrons oscillating back and forth are in a sense 'trapped' by the quasi-potential. They maintain, on average, a steady position in the moving frame, and therefore move together in the laboratory with exactly the phase velocity of the wave. There are always 'untrapped' electrons also, which have a high enough energy in the moving frame to slide along over hills and troughs of the potential curve. Their average velocities in the laboratory frame are all different, and depend on the initial velocity and phase of each.

According to the above picture, when the electrons are injected with a velocity close to the phase velocity of the wave a fraction of them, the trapped ones, will adapt their speed to the wave's, increasing their energy if their initial velocity was slightly less than the wave velocity and decreasing it if slightly more. This is the basic energy-transfer mechanism in the non-linear interaction regime of several devices, amongst which is the undulator. As is seen, energy may be transferred in either direction, from the electrons to the wave or vice versa, depending on the sign of the relative initial velocity between electron and wave. With increasing relative velocity the amount of energy transferred per electron increases, but at the same time the number of trapped electrons decreases, so that an optimum exists at some intermediate point.

The foregoing qualitative picture fits the interaction between electrons and wave in an FEL well, provided that the duration of the interaction is large compared to the period of synchrotron oscillations, which will be seen to be the condition characterizing the non-linear regime. If, on the other hand, the interaction time is small, so that each electron moves only by a fraction of a synchrotron period with respect to the wave, the phenomenon is basically linear and is more aptly described by the models presented in the previous chapters.

Let us now examine synchrotron oscillations in greater detail. Assuming, for the scope of this section only, that non-relativistic mechanics is valid in the moving frame, we may write the equation of motion for the electron phase $\xi = \beta' z'$ under the action of a sinusoidally varying force explicitly as

$$\mathrm{d}^2\xi/\mathrm{d}t'^2 = -\Omega^2 \sin \xi, \tag{7.7}$$

where $\Omega^2 = \beta' F'/m_e$, which may be recognized as the equation of motion of a simple pendulum.[1]

Let us recall the main properties of this equation for later use. By multiplying both sides of eqn (7.7) by $\mathrm{d}\xi/\mathrm{d}t'$ and integrating once, the energy first integral is obtained:

$$\mathrm{d}\xi/\mathrm{d}t' = \Omega(4m - 2 + 2\cos\xi)^{1/2}, \tag{7.8}$$

where $4m - 2$ is an arbitrary constant proportional to the energy of the pendulum, which is determined by the initial conditions. Owing to the trigonometric identity $\cos\xi = 1 - 2\sin^2(\xi/2)$, eqn (7.11) may alternatively be written as

$$\mathrm{d}\xi/\mathrm{d}t' = 2\Omega[m - \sin^2(\xi/2)]^{1/2}. \tag{7.9}$$

The main properties of pendulum motion are illustrated by the phase plot of Fig. 1.12, which shows the curves described by eqn (7.9) in the $(\xi, \dot{\xi}/\Omega)$-plane for different values of m. Two kinds of motion are possible: a bounded motion for $0 < m < 1$, in which ξ cannot escape the interval $(-2\arcsin m^{1/2}, 2\arcsin m^{1/2})$, and an unbounded motion for $m > 1$. The first case corresponds to a pendulum oscillating back and forth and to a trapped electron; the second to a continuously rotating pendulum and to an untrapped electron. When the pendulum oscillates back and forth ξ is a periodic function of t' (an elliptical function), with a period T'_p which, by a suitable change of variables, may be put in the form

$$T'_p = \frac{4}{\Omega}\int_0^1 \frac{d\zeta}{[(1-\zeta^2)(1-m\zeta^2)]^{1/2}} = \frac{4}{\Omega}\mathrm{K}(m), \tag{7.10}$$

where $\mathrm{K}(m)$ is the first complete elliptical integral. $\mathrm{K}(m)$ is an increasing function of m, which equals $\pi/2$ for $m = 0$ and ∞ for $m = 1$. For $m \to 0$, T'_p tends to $2\pi/\Omega$. It can thus be recognized that Ω is the angular frequency of small oscillations, as could also be directly seen from eqn (7.7) on observing that, for small ξ, $\sin\xi \simeq \xi$.

In the next section we shall see how a potential of the form of eqn (7.6) arises in an undulator from the beat between the undulator field and the input electromagnetic wave, and how the non-linear regime of the FEL may be described with the help of a pendulum equation.

7.3 Relativistic equations of motion of an electron in circularly polarized undulator and wave fields

In order to apply the foregoing model to the undulator we must introduce the relativistic, rather than classical, equations of motion and take the simultaneous presence of the undulator and wave fields into account. Even if the model may then seem to have become much more complicated than the one considered in the previous section, it will be seen that it can still be exactly reduced to a simple pendulum equation when both the undulator and wave fields are circularly polarized.

We shall, once again, start from a Hamiltonian formulation. Although eventually, just as in the previous chapter, we shall use the ultrarelativistic Hamiltonian (3.15), it is interesting to observe that a closed-form solution may equally well be obtained from the complete Hamiltonian (3.6). Therefore we shall initially allow for both choices.

In order to study the motion of an electron in a one-dimensional undulator traversed by an electromagnetic plane wave, let us reconsider the dimensionless vector potential (5.7),

$$\boldsymbol{a}_\perp = \mathrm{Re}[\boldsymbol{A}_u e^{ik_u z} + \boldsymbol{A}_w e^{ik(t-z)}], \tag{7.11}$$

without, this time, treating $\boldsymbol{A}_w$ as a perturbation. a_z and φ are still assumed to be zero. Since the vector potential of eqn (7.11) does not depend on transverse co-ordinates, the transverse part $\boldsymbol{P}_\perp$ of the momentum vector is still a constant of motion. Therefore, it is convenient to introduce the quantity

$$M_{eff}^2 = 1 + (\boldsymbol{P}_\perp - \boldsymbol{a}_\perp)^2, \tag{7.12}$$

so that the relativistic Hamiltonian (axial momentum) $\mathscr{K}$ of eqn (3.6) may be written as

$$\mathscr{K}(\gamma, t, z) = [\gamma^2 - M_{eff}^2(t, z)]^{1/2} \tag{7.13}$$

and the ultrarelativistic one of eqn (3.15), which is given by the first two terms of the expansion of eqn (7.13) in powers of $1/\gamma$, as

$$\mathscr{K}(\gamma, t, z) = \gamma - M_{eff}^2/2\gamma. \tag{7.14}$$

The Hamiltonian $\mathscr{K}$ of eqns (7.13) and (7.14) is analogous to the Hamiltonian of a particle of mass M_{eff} free from electromagnetic fields, except that M_{eff} is not a constant. The quantity M_{eff} is an effective mass like the one introduced in Chapter 3, and in Appendix B, and plays the same role in relativistic mechanics that potential energy plays in classical mechanics. Notice, however, that eqn (7.13) contains no approximation, and the appearance of an effective mass is due to the assumption of a one-dimensional (i.e. depending on only one co-ordinate) electromagnetic field, whereas the effective mass introduced in Appendix B is the result of a two-timing approximation in a three-dimensional field.

The advantage gained by introducing the effective mass is that all the information regarding the electromagnetic field has been condensed into the scalar quantity M_{eff}^2. In particular, for the vector potential (7.11), eqn (7.12) evaluates to (see footnote 2 in Section 3.4 for a quick way of squaring the real part of a complex vector):

$$\begin{aligned} M_{eff}^2 &= 1 + A_u^2/2 + A_w^2/2 + \mathrm{Re}\{(\boldsymbol{A}_u^2/2)\mathrm{e}^{2ik_u z} + (\boldsymbol{A}_w^2/2)\mathrm{e}^{2ik(t-z)} \\ &\quad + \boldsymbol{A}_u^* \cdot \boldsymbol{A}_w \mathrm{e}^{i[kt-(k+k_u)z]} + \boldsymbol{A}_u \cdot \boldsymbol{A}_w \mathrm{e}^{i[kt-(k-k_u)z]}\} \\ &\quad + P_\perp^2 - 2\boldsymbol{P}_\perp \cdot \mathrm{Re}[\boldsymbol{A}_u \mathrm{e}^{ik_u z} + \boldsymbol{A}_w \mathrm{e}^{ik(t-z)}]. \end{aligned} \tag{7.15}$$

As already noted with reference to eqn (3.21), when both fields are circularly polarized $\boldsymbol{A}_u^2 = \boldsymbol{A}_w^2 = 0$ and, depending on the relative orientation of the two polarizations, either $\boldsymbol{A}_u^* \cdot \boldsymbol{A}_w = 0$ or $\boldsymbol{A}_u \cdot \boldsymbol{A}_w = 0$. If, in addition, we assume that $\boldsymbol{P}_\perp = 0$, i.e. that the electron trajectory is perfectly aligned with the z axis, only one complex exponential in eqn (7.15) is left with a non-zero coefficient, and M_{eff}^2 turns out to depend on z and t only through a linear combination of them. M_{eff}^2 then reproduces the form (7.6) of the quasi-potential.

In particular, when orientations are such that $\boldsymbol{A}_u^* \cdot \boldsymbol{A}_w$ is the non-zero coefficient (we may, with no loss of generality, assume it to be real and negative), the expression of M_{eff}^2 reduces to

$$\begin{aligned} M_{eff}^2 &= 1 + A_u^2/2 + A_w^2/2 + \mathrm{Re}\{\boldsymbol{A}_u^* \cdot \boldsymbol{A}_w \mathrm{e}^{i[kt-(k+k_u z)]}\} \\ &= M^2 - A_u A_w \cos[kt - (k + k_u)z], \end{aligned} \tag{7.16}$$

where we have given the name M^2 to the constant part $1 + A_u^2/2 + A_w^2/2$ of M_{eff}^2; when $A_u \cdot A_w$ is the non-zero coefficient (which we again assume to be real and negative), one has

$$M_{eff}^2 = M^2 - A_u A_w \cos[kt - (k - k_u)z]. \tag{7.17}$$

7.4 The synchronous frame

As shown in the previous section, the combination of the vector potentials of undulator and wave into the effective mass M_{eff}, which is exactly equivalent to them in a one-dimensional model, gives rise, when both fields are circularly polarized, to an effective mass with the form of a beat travelling wave having a wavenumber equal to the sum or difference of the wavenumbers of the undulator and input-wave fields. This effective mass plays in relativistic theory the role that the quasi-potential, U, of eqn (7.6) has in the non-relativistic one.

Let us first consider the case in which $A_u^* \cdot A_w$ is the non-zero coefficient, and the expression of the effective mass is given by eqn (7.16). Following a procedure similar to that adopted in Section 7.2, we may introduce a new reference frame, the synchronous frame, moving with the phase velocity $U = k/(k + k_u)$ of the beat wave. Calling time and the axial co-ordinate in the new frame t' and z', the Lorentz transformation equations give

$$z = \frac{z' + Ut'}{(1 - U^2)^{1/2}} = \frac{(k + k_u)z' + kt'}{(2kk_u + k_u^2)^{1/2}}, \tag{7.18a}$$

$$t = \frac{t' + Uz'}{(1 - U^2)^{1/2}} = \frac{(k + k_u)t' + kz'}{(2kk_u + k_u^2)^{1/2}}. \tag{7.18b}$$

In the new frame, eqn (7.11) becomes

$$A = \mathrm{Re}\left\{A_u \exp \mathrm{i}\left[\frac{kk_u t' + (kk_u + k_u^2)z'}{(2kk_u + k_u^2)^{1/2}}\right] + A_w \exp \mathrm{i}\left[\frac{kk_u t' - kk_u z'}{(2kk_u + k_u^2)^{1/2}}\right]\right\}. \tag{7.19}$$

and eqn (7.16) becomes

$$M_{eff}^2 = M^2 - A_u A_w \cos[(2kk_u + k_u^2)^{1/2} z']. \tag{7.20}$$

Notice from eqn (7.19) that, owing to Doppler shift, the undulator and wave fields have acquired the same frequency, namely $kk_u/(2kk_u + k_u^2)^{1/2}$. It follows that in the synchronous frame the effective mass M_{eff} is independent of time t', as may be seen in eqn (7.20).

In the synchronous frame the relativistic Hamiltonian is still given by an equation similar to eqn (7.13), which is valid in any inertial frame,

$$\mathscr{H}' = [\gamma'^2 - M_{eff}^2(z')]^{1/2} \tag{7.21}$$

with a new energy γ' and momentum $\mathscr{H}'$ but the same value of M_{eff}^2, because M_{eff}^2 is defined in terms of transverse vectors which are left unaltered by a Lorentz transformation in the axial direction.

Moreover, it turns out that, as was shown in Section 3.3, the ultrarelativistic Hamiltonian (7.14) also transforms into eqn (7.21), provided that the transformation equations are suitably adapted to take the ultrarelativistic limit into account. The appropriate transformation equations are eqns (3.17), which may be written

$$z \simeq (k/2k_u)^{1/2}(t' + z'), \qquad t - z \simeq (2k_u/k)^{1/2}(t' - z'). \qquad (7.22\text{a,b})$$

Therefore the exact relativistic and the ultrarelativistic version of the problem share the same formulation in the synchronous frame (but, as will be seen in detail later, with a different meaning of constant parameters).

To solve the equations of motion in the synchronous frame it is sufficient to observe that, since $\mathscr{H}'$ is independent of time t', energy γ' is a constant of motion. The Hamilton equation $\mathrm{d}t'/\mathrm{d}z' = \partial\mathscr{H}'/\partial\gamma'$ (the second of eqns (3.4a)) then reads

$$\mathrm{d}t'/\mathrm{d}z' = \gamma'/[\gamma'^2 - M_{eff}^2(z')]^{1/2}, \qquad (7.23\text{a})$$

i.e.

$$\frac{\mathrm{d}z'}{\mathrm{d}t'} = \left\{\frac{\gamma'^2 - M^2}{\gamma'^2} + \frac{A_w A_u}{\gamma'^2}\cos[(2kk_u + k_u^2)z']\right\}^{1/2}, \qquad (7.23\text{b})$$

and may be seen to be equivalent to the energy integral (7.8) of the non-relativistic equation of motion of a pendulum, provided that the oscillation phase ξ and the two parameters m and Ω are redefined as

$$\xi = (2kk_u + k_u^2)^{1/2}z' = (k + k_u)z - kt, \qquad (7.24\text{a})$$

$$m = \tfrac{1}{2} + (\gamma'^2 - M^2)/2A_w A_u, \qquad (7.24\text{b})$$

$$\Omega = [(2kk_u + k_u^2)A_w A_u/2\gamma'^2]^{1/2}. \qquad (7.24\text{c})$$

Therefore, the non-relativistic picture of electron–wave interaction in the synchronous frame given in Section 7.2 directly applies to the relativistic circular-polarization FEL as well.

It is somewhat puzzling to see that eqn (7.23) gives the exact solution of the complete relativistic equations of motion of the electron and yet has the same functional form, albeit with differently defined parameters, as the non-relativistic result. The reason for this may be better understood by looking at the way in which the non-relativistic limit of the Hamiltonian (7.21) is attained. The non-relativistic limit corresponds to the case in which both M_{eff} and γ' are close to 1. Then, one may set $\gamma' = 1 + \mathscr{E}_c/m_e c^2$ and $M_{eff} = 1 + U/m_e c^2$, where U and $\mathscr{E}_c$ shall be recognized to be the classical potential and total energies; inserting these expressions into eqn (7.21) and remembering that $\mathscr{K}'$ coincides with the dimensionless axial component $P'_{z'}/m_e c$ of momentum, one has, at first order in $\gamma' - 1$ and $M_{eff} - 1$,

$$P'_{z'} = (2m_e)^{1/2}[\mathscr{E}_c - U(z')]^{1/2}, \tag{7.25}$$

which is the standard non-relativistic expression of momentum. It is clearly seen that, leading-order terms cancelling out, eqn (7.25) has retained the same form as eqn (7.21), except for a different naming of constants. That is why the law of motion has the same form in the relativistic and non-relativistic cases.

Let us now briefly examine the other possibility of having a non-zero coefficient in eqn (7.15); that is, when $\boldsymbol{A}_u^* \cdot \boldsymbol{A}_w = 0$ and $\boldsymbol{A}_u \cdot \boldsymbol{A}_w \neq 0$, so that M_{eff}^2 is expressed by eqn (7.17). In this case M_{eff}^2 again depends only on a linear combination of t and z, namely $kt - (k - k_u)z$, but this combination cannot be interpreted as a co-ordinate in a moving frame because the resulting speed $U = k/(k - k_u)$ would be greater than the speed of light (unless $2k < k_u$). However, it can be interpreted as time, in a reference frame moving with the speed $U = (k - k_u)/k$ (which is less than the speed of light provided that $2k > k_u$). In fact, the equations of a Lorentz transformation with the latter speed U yield

$$z = [kz' + (k - k_u)t']/(2kk_u - k_u^2)^{1/2}, \tag{7.26a}$$

$$t = [kt' + (k - k_u)z']/(2kk_u - k_u^2)^{1/2}, \tag{7.26b}$$

so that $kt - (k - k_u)z = (2kk_u - k_u^2)t'$, and in the new frame M_{eff} is a function of t' alone. Equation (7.21) still applies, but it is now more convenient to use the ordinary Hamiltonian $\mathscr{H}' = [P'^2_{z'} + M_{eff}^2]^{1/2}$. Since in the moving frame the Hamiltonian is independent of z', $P'_{z'}$ is a constant of motion, and from the Hamilton equation $\mathrm{d}z'/\mathrm{d}t' = \partial\mathscr{H}'/\partial P'_{z'}$ the explicit quadrature solution can be obtained:

$$z' = \int \frac{P'_{z'}\,\mathrm{d}t'}{[P'^2_{z'} + M_{eff}^2(t')]^{1/2}}. \tag{7.27}$$

In this case $z'(t')$ is always monotonic (because the square root never changes sign) and no trapping is possible. Since trapping is fundamental to the interaction mechanism presented in Section 7.2, we shall not spend much time on this second solution.

7.5 Description of the electron motion in the laboratory frame

While the foregoing analysis has served to illustrate the non-linear interaction mechanism between electrons and waves in an undulator, it still does not allow us to calculate FEL gain and energy extraction. For this purpose we need to determine, as shown in Section 7.1, the variation of the electron energy γ in the laboratory frame after going through the undulator. To do so only requires transforming the solution obtained for the electron motion in the synchronous frame back to the laboratory frame but, as will now be seen, the exact result thus obtained is not easily manageable, and it turns out to be useful to introduce approximations of the same kind as those adopted in the linear analysis of the previous chapters.

The law of variation of electron energy $\gamma(z)$ along the undulator axis may be calculated in two steps, by first expressing γ as a function of time t' in the synchronous frame and then t' as a function of z. To perform the first step, it is necessary to remember that energy changes under a Lorentz transformation as time and axial momentum as the opposite of the axial co-ordinate. Therefore γ is given by a formula analogous to eqn (7.18b), namely

$$\gamma = \frac{(k + k_u)\gamma' - kP'_{z'}}{(2kk_u + k_u^2)^{1/2}}. \tag{7.28}$$

On replacing $P'_{z'}$ by the Hamiltonian (7.21) and expressing γ' (which is a constant of motion) in terms of the parameter m by eqn (7.24b), eqn (7.28) becomes

$$\gamma = \frac{(k + k_u)[M^2 + (2m - 1)A_u A_w]^{1/2}}{(2kk_u + k_u^2)^{1/2}} \mp \frac{k(2A_u A_w)^{1/2}[m - \sin^2(\xi/2)]^{1/2}}{(2kk_u + k_u^2)^{1/2}}. \tag{7.29}$$

For the second step, denoting by $\xi_m(\Omega t')$ the solution of the pendulum equation (7.9) and remembering the definition (7.24a) of ξ, one has from the Lorentz transformation equation (7.18a):

$$z = \frac{k + k_u}{2kk_u + k_u^2}\xi_m(\Omega t') + \frac{k}{(2kk_u + k_u^2)^{1/2}}t', \tag{7.30}$$

which is an equation relating implicitly time t' in the synchronous frame to the axial co-ordinate z in the laboratory frame.

Equations (7.29) and (7.30) formally solve the problem of determining $\gamma(z)$, but they are not easy to use, mainly because they are in implicit form and contain several independent parameters. Fortunately, in practice some of these parameters always take on very large or very small values, and thus approximations are possible.

A first simplification comes from the ultrarelativistic approximation, routinely adopted in the previous chapters, which in this case consists of replacing the Lorentz transformation formulae (7.18) by eqns (7.22), and may be verified to be equivalent to expanding eqns (7.29) and (7.30) at leading order in the ratio k_u/k (because it is when this ratio is small that the speed of motion of the synchronous frame is ultrarelativistic). In the ultrarelativistic limit, eqn (7.29) becomes

$$\gamma = (k/2k_u)^{1/2}\{[M^2 + (2m - 1)A_u A_w]^{1/2} \mp [2A_u A_w[m - \sin^2(\xi/2)]]^{1/2}\} \tag{7.31}$$

and eqn (7.30):

$$z = (1/2k_u)\xi_m(\Omega t') + (k/2k_u)^{1/2}t'. \qquad (7.32)$$

At the same time eqns (7.24a) and (7.24c) may be simplified into

$$\xi = (2kk_u)^{1/2}z', \qquad \Omega = (kk_u A_w A_u/\gamma'^2)^{1/2}. \qquad (7.33a,b)$$

A second simplification may be introduced when the spatial period of synchrotron oscillations L_s is much longer than the undulator period.[2] L_s may be obtained from eqns (7.32) and (7.10) as

$$L_s = (k/2k_u)^{1/2}T'_p = (k/2k_u)^{1/2}4\mathrm{K}(m)/\Omega \qquad (7.34)$$

which, on using eqn (7.33b) to express Ω and eqn (7.24b) to eliminate γ', becomes

$$L_s = [4\mathrm{K}(m)/k_u](M^2/2A_u A_w + m - \tfrac{1}{2})^{1/2}. \qquad (7.35)$$

The minimum of $\mathrm{K}(m)$ is attained when $m = 0$, and is $\pi/2$. Since $2\pi/k_u = l_u$, it is then seen that L_s/l_u is large when $M^2/A_w A_u$ is. Under the assumption that L_s/l_u is large we may neglect the first term of eqn (7.32), of order l_u when bounded oscillations of ξ take place, with respect to the second, of order L_s, and treat time t' as approximately proportional to the laboratory axial co-ordinate z:

$$t' \simeq (2k_u/k)^{1/2}z. \qquad (7.36)$$

As $M^2/A_u A_w$ must also be large, we may neglect, within the same order of approximation, the term $(2m - 1)A_u A_w$ with respect to M^2 in eqn (7.31), thus leaving

$$\gamma = (k/2k_u)^{1/2}(M \mp (2A_u A_w)^{1/2}\{m - \sin^2[\xi_m(\Omega t')/2]\}^{1/2}). \qquad (7.37)$$

Equations (7.36) and (7.37) finally give the simplified expression of $\gamma(z)$ we need in order to calculate FEL gain and energy extraction under saturation conditions. Notice that when $m = 0$, i.e. when the electron sits still at the bottom of the potential well, eqn (7.37) gives $\gamma = \gamma_s$, where

$$\gamma_s = (M^2 k/2k_u)^{1/2}. \qquad (7.38)$$

This is the value of γ at exact synchronism between electron and wave. As may be seen by comparison with eqn (4.42), γ equals γ_s when k equals the resonance value k_r that was seen in the previous chapters to characterize FEL interaction.

7.6 Direct approximate solution of the equations of motion in the laboratory frame

Although the foregoing analysis enabled us to display several facets of the non-linear interaction between electrons and waves in an undulator, we arrived at eqns (7.36) and (7.37) through a rather long route. There is also a quicker way to obtain this result by working directly on the Hamilton equations in the laboratory frame, which we are now going to illustrate. We have not taken this approach in the first place because it hides the fact that a pendulum equation holds exactly in the synchronous frame and is not a construct of approximations.

In order to obtain the approximate solution (7.36) and (7.37) directly in the laboratory frame, we shall modify the Hamiltonian so that the role of dependent variables is taken by the deviations from synchronous motion, and then expand it in powers of these deviations, just as is generally done in presenting particle-accelerator theory. With this aim, let us consider the Hamilton equations

$$\mathrm{d}t/\mathrm{d}z = \partial\mathscr{H}/\partial\gamma, \qquad \mathrm{d}\gamma/\mathrm{d}z = -\partial\mathscr{H}/\partial t \tag{7.39, 7.40}$$

for the ultrarelativistic Hamiltonian (7.14). We want the electron phase $\xi = (k + k_u)z - kt$ to be one of the new variables. The derivative of ξ with respect to z is given by $\mathrm{d}\xi/\mathrm{d}z = k + k_u - k\,\mathrm{d}t/\mathrm{d}z$, i.e.

$$\mathrm{d}\xi/\mathrm{d}z = k + k_u - k\,\partial\mathscr{H}/\partial\gamma. \tag{7.41}$$

On the other hand, $\partial\mathscr{H}/\partial t = -k\,\partial\mathscr{H}/\partial\xi$. Therefore, eqns (7.40) and (7.41) may together be given the form of Hamilton equations for the variables ξ and γ, i.e.

$$\mathrm{d}\xi/\mathrm{d}z = \partial\mathscr{H}'/\partial\gamma, \qquad \mathrm{d}\gamma/\mathrm{d}z = -\partial\mathscr{H}'/\partial\xi, \tag{7.42}$$

simply by letting $\mathscr{H}' = (k + k_u)\gamma - k\mathscr{H}$, as may easily be verified. From eqn (7.14), the explicit expression of $\mathscr{H}'$ turns out to be

$$\mathscr{H}' = k_u\gamma + kM_{eff}^2/2\gamma. \tag{7.43}$$

Let us now introduce a function $U = -A_uA_w\cos\xi$, so that the square of the effective mass M_{eff} may be written as $M_{eff}^2 = M^2 + U$, and let us set $\gamma = \gamma_s + \delta\gamma$, with γ_s a constant to be determined, and expand eqn (7.43) to second order in $\delta\gamma$. We obtain

$$\begin{aligned}\mathscr{H}' \simeq\ & k_u\gamma_s + kM^2/2\gamma_s + kU/2\gamma_s - (kM^2/2\gamma_s^2 - k_u)\delta\gamma \\ & - kU\delta\gamma/2\gamma_s^2 + (kM^2/2\gamma_s^3)\delta\gamma^2.\end{aligned} \tag{7.44}$$

By letting $\gamma_s = (kM^2/2k_u)^{1/2}$, which happens to be the synchronism condition (7.38), we can get rid of terms of first order in $\delta\gamma$. If it is then assumed that $U = O(\delta\gamma^2)$, as can be verified *a posteriori*, the term containing the product $U\delta\gamma$ turns out to be $O(\delta\gamma^3)$ and may be neglected. The constant term in eqn

(7.44), moreover, may be omitted since an arbitrary constant may always be added to a Hamiltonian. Equation (7.44) thus reduces to

$$\mathscr{H}' \simeq (kM^2/2\gamma_s^3)\delta\gamma^2 + kU/2\gamma_s, \tag{7.45}$$

which looks like the non-relativistic Hamiltonian of a system of momentum proportional to $\delta\gamma$ and potential energy U proportional to $\cos\xi$, i.e. a pendulum.

In order to solve the equations of motion associated with the Hamiltonian (7.45) one only needs to observe that, as the independent variable z does not appear explicitly in the expression of the Hamiltonian, $\mathscr{H}'$ itself is a constant of motion. Inverting eqn (7.45) with respect to $\delta\gamma$ and inserting the explicit expression of U in the form $U = A_u A_w[2\sin^2(\xi/2) - 1]$ gives

$$\delta\gamma = (2\gamma_s^2 A_u A_w/M^2)^{1/2}[\gamma_s \mathscr{H}'/kA_u A_w + \tfrac{1}{2} - \sin^2(\xi/2)]^{1/2} \tag{7.46}$$

which, on replacing γ_s^2/M^2 by $k/2k_u$ according to eqn (7.38) and defining the parameter m as $\gamma_s\mathscr{H}'/kA_u A_w + \frac{1}{2}$, may be recognized to be equivalent to eqn (7.37). In addition, differentiating eqn (7.45) with respect to $\delta\gamma$ gives

$$\frac{\mathrm{d}\xi}{\mathrm{d}z} = \frac{\partial\mathscr{H}'}{\partial\delta\gamma} = (kM^2/\gamma_s^3)\delta\gamma \tag{7.47}$$

which, with the expression (7.46) of $\delta\gamma$, may be recognized to be the pendulum equation (7.9). Note that exactly the same result as before is obtained.

7.7 Energy extraction of the FEL at complete saturation

Having solved the non-linear equations of motion of an electron in the superposition of the undulator and input-wave fields, we may now go back to the problem posed in Section 7.1, of calculating the energy extraction of the FEL. In general, this requires numerical computations, which give numbers but little insight. A better understanding of the saturation mechanism of the free-electron laser and of the order of magnitude of the parameters involved may be gained by analysing the condition of complete saturation first, the one where the input field is so strong, or the undulator so long, that output power does not increase further with undulator length. Although difficult to reach in practice, this condition is much easier to treat and its study gives the right orders of magnitude. We shall also continue to limit our consideration to the case of circularly polarized fields. In the next sections it will be seen what the numerical calculations predict for intermediate saturation levels, and how the results can be extended to general polarization.

Let us start out by determining the condition under which saturation occurs. This has a simple expression in terms of the minimum period of synchrotron oscillations, given by eqn (7.35) for $m = 0$, i.e.

$$L_{s,\min} = (M^2/2A_uA_w)^{1/2} l_u \tag{7.48}$$

(we may neglect the term $\frac{1}{2}$ in eqn (7.35) for we shall always assume that $M^2/2A_uA_w \gg 1$). As long as the undulator length L is small compared to $L_{s,\min}$, only a fraction of a synchrotron oscillation takes place in the whole undulator length, and the perturbative solutions considered in Chapter 5, which may be seen as the result of an expansion of eqn (7.37) in powers of A_w, are good approximations. In terms of input-field amplitude the condition that $L \ll L_{s,\min}$ may be restated as

$$A_w \ll M^2/2A_uN^2, \tag{7.49}$$

$N = L/l_u$ being, as usual, the number of periods of the undulator.

Equation (7.49), therefore, qualifies the range of the small-signal regime. In the opposite case, when

$$A_w \gg M^2/2A_uN^2, \tag{7.50}$$

complete saturation is attained. In this case many synchrotron oscillations take place before each electron exits the undulator.

Notice that the FEL remains in the linear amplification regime up to an input power level which is very high from a practical standpoint. For instance, $A_w = 10^{-4}$, which is the saturation level for $A_u \simeq 1$ and $N \simeq 100$, corresponds to a power density of the order of 10^{10} W/cm^2 in the visible.

Let us now calculate the energy extraction of the FEL at saturation. We shall consider, for the sake of simplicity, the case of a monoenergetic input electron beam of energy γ_i. Since at the undulator entrance, $z = 0$, the electrons are uniformly distributed in time t_i, it follows from eqn (7.24a) that they are also uniformly distributed in initial phase ξ_i. The constant parameter m is determined for each electron by the condition that eqn (7.37), evaluated at $z = 0$, should yield the correct initial value γ_i of γ. This condition gives

$$m = \sin^2(\xi_i/2) + k_u(\gamma_i - \gamma_s)^2/kA_uA_w, \tag{7.51}$$

where γ_s is given by eqn (7.38). On introducing the parameter ν as

$$\nu = (k_u/kA_uA_w)^{1/2}(\gamma_s - \gamma_i), \tag{7.52}$$

eqn (7.51) may be rewritten as

$$m = \nu^2 + \sin^2(\xi_i/2). \tag{7.53}$$

In eqn (7.52) ν is proportional to the difference between the synchronous γ and the actual initial γ of the electron beam. However, since γ_s is a function of the input wavenumber, ν may also be re-expressed in terms of the difference between the input wavenumber and the resonance wavenumber k_r associated with the initial γ, γ_i. In particular, remembering that $k_r = 2\gamma_i^2 k_u/M^2$ and linearizing the expression (7.38) of γ_s with respect to $k - k_r$ gives $\gamma_s - \gamma_i \simeq (M^2/2kk_u)^{1/2}(k - k_r)/2$, and thus

$$\nu = (M^2/2A_u A_w)^{1/2}(k - k_r)/2k_r. \tag{7.54}$$

As ν turns out to be proportional to the difference between actual and resonance wavenumber, it may be called a detuning parameter, and in fact eqn (7.54) is the definition of the detuning parameter appropriate to the saturation regime.

The value of ν determines the fraction of the electrons that is trapped by the quasi-potential. According to eqn (7.53), those electrons having an initial phase such that

$$0 < |\xi_i| < 2 \arcsin (1 - \nu^2)^{1/2} = 2 \arccos \nu$$

have $m < 1$ and therefore are trapped, whereas those with

$$2 \arccos \nu < |\xi_i| < \pi$$

are not. Since ξ_i is uniformly distributed, we may conclude that a fraction

$$f_t = (2/\pi)\arccos \nu \tag{7.55}$$

of the electrons are trapped and the remaining $1 - f_t$ are not, provided, of course, that $|\nu| < 1$. If $|\nu| > 1$ no electrons are trapped.

In order to calculate the energy extraction according to eqn (7.3) we now need the average electron γ at the end of the undulator, that is, the average of eqn (7.37) evaluated at $z = L$ over a uniform ξ_i distribution. In general, calculating this average requires evaluating $\xi_m(\Omega t')$ as a function of both m and t', which is a known but complicated function. The case of complete saturation, however, corresponds to the limit for $L \to \infty$. In this limit the average γ becomes coincident,[3] for electrons falling in every small interval m, $m + \mathrm{d}m$ of the parameter m, with the temporal average over time t'.

For trapped electrons the average of the second term of eqn (7.37) is evidently zero, because this term changes sign periodically while performing a symmetric oscillation, so for trapped electrons $\langle\gamma_f\rangle = \gamma_s$.

For untrapped electrons, which have $m > 1$, the second term of eqn (7.37) can never vanish and change sign, and therefore its average is non-zero. In fact, the average γ of untrapped electrons can be expressed in terms of elliptical integrals with relative ease, and it can be shown that with increasing m, when the motion tends to approach a sinusoidal oscillation around the trajectory that the electron would follow in the absence of the electromagnetic wave, $\langle\gamma\rangle$ becomes more and more close to γ_i. For the purpose of obtaining a simple estimate of energy extraction we may roughly assume that for all untrapped electrons $\langle \gamma_f \rangle = \gamma_i$. The overall average of γ_f is then simply given by

$$\langle \gamma_f \rangle \simeq f_t \gamma_s + (1 - f_t)\gamma_i, \tag{7.56}$$

the first term referring to trapped and the second to untrapped electrons. With the relation (7.52) between $\gamma_s - \gamma_i$ and ν, eqn (7.3) yields

$$\eta_{\text{sat}} = f_t \frac{\gamma_i - \gamma_s}{\gamma_i - 1} = -\frac{(A_u A_w k/k_u)^{1/2}}{\gamma_i - 1} \frac{\nu \arccos \nu}{\pi/2}. \tag{7.57}$$

On neglecting 1 with respect to γ_i, as may be done in the ultrarelativistic limit, and approximately replacing k by $k_r = M^2 k_u / 2\gamma_i^2$, eqn (7.57) may be rewritten as

$$\eta_{\text{sat}} \simeq -\left(\frac{2A_u A_w}{M^2}\right)^{1/2} \frac{\nu \arccos \nu}{\pi/2}, \tag{7.58}$$

which is the approximate analytical expression of energy extraction that we were seeking.

A plot of the function $-(\nu \arccos \nu)/(\pi/2)$ may be seen in Fig. 7.1, together with a plot of the result obtained from an exact numerical evaluation of the contribution of untrapped electrons to $\langle \gamma_f \rangle$. It may be noticed that, just as in the small-signal regime, the curve is antisymmetric about $\nu = 0$, so that gain in radiation power is obtained for $\nu < 0$ and loss for $\nu > 0$. Maximum energy extraction occurs for $\nu = -0.697$ and is given by

$$\eta_{\text{sat, max}} = 0.302(2A_u A_w/M^2)^{1/2}. \tag{7.59}$$

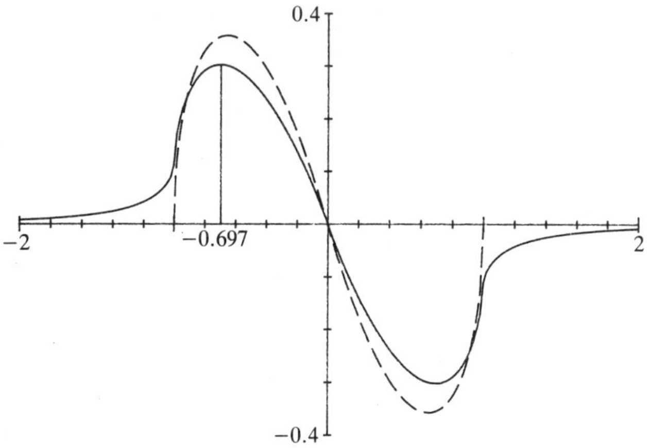

FIG. 7.1. Normalized energy extraction $\bar{\eta} = (M^2/2A_u A_w)^{1/2} \eta_{\text{sat}}$ at complete saturation as a function of the detuning parameter ν. Solid line, numerical result; broken line, approximate formula (7.58).

The approximate equation (7.58) leads to overestimating $\eta_{\text{sat, max}}$ by 17 per cent.

The full width at half maximum of the energy-extraction curve of Fig. 7.1 is $\delta\nu_{fwhm} = 0.76$, which by setting $k \simeq k_r = 2\gamma_i^2 k_u/M^2$ in eqn (7.52) can be seen to correspond to a relative input energy acceptance given by

$$\delta\gamma_{fwhm}/\gamma_i = 0.76(2A_u A_w/M^2)^{1/2}. \tag{7.60}$$

7.8 Approach to saturation

The method of calculating energy extraction described in the previous section applies under the limiting condition of $L \gg L_{s,\text{min}}$, whereas the linear, small-signal analysis of Chapter 5 applies for $L \ll L_{s,\text{min}}$. In the intermediate range where $L \sim L_{s,\text{min}}$ there is hardly any choice other than numerical computation, which in the simplest case consists of following a number of electrons with different, suitably chosen, initial phases in their motion through the undulator with the help of an ordinary differential equation solver (for instance, of the Runge–Kutta type) and then reckoning the average of the values of γ assumed by this group of electrons at the end of the undulator.

Although a large variety of physical effects and corrections are easily incorporated into a numerical model, and that is why numerical models are so useful for simulating particular instances of reality, for the purpose of gaining a general understanding of the properties of undulators we need a model containing the least possible number of free parameters. The one-dimensional, monoenergetic model considered in Sections 7.5 and 7.6 with the approximations coming from the ultrarelativistic limit and the assumption of $L_{s,\text{min}} \gg l_u$ can be described in terms of only two parameters; the detuning parameter ν of eqn (7.52), or (7.54), and the saturation parameter $L/L_{s,\text{min}}$.

Indeed, upon introducing a dimensionless longitudinal co-ordinate as $\bar{z} = 2\pi z/L_{s,\text{min}}$, so that the period of small oscillations becomes 2π and their angular frequency 1, the pendulum equation (7.7) for synchrotron oscillations may be rewritten as[4]

$$\mathrm{d}^2\xi/\mathrm{d}\bar{z}^2 = -\sin\xi. \tag{7.61}$$

This equation must be integrated numerically for $\bar{z}$ varying from 0 to $2\pi L/L_{s,\text{min}}$ for a number N_e of electrons with different initial phases. The initial phase $\xi_n(0)$ for the nth electron may be conveniently chosen to be $(n/N_e)2\pi$; in order to find an initial condition for the derivative $\mathrm{d}\xi_n/\mathrm{d}\bar{z}$ eqn (7.47) may be used, which on multiplying both sides by $L_{s,\text{min}}/2\pi$ becomes

$$\frac{\mathrm{d}\xi}{\mathrm{d}\bar{z}} = \frac{kM^2 L_{s,\text{min}}}{2\pi\gamma_s^3}\,\delta\gamma = \left(\frac{4k_u}{kA_u A_w}\right)^{1/2}(\gamma - \gamma_s). \tag{7.62}$$

By comparing eqn (7.62) for $\gamma = \gamma_i$ with the definition (7.52) of ν, the required initial condition may be seen to be $(d\xi_n/d\bar{z})_0 = -2\nu$.

Once eqn (7.61) has been integrated for all the electrons up to a certain abscissa $\bar{z}$, the average energy $\langle\gamma\rangle$ may be calculated using eqn (7.62) again, as

$$\langle\gamma\rangle = \frac{1}{N_e}\sum_{n=1}^{N_e}\left[\gamma_s + \left(\frac{kA_uA_w}{4k_u}\right)^{1/2}\frac{d\xi_n}{d\bar{z}}\right] \tag{7.63}$$

whence, by performing the same steps that led from eqn (7.56) to eqn (7.58), energy extraction may be obtained as

$$\eta = -\left(\frac{2A_uA_w}{M^2}\right)^{1/2}\frac{1}{N_e}\sum_{n=1}^{N_e}\left(\nu + \frac{1}{2}\frac{d\xi_n}{d\bar{z}}\right). \tag{7.64}$$

The result of the just-described numerical calculation is shown, for the value -0.697 of ν which was seen to correspond to the maximum gain at complete saturation, in Fig. 7.2.

It may be noticed that the approach to the complete-saturation limit is slow and oscillatory, so that the conditions envisaged in the preceding section are hardly ever realized in practice. Nevertheless, eqns (7.58) and (7.59) give a correct estimate of the order of magnitude of energy extraction in a saturated FEL, representing, by the way itself in which they were calculated, the mean value around which η oscillates; in particular, Fig. 7.1 may be assumed to

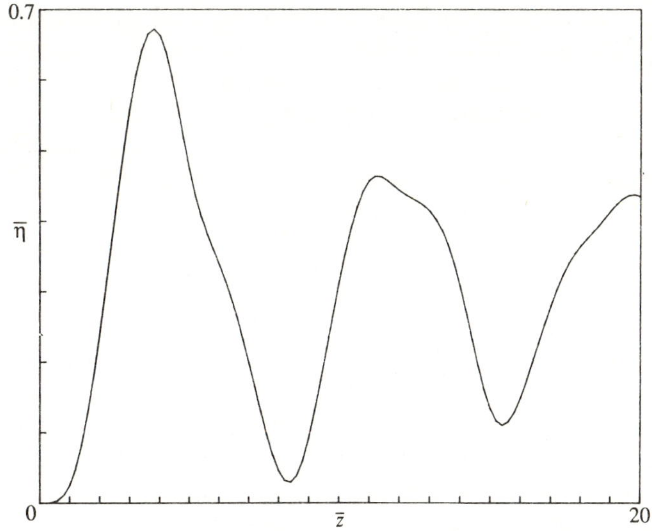

FIG. 7.2. Normalized energy extraction $\bar{\eta} = (M^2/2A_uA_w)^{1/2}\eta$ vs. normalized co-ordinate $\bar{z} = 2\pi L/L_{s,\min} = 2\pi(2A_uA_w/M^2)^{1/2}N$, for $\nu = -0.697$.

represent the dependence of energy extraction on detuning reasonably well, even before complete saturation.

7.9 Maximum energy extraction

It may be seen from Fig. 7.2 that the first, and highest, overshoot of the energy-extraction curve occurs for $\bar{z} \simeq 3.8$, i.e. $L \simeq 0.6L_{s,\min}$. The corresponding value of η is given by

$$\eta = 0.67(2A_u A_w/M^2)^{1/2} \tag{7.65}$$

and is roughly twice the one predicted by eqn (7.59). Notice that since $(2A_u A_w/M^2)^{1/2} = l_u/L_{s,\min}$ (from eqn (7.48)) and $L_{s,\min} \simeq L$, the value of η given by eqn (7.65) is of the order of $l_u/L = 1/N$.

The energy extraction of eqn (7.65) is maximum with respect to the undulator length for a given input field; however, one may more frequently want to know the maximum energy extraction obtainable with varying input field for a given undulator length. This maximum can be found to occur for a slightly different value of $\bar{z}$, namely 4.28 according to numerical results, which corresponds to $L/L_{s,\min} = 0.68$, and is given by

$$\eta_{\max} = 0.44/N. \tag{7.66}$$

Numerical calculations also show that maximum energy extraction is attained for the value -0.62 of v, slightly different from the one which grants maximum energy extraction at full saturation ($v = -0.697$). The result expressed by eqn (7.66) has been computed for $v = -0.62$.

The input field level which grants maximum energy extraction, and therefore maximum output power, may be usefully re-expressed in terms of the corresponding gain. From eqn (7.48), for $L/L_{s,\min} = 0.68$ we obtain $A_w = 0.23M^2/A_u N^2$. Putting this value of A_w, together with the resonant value k_r of k, into eqn (7.4) gives the desired expression of gain at maximum output,

$$g_{\text{maxout}} = 0.14N^3 A_u^2 \gamma_i^{-3} \pi^2 l_u^2 j_b/I_A, \tag{7.67}$$

which turns out to be about one quarter of the small-signal gain of eqn (5.24). It follows, for instance, that the designer of an FEL in oscillator configuration ought to choose a value of small-signal gain exceeding the losses of the feedback optical cavity by a factor of four if he wants to obtain the maximum energy extraction predicted by eqn (7.66). Both lower and higher values of gain would give a lesser output.

7.10 Fields of any polarization

Within the assumption of $L_{s,\min} \gg l_u$ it is possible to extend all the results of this chapter to undulator and wave fields of general polarization. We shall show that, just as in the small-signal regime, when the polarization of the undulator and/or laser fields is other than circular the behaviour of a free-electron laser is qualitatively similar to that corresponding to circular polarization, the only difference being that operation on higher harmonics becomes possible, and the effective undulator vector potential $A_{ueff,n}$ of eqn (4.36) must be substituted for A_u in the final results (but not in the effective mass).

Under the assumption that $L_{s,\min} \gg l_u$ two very different periods of oscillation exist in the electron motion, and therefore it is possible to apply a two-timing approximation. This approximation may be introduced in several ways, one of which is to consider it as a special case of the effective-mass approximation described in Appendix B.

Let us then introduce the guiding centre, just as was done in the study of betatron oscillations in Section 3.8, the motion of which is described in the present one-dimensional context by the time $t_g(z)$ of passage at the position z only. The relation between the time $t(z)$ when the electron passes at a given position z and the time $t_g(z)$ when the guiding centre goes past the same position is given by the equation of motion of the electron in the undulator field alone, and is the same as eqn (3.42b), but with $\boldsymbol{P}_\perp = 0$:

$$t(z) = t_g(z) + \mathrm{Re}\left(\frac{A_u^2}{8i\gamma^2 k_u}\,\mathrm{e}^{2ik_u z}\right). \tag{7.68}$$

$\mathrm{d}t_g/\mathrm{d}z$ and γ are now assumed to be functions of z varying slowly on the scale of the undulator period l_u. According to eqn (B.8), the effective mass is given by $M_{eff}^2 = 1 + \langle \boldsymbol{a}^2 \rangle$, where the average is to be calculated, on a scale of the order of l_u, upon inserting eqn (7.68) into the expression (7.11) of the vector potential. Thus one obtains, in place of eqn (7.15),

$$\begin{aligned} M_{eff}^2 = 1 + \frac{A_u^2}{2} + \frac{A_w^2}{2} + \Bigg\langle \mathrm{Re}\Bigg\{ &\frac{A_u^2}{2}\,\mathrm{e}^{2ik_u z} \\ &+ (\boldsymbol{A}_u^*\mathrm{e}^{-ik_u z} + \boldsymbol{A}_u\mathrm{e}^{ik_u z})\cdot \boldsymbol{A}_w \exp i\left[kt_g(z) - kz + \mathrm{Re}\left(\frac{kA_u^2}{8i\gamma^2 k_u}\,\mathrm{e}^{2ik_u z}\right)\right] \\ &+ \frac{A_w^2}{2}\exp 2i\left[kt_g(z) - kz - \mathrm{Re}\left(\frac{kA_u^2}{8i\gamma^2 k_u}\,\mathrm{e}^{2ik_u z}\right)\right]\Bigg\}\Bigg\rangle. \end{aligned} \tag{7.69}$$

In general, the average of the oscillating terms enclosed in curly brackets in eqn (7.69) is zero, and the effective mass is a constant. It may happen, however, that the beat between the undulator and the laser oscillations generates a slowly varying contribution to M_{eff}^2 which is not eliminated by an average on the scale of l_u. This eventuality may be seen to occur when $k(\mathrm{d}t_g/\mathrm{d}z - 1)$ is close to an integer multiple of k_u, say nk_u.

Since $\mathrm{d}t_g/\mathrm{d}z$ is by assumption a slowly varying quantity, if this condition is verified at the beginning of an undulator period it will be satisfied for the whole period; $(\mathrm{d}/\mathrm{d}z)(kt_g - kz - nk_u z)$ is then a small quantity throughout the period and we may take $\mathrm{e}^{\mathrm{i}(kt_g - kz - nk_u z)}$ out of the average, treating it as a constant for the duration of one period. Equation (7.69) then becomes

$$M_{eff}^2 = 1 + \frac{A_u^2}{2} + \frac{A_w^2}{2} + \mathrm{Re}\Big\{\mathrm{e}^{\mathrm{i}[kt_g - (k + nk_u)z]} \times \Big\langle (A_u^* \mathrm{e}^{-\mathrm{i}k_u z} + A_u \mathrm{e}^{\mathrm{i}k_u z}) \cdot A_w \exp \mathrm{i}\Big[nk_u z + \mathrm{Re}\Big(\frac{kA_u^2}{8\mathrm{i}\gamma^2 k_u}\,\mathrm{e}^{2\mathrm{i}k_u z}\Big)\Big]\Big\rangle\Big\}, \tag{7.70}$$

where the first and the last term inside the angular brackets, the average of which is always zero, have been dropped. Equation (7.70) may also be written as

$$M_{eff}^2 = M^2 + \mathrm{Re}\{A_{ueff,n}^* \cdot A_w \mathrm{e}^{\mathrm{i}[kt_g - (k + nk_u)z]}\}, \tag{7.71}$$

where M^2 denotes, as before, the constant part $1 + A_u^2/2 + A_w^2/2$ of the effective mass, and $A_{ueff,n}$, the result of the averaging denoted by angular brackets in eqn (7.70), is given by

$$A_{ueff,n} = \frac{1}{l_u}\int_0^{l_u} (A_u \mathrm{e}^{\mathrm{i}k_u z} + A_u^* \mathrm{e}^{-\mathrm{i}k_u z}) \times \exp -\mathrm{i}\Big[nk_u z + \mathrm{Re}\Big(\frac{kA_u^2}{8\mathrm{i}\gamma^2 k_u}\,\mathrm{e}^{2\mathrm{i}k_u z}\Big)\Big]\,\mathrm{d}z. \tag{7.72}$$

It should have been noticed by now that eqn (7.72) is identical to eqn (4.36), apart from the angular dependence which is not considered in this chapter (i.e. we now set $\vartheta_{we} = 0$). $A_{ueff,n}$, it will be remembered, is different from zero for n odd only, and is expressed in terms of Bessel functions by eqn (4.49).

Equation (7.71) differs from eqn (7.16) only by the substitution of t_g for t, $A_{ueff,n}$ for A_u, and nk_u for k_u. Once eqn (7.16) is replaced by eqn (7.71) all the calculations performed in the previous sections may be repeated with no modification, and all of the foregoing results can be immediately extended to free-electron lasers of general polarization working on their fundamental resonance frequency or on its odd harmonics, simply by substituting nk_u for k_u and $A_{ueff,n}$ for A_u, provided that the condition of $L_{s,\min} \gg l_u$, i.e. $2A_u A_w \ll M^2$, is satisfied.

As the present extension of saturation FEL theory to any polarization is the result of an approximation, one may wonder how good an approximation it is. Figure 7.3 shows the exact (numerically computed) electron trajectories

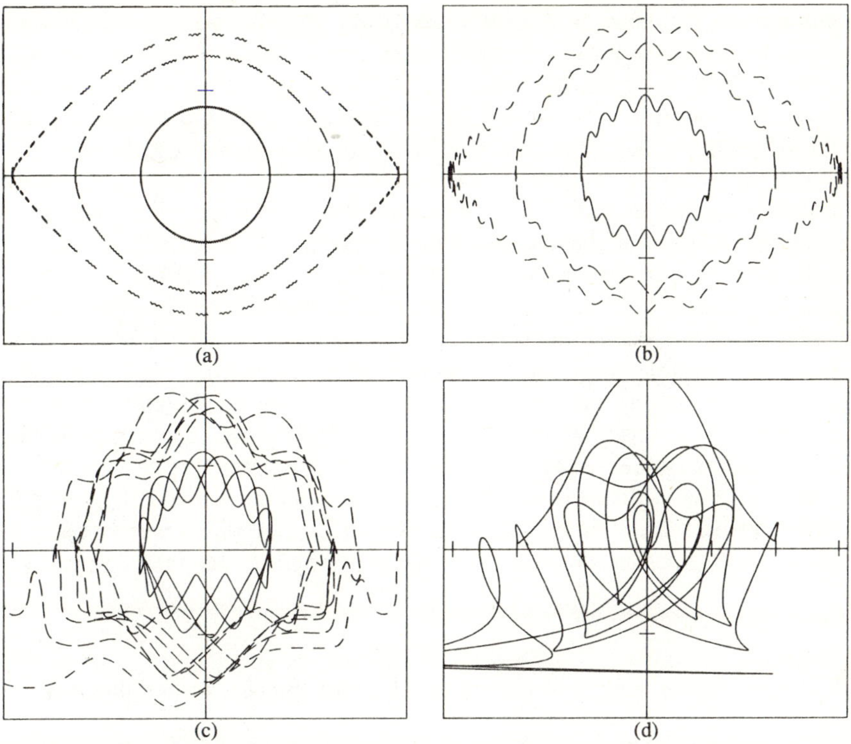

FIG. 7.3. Numerically computed phase-space trajectories of the guiding centre in a linearly polarized FEL, for (a) $2A_uA_w/M^2 = 10^{-4}$, (b) $2A_uA_w/M^2 = 10^{-2}$, (c) $2A_uA_w/M^2 = 0.1$, and (d) $2A_uA_w/M^2 = 0.6$. Different line types correspond to initial conditions as follows: solid line, $\xi_g = 1$; broken line, $\xi_g = 2$; dashed line, $\xi_g = 3$, imposed at the point where $d\xi_g/dz = 0$. The horizontal scale represents exactly one period of pendulum phase from $\xi_g = -\pi$ to $\xi_g = \pi$; the vertical scale is in arbitrary units.

in the phase space of the variables $\xi_g = (k + k_u)z - kt_g$ and $(k_u/kA_uA_w)^{1/2}\delta\gamma$ for different values of the ratio $2A_uA_w/M^2 = (l_u/L_{s,\min})^2$ in the limiting condition of $A_u \to \infty$. (The trajectory depends on the exact value of A_u, and a value had to be chosen for the sake of exemplification. The limit of $A_u \to \infty$ is practically significant because FEL operation is favoured by a large value of A_u.)

For $2A_uA_w/M^2 = 10^{-4}$, which is the saturation limit for an undulator with 100 periods, the trajectory shown in Fig. 7.3(a) is indistinguishable from the one provided by the effective-mass approximation; that is, the phase trajectory of a pendulum represented in Fig. 1.12, except for an almost imperceptible modulation of $\delta\gamma$ with the undulator period. For $2A_uA_w/M^2 = 10^{-2}$ (Fig. 7.3(b)), the saturation limit of a 10-period undulator, the energy modulation has become definitely significant but the underlying pendulum trajectory is still respected. For $2A_uA_w/M^2 = 0.1$ signs of instability begin to appear: only electrons moving deep inside the potential well (solid line) maintain a bounded orbit, while those which get closer to the edges (broken lines) are detrapped after a few oscillations. Finally, for $2A_uA_w/M^2 = 0.6$ (Fig. 7.3(d)) instability has set in completely, electrons move on chaotic trajectories, and even the one whose motion is represented in the figure, which starts right at the bottom of the potential well, eventually escapes.

In fact, it can be shown theoretically (Luchini and Motz 1984), through an analysis based on the known mathematical stability properties of the solutions of the Mathieu and Hill equations, that the motion of the electron becomes unstable, even at the bottom of the potential well, when the parameter $2A_uA_w/M^2$ grows beyond a certain threshold. It is also interesting to observe that the unpredictable behaviour of the electron exemplified by Fig. 7.3(d) is a consequence of the loss of sharpness of the undulator resonance due to the near equality of the characteristic lengths $L_{s,\mathrm{min}}$ and l_u; in this respect the phenomenon is analogous to that which was foreseen to occur, in the previous chapter, when the gain length L_g becomes comparable to l_u.

7.11 Saturation regime of the optical klystron: frequency multiplication

The optical klystron, introduced in Section 5.14, is a modified undulator structure which consists of two short undulators separated by a dispersive section where a more or less uniform magnetic field is present. This device delegates to physically separate stages the three functions of modulating the electron beam energy at the frequency of the input wave, converting the energy modulation into position modulation, and extracting radiation from the position-modulated (or, as is often said, 'bunched') beam.

Because of the more efficient conversion of energy modulation into position modulation, the optical klystron exhibits a higher small-signal gain than a uniform undulator of the same length, as was shown in Section 5.14, at the cost of a reduced tolerance to electron beam energy spread. In addition, as we are going to show, the optical klystron enters the non-linear regime at a lower power level than a uniform undulator. For this reason, the optical klystron is not fit for generating very high power, but it has an advantage for those applications which explicitly need a non-linear component, such as frequency multiplication.

The non-linear regime of the optical klystron is somewhat simpler to describe than that of the uniform-undulator FEL, just because of the above-mentioned separation of functions. It is, in fact, sufficient to take non-linearity into account only in the central dispersive section, where the motion is relatively simple. In the following we shall illustrate how the non-linear analysis of the optical klystron proceeds, assuming circular polarization. It can be shown that, just as for a uniform undulator, extension of the results to any polarization only requires substituting $\boldsymbol{A}_{ueff,n}$ for $\boldsymbol{A}_u$.

According to the approximate description of undulator interaction given in Section 5.4, in the first undulator section of an optical klystron each electron acquires a small energy deviation γ_1, which at first order in the input-wave amplitude is given by eqn (5.13),

$$\gamma_1 = \frac{k}{2\mu\gamma_i} \mathrm{Re}[\boldsymbol{A}_u^* \cdot \boldsymbol{A}_w \mathrm{e}^{ikt_i}(\mathrm{e}^{i\mu L_1} - 1)], \tag{7.73}$$

where μ is still the beat wavenumber defined by eqn (5.14), but L_1 is now the length of the first undulator section. Thereafter, γ remains approximately constant and equal to $\gamma_i + \gamma_1$ in the dispersive section, where no resonant interactions with the electromagnetic field take place. On the other hand, the deviation in position, or equivalently in time $t(z)$ of passage by a given position, acquired by the electron in the dispersive section, is much larger than that acquired in the undulator section, so that the latter may be neglected and the distribution of electrons arriving at the beginning of the dispersive section may be considered to be uniform. In other words, we may use the unperturbed value of time t_1 of arrival at the end of the first undulator section, as if the wave field were not present, i.e. $t_i + \mu L_1$. In terms of this t_1 eqn (7.73) may be rewritten as

$$\gamma_1 = \frac{k}{2\mu\gamma_i} \mathrm{Re}[\boldsymbol{A}_u^* \cdot \boldsymbol{A}_w \mathrm{e}^{ikt_1}(1 - \mathrm{e}^{i\mu L_1})]. \tag{7.74}$$

In the dispersive section, the time $t(z)$ of passage at the position z obeys the equation of motion (5.11c):[5]

$$\frac{\mathrm{d}t}{\mathrm{d}z} = 1 + \frac{1 + \boldsymbol{a}_\perp^2}{2\gamma^2}. \tag{7.75}$$

Since γ may be considered constant and equal to $\gamma_i + \gamma_1$, the integration of eqn (7.75) from time t_1 of exit from the first undulator to time t_2 of entry into the second is straightforward, and gives, at first order in γ_1,

$$t_2 = t_1 + \left[1 + \frac{\langle a_d^2 \rangle}{2(\gamma_i + \gamma_1)^2}\right] L_d \simeq t_1 + \left(1 + \frac{\langle a_d^2 \rangle}{2\gamma_i^2}\right) L_d - \frac{\langle a_d^2 \rangle L_d}{\gamma_i^3} \gamma_1, \tag{7.76}$$

where, as in Section 5.14, $\langle a_d^2 \rangle$ and L_d denote the mean square vector potential and the length of the dispersive section.

Once t_2 has been calculated as a function of t_1, the probability density $f_2(t_2)$ representing the distribution in arrival time of the electrons at the entrance of the second undulator may simply be obtained as $f_1(t_1)/(\mathrm{d}t_2/\mathrm{d}t_1)$, which, on using eqns (7.74)–(7.76) and remembering that f_1 is constant and may be taken to be unity, gives

$$f_2 = \frac{1}{1 + q\cos(kt_1 - \varphi_i)}, \tag{7.77}$$

where

$$q = \frac{\langle a_d^2\rangle k^2 L_d L_1}{4\gamma_i^4} \frac{\sin(\mu L_1/2)}{\mu L_1/2} A_u A_w \tag{7.78}$$

and φ_i is an irrelevant initial phase term.

On a qualitative basis, eqn (7.77) clearly shows that when q is small the electron distribution f_2 differs from a uniform one only by a shallow sinusoidal modulation of amplitude proportional to q (because eqn (7.77) itself may be linearized as $f_2 \sim 1 - q\cos(kt_1 - \varphi_i)$), whereas when q is of order unity or greater f_2 has a largely non-sinusoidal behaviour and a manifestly non-linear dependence on q and, consequently, on input field amplitude A_w. Therefore, the parameter controlling non-linearity in an optical klystron is q, and when q is not small the dependence of all relevant quantities on input field becomes non-linear. As a side-effect, harmonics higher than the first may be produced.

Using the expression (7.78) of q, we can now prove the claim made at the beginning, that the optical klystron enters the non-linear regime at a lower power level than a uniform undulator. On assuming, for the sake of simplicity, that μ has the value which maximizes q (i.e. 0) and that $k \simeq k_r = 4\pi\gamma^2/M^2 l_u$, eqn (7.78) may be rewritten as

$$q = \frac{4\pi^2\langle a_d^2\rangle L_d L_1}{M^4 l_d^2} A_u A_w. \tag{7.79}$$

The corresponding parameter for a uniform undulator of length $2L_1 + L_d$ is given by

$$\left(\frac{2L_1 + L_d}{L_{s,\min}}\right)^2 = \frac{(2L_1 + L_d)^2}{l_u^2} \frac{2A_u A_w}{M^2} \tag{7.80}$$

(according to eqn (7.48)). It may easily be seen that whenever $\langle a_d^2\rangle \gg M^2 = 1 + \langle a_1^2\rangle$, as is necessary in any case for the gain of the optical klystron to be sensibly higher than that of a uniform undulator, q exceeds the value given by eqn (7.80).

Quantitatively, the radiation emitted in the second stage of an optical klystron by the beam bunched according to eqn (7.77) can be calculated by the classical radiation formulae recalled in Chapter 4. If further action of the input wave upon the electrons inside the second stage is neglected, just as time shift was in the first stage where energy shift is the dominant effect, the calculation proceeds exactly as in Section 4.4, the only difference being that at the moment of summing eqn (4.38) over the electron beam distribution, as was done to obtain eqn (4.46), the electron distribution must no longer be assumed uniform. The result of this calculation, the details of which we omit, is that if the second undulator is tuned on a harmonic of the input wavenumber, say the nth, radiation is extracted on this harmonic with an amplitude proportional to the corresponding component $F_{2,n}$ of the discrete Fourier spectrum of the electron density f_2 of eqn (7.77), given by

$$F_{2,n} = \frac{1}{\lambda} \int_0^{\lambda} \frac{\mathrm{e}^{-inkt_2}}{1 + q\cos(kt_1 - \varphi_i)} \, \mathrm{d}t_2. \tag{7.81}$$

The integral of eqn (7.81) can be calculated by first recasting it as an integral in $\mathrm{d}(kt_1)$, namely as

$$F_{2,n} = \frac{1}{2\pi} \int_0^{\pi} \mathrm{e}^{-in[kt_1 + q\sin(kt_1 - \varphi_i)]} \, \mathrm{d}(kt_1) \tag{7.82}$$

which, according to eqn (4.48), evaluates to

$$F_{2,n} = J_n(-nq)\mathrm{e}^{in\varphi_i}. \tag{7.83}$$

When q is of order unity or larger, $F_{2,n}$ decreases very slowly with increasing n, as may be shown by means of the asymptotic formula (4.53). The optical klystron, therefore, may be used as an efficient frequency multiplier in a broad wavelength range.

Notice that, despite a superficial similarity, frequency multiplication in an optical klystron is a phenomenon that is completely different from the generation of higher harmonics in the spontaneous radiation spectrum of an undulator, or the amplification of radiation at harmonics of the resonance wavenumber by an FEL, and should not be confused with them. In fact, in the latter two cases, operation of the undulator, either as a spontaneous radiator or as an amplifier, is obtained on a wavelength shorter than the *resonant* wavelength with more or less unchanged characteristics, much as though the undulator period had been shortened. In the case of frequency multiplication, the optical klystron must instead be driven by an external laser, and radiation is generated at a frequency which is an integral multiple of the *input* frequency and has the same spectral structure and coherence properties as the input.

This phenomenon provides a possible alternative route to the generation of coherent radiation at far-uv or X-ray wavelengths, where sufficiently good mirrors may not be available to build a (conventional or FEL) laser oscillator: frequency-multiplying a visible laser beam of longer wavelength.

Experiments on frequency multiplication of visible radiation by an optical klystron were performed by Girard *et al.* on the storage ring ACO in 1984. Further details on this application may be found in their paper and references therein.

Notes

1. For a simple pendulum of mass m_p and length l, the tangential component of the gravitational force is $-m_p g \sin \xi$, ξ being the angular displacement from the bottom position, and the tangential component of acceleration is $l\,d^2\xi/dt'^2$, whence eqn (7.7) follows with $\Omega^2 = g/l$.
2. Since the period of synchrotron oscillations is the typical longitudinal scale of variation in the non-linear regime, the approximation of neglecting terms of order l_u/L_s in the calculation of gain is analogous to that of neglecting terms of order l_u/L, as done in Chapter 5, or terms of order l_u/L_g as in Chapter 6.
3. A formal proof of this statement would lead us too far. The basic idea is that, γ being a periodic function of t' with a frequency varying with m, electrons characterized by a slightly different m acquire with time an increasing phase difference between them, and therefore in the limit for t' tending to infinity the electrons which are contained in any small interval m, $m + dm$, whatever their initial phases, end up being uniformly spread over the whole phase range $(0, 2\pi)$. It follows that the ensemble and time averages of γ coincide.
4. Notice that the first-order pendulum equation (7.8), although simpler from the analytical viewpoint, is not preferable to the second-order equation (7.7) in numerical calculations because of the non-analytic behaviour of the square root and the added complication of handling its periodic change of sign.
5. The above discussion does not account for the fact that the direction of the electron trajectory in the dispersive section may deviate significantly from that of the z-axis, making the ultrarelativistic equations of motion in the form (5.11) invalid. Adaptation of the results to this more general situation, however, involves hardly anything more than replacing dt/dz in eqn (7.75) by the derivative of time with respect to the curvilinear abscissa along the electron path. In order not to complicate the formalism, we restrict the discussion here to small angular deviations from the direction of the z-axis.

References

Courant, E. D. and Snyder, S. (1958). Theory of the alternating-gradient synchrotron. *Ann. Phys.* **3**, 1–48.

Girard, B., Lapierre, Y. Ortega, J. M., Bazin, C., Billardon, M., Elleaume, P., Bergher, M., Velghe, M., Petroff, Y. (1984). Optical frequency multiplication by an optical klystron on the ACO storage ring. *Phys. Rev. Lett.* **53**, 2405–8.

Lawson, J. D. (1980). Coherent interaction between waves and particle streams. *Particle Accelerators* **10**, 73–81.

Lawson, J. D. (1988). *The physics of charged particle beams.* Second edition. Clarendon Press, Oxford.

Luchini, P. and Motz, H. (1984). Instability of an equation arising in the theory of free-electron lasers. *Phys. Lett.* **103A**, 110–12.

Sands, M. (1970). The physics of electron storage rings: an introduction. *Report SLAC-121.* Stanford Linear Accelerator Center.

Webber, S. E. (1957). *IRE Trans. Electron. Dev.* **ED-4**, 87.

8

TAPERED UNDULATORS

In a large-gain free-electron laser the effect of amplitude variations of the laser field upon electron motion cannot be neglected. If, in addition to large gain, there is a substantial power extraction, that is if the laser power is a non-negligible fraction of the power transported by the electron beam, the effect of power extraction upon electron energy cannot be neglected. Under these conditions the laser power and the energy of the electron beam, i.e. its mean γ, can no longer be considered constant along the undulator. Since variations in γ affect the synchronism condition which is essential for FEL operation, it becomes necessary to allow the undulator parameters (period and/or field amplitude) to vary as well if synchronism is to be maintained. The undulator becomes 'tapered'.

The topic of this chapter is this non-linear regime, in which the wave field amplitude, undulator field amplitude and period, and mean electron γ are all allowed to change smoothly along the undulator. In this regime a completely new gain mechanism becomes dominant, one in which electrons trapped by synchrotron oscillations are driven by the slow variation of undulator and wave parameters up or down in energy, as though in a bucket chain or moving staircase, thus exchanging energy with the laser wave. This is the mechanism by which the greatest energy transfer can be achieved, in either direction, and therefore forms the basis of very-high-power lasers and laser particle accelerators.

8.1 Motion of an electron in a tapered wave structure

According to the interaction mechanism described in the previous chapter, when an electron beam and a travelling electromagnetic wave move with a sufficiently similar speed some of the electrons become trapped by the quasi-potential existing in the synchronous frame and begin to move with exactly the same speed as the wave. It follows naturally that a way to accelerate or slow down these electrons is simply to gradually change the velocity of the wave. This is the basic principle of operation of travelling-wave linear accelerators (linacs), where a microwave travelling with changing speed in a tapered waveguide is used to accelerate a bunch of particles, and of high-power travelling-wave tubes, where it is used to slow them down. The same principle is put to work in a tapered undulator, where a quasi-potential pattern with changing speed is generated by the superposition of an electromagnetic wave with a static magnetic field of changing period and/or amplitude.

The basic mechanism of interaction between an electron and a tapered wave structure was explained by Slater (1948) in connection with linacs. We now want to show, on similar lines, how the general picture of the non-linear interaction between electron and wave given in Section 7.2 may be generalized to include the possibility of tapering.

We shall initially consider a wave the phase of which varies non-linearly in *time*, although we are eventually interested in a wave the speed of which changes in *space*. It may be intuitively imagined, and will be shown later, that a particle travelling along with the wave sees little difference between these two situations.

In order to study the effects of a wave with changing speed, let us replace the expression (7.5) of the force acting upon the electron by

$$f_z = F \sin[\psi(t) - \beta z], \tag{8.1}$$

where $\psi(t)$ is a non-linear function and β a constant. This expression represents a travelling wave having a phase velocity $v_{ph} = \omega(t)/\beta$, where $\omega = \mathrm{d}\psi/\mathrm{d}t$, which changes with time. Wishing to proceed in analogy with Section 7.2, we should look for a reference frame in which the force pattern of eqn (8.1) is stationary. However, in a tapered wave structure such a reference frame is non-inertial, and suitable inertia forces must be introduced in order to preserve the validity of the equations of motion.

Moving at first within the framework of non-relativistic mechanics, we may introduce a moving co-ordinate z' simply as $z' = z - \psi(t)/\beta$. A point characterized by a constant value of z' moves with the velocity v_{ph} and the acceleration $\mathrm{d}v_{ph}/\mathrm{d}t$; therefore, in the primed frame an electron experiences the force $f'_{z'} = f_z - m_e\, \mathrm{d}v_{ph}/\mathrm{d}t$.

Let us consider in particular what happens when the acceleration $\mathrm{d}v_{\varphi}/\mathrm{d}t$ is constant. In this case the force $f'_{z'}$ is again a function of z' only, and may be derived from a potential given by

$$U = -(F/\beta)\cos(\beta z') + (m_e\, \mathrm{d}v_{ph}/\mathrm{d}t)z', \tag{8.2}$$

i.e., in terms of the electron phase $\xi = \beta z - \psi(t) = \beta z'$,

$$U = -(F/\beta)(\cos \xi + p\xi), \tag{8.3}$$

where $p = -(\beta m_e/F)\, \mathrm{d}v_{ph}/\mathrm{d}t$. The potential of eqn (8.3) differs from that of eqn (7.6) for the term $p\xi$, and may or may not have wells depending on the value of the quantity p, which we shall call the deceleration parameter. When $|p| > 1$ the function $U(\xi)$ is monotonic, and has no local minima or maxima. For $|p| < 1$, a plot of the potential $U(\xi)$ is instead shaped as in Fig. 8.1, and presents a sequel of wells in which electrons may be trapped, if the initial conditions allow it.

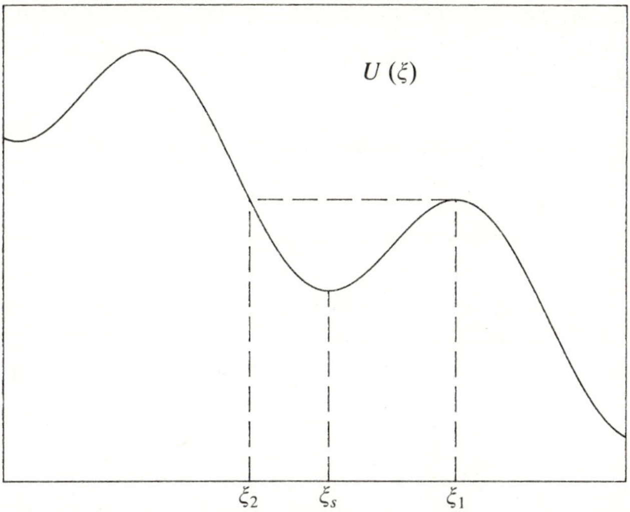

FIG. 8.1. Quasi-potential created in the synchronous frame by a decelerating wave, for $p = 0.3$. ξ_s is the synchronous phase; ξ_1, ξ_2 are the end points of the potential well.

Just as in Section 7.2, the bottom of each well defines the 'synchronous' electron phase ξ_s, which the electron can maintain indefinitely if it starts out with that phase and zero velocity in the synchronous frame. Simple calculations show that the synchronous phase ξ_s, a local minimum of the function U, is given by a solution of the equation $\sin \xi_s = p$, where $\cos \xi_s > 0$, i.e. by the principal value of $\arcsin p$ bound between $-\pi/2$ and $\pi/2$ (plus any integral multiple of 2π). The endpoints ξ_1 and ξ_2 of the potential well are given by the equations $\sin \xi_1 = p$ (with $\cos \xi_1 < 0$) and $\cos \xi_2 + p\xi_2 = \cos \xi_1 + p\xi_1$ (with $\xi_2 \neq \xi_1$). A little trigonometry allows an equation for the well width, $\Delta\xi = \xi_2 - \xi_1$, to be written directly in terms of p as

$$\frac{1 - \cos \Delta\xi}{\Delta\xi - \sin \Delta\xi} = \frac{p}{(1 - p^2)^{1/2}}. \tag{8.4}$$

The non-relativistic motion of an electron in the potential of eqn (8.3) is governed by the Hamiltonian

$$\mathscr{H} = P_{z'}^{\prime 2}/2m_e - (F/\beta)(\cos \xi + p\xi), \tag{8.5}$$

where $P'_{z'} = m_e\, \mathrm{d}z'/\mathrm{d}t' = (m_e/\beta)\, \mathrm{d}\xi/\mathrm{d}t'$. Just as in Section 7.2, this motion may be easily determined with the aid of the energy first integral, which follows from the constancy of $\mathscr{H}$ and allows the law of motion $\xi(t)$ to be implicitly expressed as

$$\pm \int \frac{\mathrm{d}\xi}{(\beta\mathscr{H}/F + \cos \xi + p\xi)^{1/2}} = \left(\frac{2\beta F}{m_e}\right)^{1/2} t. \tag{8.6}$$

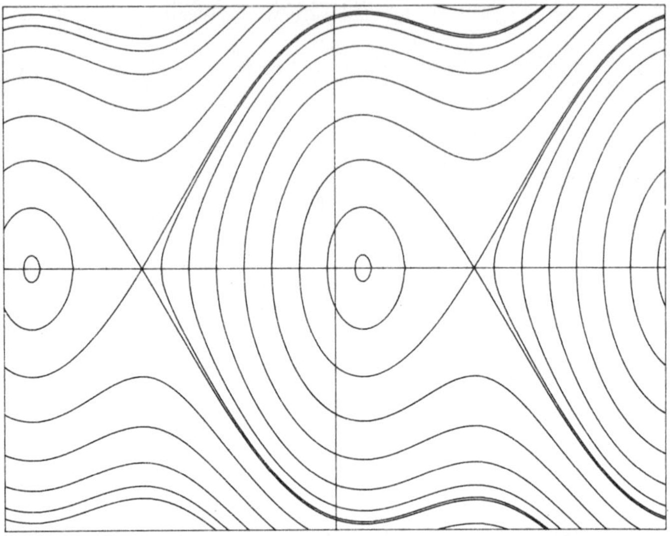

FIG. 8.2. Phase-space trajectories of particles moving in the potential of Fig. 8.1.

The trajectories in phase space (constant-$\mathscr{H}$ lines) described by eqn (8.5) may be seen in Fig. 8.2: those trajectories crossing the ξ-axis within the range (ξ_1, ξ_2) around each value of the synchronous phase are closed and the motion of the electron is bounded; the other trajectories are open and the motion is unbounded. As before, in the first case we say that the electron is trapped by the wave quasi-potential and in the second that it is not.

Notice that, since the moving frame is non-inertial, a trapped electron moves of accelerated motion in the laboratory frame, and therefore continuously gains (if $p < 0$) or loses (if $p > 0$) energy which must come from, or go to, the electromagnetic field.

8.2 Solution of the ultrarelativistic equations of motion of an electron in a tapered undulator

It has just been shown that within the framework of non-relativistic mechanics the motion of an electron in a tapered wave structure may still be described with the aid of a quasi-potential when the acceleration of the synchronous frame, i.e. the derivative of the phase velocity of the wave, is constant. We shall indicate later a suitable approximation to deal with the case in which the acceleration is not constant but slowly varying; in this section we want to extend the foregoing analysis into the domain of relativistic mechanics, considering an undulator structure the properties of which are tapered in space rather than in time. It will be seen that, at least in the ultrarelativistic case, it is still possible to find a particular choice of the law of variation of the wave parameters which allows a closed-form solution of the equations of motion, and that this solution can be expressed in terms of an effective mass having the same form as the quasi-potential, U, of eqn (8.3).

In order to introduce the possibility of a tapered undulator structure, i.e. one with varying parameters, let us generalize the electromagnetic vector potential considered in the previous chapter, eqn (7.11), to

$$\boldsymbol{a}_\perp = \mathrm{Re}[\boldsymbol{A}_u(z)\mathrm{e}^{\mathrm{i}\psi_u(z)} + \boldsymbol{A}_w(z)\mathrm{e}^{\mathrm{i}k(t-z)}], \tag{8.7}$$

where $\boldsymbol{A}_u(z)$, $\psi_u(z)$, and $\boldsymbol{A}_w(z)$ are unspecified functions. We shall also find it convenient to introduce a local undulator wavenumber, defined as $k_u(z) = \mathrm{d}\psi_u(z)/\mathrm{d}z$.

The vector potential of eqn (8.7), just as the one of eqn (7.11), does not depend on transverse co-ordinates, and therefore the transverse part $\boldsymbol{P}_\perp$ of the momentum vector is still a constant of motion. The motion of the electron along the z-axis is determined by the Hamiltonian of eqn (7.13), or, in the ultrarelativistic approximation, by that of eqn (7.14),

$$\mathscr{H}(\gamma, t, z) = \gamma - M_{eff}^2(t, z)/2\gamma, \tag{8.8}$$

where M_{eff}^2 is defined by eqn (7.12): $M_{eff}^2 = 1 + (\boldsymbol{P}_\perp - \boldsymbol{a}_\perp)^2$. For simplicity we shall only consider the case of $\boldsymbol{P}_\perp = 0$. The explicit expression of M_{eff}^2, analogous to eqn (7.15), is then

$$\begin{aligned} M_{eff}^2 = 1 + A_u^2/2 + A_w^2/2 + \mathrm{Re}\{(A_u^2/2)\mathrm{e}^{2\mathrm{i}\psi_u} + (A_w^2/2)\mathrm{e}^{2\mathrm{i}k(t-z)} \\ + \boldsymbol{A}_u^* \cdot \boldsymbol{A}_w \mathrm{e}^{\mathrm{i}[k(t-z)-\psi_u]} + \boldsymbol{A}_u \cdot \boldsymbol{A}_w \mathrm{e}^{\mathrm{i}[k(t-z)+\psi_u]}\}, \end{aligned} \tag{8.9}$$

and on assuming again that the undulator and wave fields are circularly polarized,[1] with orientations such that $\boldsymbol{A}_u^* \cdot \boldsymbol{A}_w \neq 0$, it reduces to

$$M_{eff}^2 = 1 + A_u^2/2 + A_w^2/2 + \mathrm{Re}\{\boldsymbol{A}_u^* \cdot \boldsymbol{A}_w \mathrm{e}^{\mathrm{i}[k(t-z)-\psi_u(z)]}\}. \tag{8.10}$$

Nothing is hitherto really different from what was done in the previous chapter, where we found that the energy of the electron in a certain moving reference frame (the one in which M_{eff} is independent of time) is a constant of motion, and a Lorentz transformation to this new frame enabled us to give an explicit solution of the equations of motion. Now, however, the frame in which M_{eff} is independent of time is non-inertial, and cannot be reached through a Lorentz transformation. In order to find, even in this situation, an explicit solution of the equations of motion, the Hamilton–Jacobi method may be applied (Luchini and Motz 1985). The Hamilton–Jacobi equation corresponding to the Hamiltonian (8.8) may immediately be written by substituting $\partial S/\partial z$ for $\mathscr{H}$ and $-\partial S/\partial t$ for γ, as already done several times in the previous chapters, and is

$$\frac{\partial S}{\partial z}\frac{\partial S}{\partial t}+\left(\frac{\partial S}{\partial t}\right)^2=\frac{M_{eff}^2}{2}. \tag{8.11}$$

Aiming at simplifying eqn (8.11), we may adopt as a new space variable the argument of the exponential appearing in eqn (8.10), i.e. the electron phase, defined by

$$\xi=\psi_u(z)-k(t-z). \tag{8.12}$$

In addition, we may choose a new time variable t' in such a manner that the cross-product of partial derivatives is eliminated from eqn (8.11). A suitable choice is

$$t'=\psi_u(z)+k(t-z). \tag{8.13}$$

On performing a change of variables from z, t to ξ, t', eqn (8.11) becomes

$$(\partial S/\partial t')^2-(\partial S/\partial \xi)^2=M_{eff}^2/2kk_u, \tag{8.14}$$

where $k_u = \mathrm{d}\psi_u/\mathrm{d}z$. It may easily be recognized that in the particular case of constant k_u eqns (8.12) and (8.13) become equivalent, up to a constant scale factor, to the ultrarelativistic Lorentz transformation (7.22), and eqn (8.14) to the Hamilton–Jacobi equation corresponding to the Hamiltonian of eqn (7.21). However, eqn (8.14), is of more general validity, for the change of variables expressed by eqns (8.12) and (8.13) has meaning whatever the function $\psi_u(z)$. In Section 7.2 the existence of quadrature solutions followed from M_{eff}^2's depending on only one of the variables ξ and t'. More generally, however, quadrature expressions of the law of motion may be seen to exist whenever the r.h.s. of eqn (8.14) can be split into the sum of a function of ξ alone and a function of t' alone.

In fact, the equation

$$(\partial S/\partial t')^2-(\partial S/\partial \xi)^2=f_1(\xi)+f_2(t') \tag{8.15}$$

admits separable solutions of the form $S = S_1(\xi) + S_2(t')$, where

$$(\mathrm{d}S_1/\mathrm{d}\xi)^2=I-f_1(\xi),\qquad (\mathrm{d}S_2/\mathrm{d}t')^2=I+f_2(t'), \tag{8.16, 8.17}$$

and I is an arbitrary constant.

By using eqn (8.10), where the undulator phase ψ_u can always be adjusted so that the product $A_u^*\cdot A_w$ is real and negative, the r.h.s. of eqn (8.14) may be rewritten as

$$M_{eff}^2/2kk_u=k^{-2}[a(z)-b(z)\cos\xi], \tag{8.18}$$

where

$$a(z)=(1+A_u^2/2+A_w^2/2)k/2k_u \tag{8.19a}$$

and

$$b(z) = A_u A_w k/2k_u. \tag{8.19b}$$

The r.h.s. of eqn (8.14) being given by eqn (8.18), separation of variables turns out to be possible when $a(z)$ is a linear function of $\psi_u(z)$ and $b(z)$ is a constant, say

$$a = a_0 + 2a_1\psi_u, \qquad b = b_0. \tag{8.20}$$

In fact, it follows from eqns (8.12) and (8.13) that $2\psi_u = t' + \xi$, and eqn (8.14) can be reduced to the form (8.15) with $f_1(\xi) = k^{-2}(a_1\xi - b_0 \cos \xi)$ and $f_2(t') = k^{-2}(a_0 + a_1 t')$. Equations (8.16) and (8.17) then give the solution:

$$kS = \pm \int (I - a_1\xi + b_0 \cos \xi)^{1/2}\, \mathrm{d}\xi - \int (I + a_0 + a_1 t')^{1/2}\, dt'. \tag{8.21}$$

On having a solution of the Hamilton–Jacobi equation such as eqn (8.21), which depends on an arbitrary parameter, the constant I, the law of motion can be determined, according to the general rules of the Hamilton–Jacobi method, by equating $\partial S/\partial I$ to a new arbitrary constant. Hence one obtains

$$\pm \int \frac{\mathrm{d}\xi}{(I - a_1\xi + b_0 \cos \xi)^{1/2}} = \int \frac{\mathrm{d}t'}{(I + a_0 + a_1 t')^{1/2}} = \frac{2}{a_1}(I + a_0 + a_1 t')^{1/2} + C. \tag{8.22}$$

Equation (8.22) gives, in implicit form, the law of motion $\xi(t')$. I and C are the two arbitrary constants needed to satisfy general initial conditions. It will be noticed that the l.h.s. of eqn (8.22) is identical, apart from the naming of constants, to that of eqn (8.6). Once again, the ultrarelativistic solution is similar in form to the non-relativistic one.

8.3 The motion in the laboratory frame

Just as in the previous chapter, in order to utilize the solution obtained in the synchronous frame for the calculation of energy transferred between electron and wave it is first necessary to transform it back to the laboratory frame, so as to obtain the electron energy γ as a function of the axial co-ordinate z.

The electron γ may easily be expressed in terms of quantities relating to the synchronous frame on remembering the relation between energy and action function, $\gamma = -\partial S/\partial t$. The formula for compound derivatives applied to eqns (8.12) and (8.13) gives

$$\gamma = -\partial S/\partial t = k(\partial S/\partial \xi - \partial S/\partial t'), \tag{8.23}$$

and thus, from eqn (8.21),

$$\gamma(t') = (I + a_0 + a_1 t')^{1/2} \pm [I - a_1\xi(t') + b_0 \cos \xi(t')]^{1/2}, \qquad (8.24)$$

where $\xi(t')$ is the function implicitly defined by eqn (8.22). This is the analogue of eqn (7.31).

In addition, eliminating t between eqns (8.12) and (8.13) gives the relation between t' and z implicitly as

$$t' + \xi(t') = 2\psi_u(z). \qquad (8.25)$$

This is the analogue of eqn (7.32). Just as in that case, eqn (8.25) may be replaced by an approximate explicit equation under the assumption that a typical electron's period of oscillation in the potential well (synchrotron oscillation period) is much longer than the undulator period. This assumption enables us to neglect the oscillating term $\xi(t')$ in eqn (8.25) and to write, approximately,

$$t' \simeq 2\psi_u(z). \qquad (8.26)$$

Therefore eqn (8.24) may be assumed, approximately, to give the electron energy γ as a function of the undulator phase ψ_u. Moreover, within the same order of approximation I may be neglected with respect to $a_0 + a_1 t'$, just as $A_u A_w(m - \frac{1}{2})$ was neglected with respect to M^2 in eqn (7.31), and eqn (8.24) is thus further simplified to

$$\gamma \simeq \gamma_s \pm [I - a_1\xi(2\psi_u) + b_0 \cos \xi(2\psi_u)]^{1/2}, \qquad (8.27)$$

where

$$\gamma_s \simeq (a_0 + 2a_1\psi_u)^{1/2} = a^{1/2}(z). \qquad (8.28)$$

The same approximate result may also be arrived at on introducing suitable approximations directly into the electron Hamiltonian $\mathscr{H}$ in the laboratory frame, in a manner similar to that used in Section 7.6. Using the expression (8.12) of the phase ξ and the Hamilton equation $\mathrm{d}t/\mathrm{d}z = \partial\mathscr{H}/\partial\gamma$ it can be easily verified that $\mathrm{d}\xi/\mathrm{d}z$ is still correctly expressed by eqn (7.41), provided that k_u is defined as $\mathrm{d}\psi_u/\mathrm{d}z$. Therefore a Hamiltonian $\mathscr{H}'$ from which the equations of motion for the variables ξ and γ can be derived is still given by eqn (7.43),

$$\mathscr{H}' = k_u\gamma + kM_{eff}^2/2\gamma, \qquad (8.29)$$

the difference with the previous chapter being that k_u is now a function of z and M_{eff} is a function of both ξ and z. Writing M_{eff}^2 in the form (8.18) we then obtain

$$\mathscr{H}' = k_u[\gamma + (a - b \cos \xi)/\gamma]. \qquad (8.30)$$

Proceeding just as in Section 7.6, we should now let $\gamma = \gamma_s + \delta\gamma$, with γ_s to be determined later, and expand the Hamiltonian in powers of $\delta\gamma$. Before doing so, however, we must perform one more change of Hamiltonian to change from the variables ξ, γ to ξ, $\delta\gamma$ (this step was not needed in Section 7.6 because γ_s was constant there). From the equation $d\gamma/dz = -\partial\mathscr{K}'/\partial\xi$ it may be seen that the Hamiltonian appropriate to the variables ξ and $\delta\gamma$ is $\mathscr{K}'' = \mathscr{K}' + (d\gamma_s/dz)\xi$. Expanding $\mathscr{K}''$ we obtain

$$\mathscr{K}'' \simeq k_u[\gamma_s + a/\gamma_s + (1 - a/\gamma_s^2)\delta\gamma - (b\cos\xi)/\gamma_s + a\delta\gamma^2/\gamma_s^3] + (d\gamma_s/dz)\xi. \tag{8.31}$$

Choosing γ_s so as to eliminate the term of first order in $\delta\gamma$ requires $\gamma_s^2 = a$, thus giving back eqn (8.28). Then, deleting the zero-order term which is a given function of the independent variable z and is irrelevant in a Hamiltonian, we are left with the second-order Hamiltonian,

$$\mathscr{K}'' = k_u[\delta\gamma^2/\gamma_s - (b\cos\xi)/\gamma_s] + (d\gamma_s/dz)\xi, \tag{8.32}$$

which is the correspondent for a tapered undulator of the pendulum Hamiltonian (7.45). Unlike that of eqn (7.45), the Hamiltonian $\mathscr{K}''$ of eqn (8.32) is not quite independent of z, because of the z dependence of k_u, γ_s, and b, and is not a conserved quantity. However, when the parameters a ($=\gamma_s^2$) and b have the special form given by eqns (8.20), so that, in particular, $2\gamma_s\, d\gamma_s/dz = 2a_1k_u$, eqn (8.32) may be rewritten as

$$\mathscr{K}'' = (k_u/\gamma_s)(\delta\gamma^2 - b_0\cos\xi + a_1\xi) \tag{8.33}$$

and $\mathscr{K}''$ can be transformed by a further change of variables into a Hamiltonian independent of z. In fact, if the independent variable is changed from z to $\zeta = f(z)$, the Hamiltonian changes from $\mathscr{K}$ to $\mathscr{K}/f'$, as may be verified either by comparing the form of the resulting Hamilton equations or by noticing that the differential $\mathscr{K}\,dz$ in the variational principle (A.15) must remain unchanged. On choosing $f'(z) = k_u/\gamma_s$ we obtain the new Hamiltonian given by

$$\mathscr{K}''' = \delta\gamma^2 - b_0\cos\xi + a_1\xi, \tag{8.34}$$

which is finally independent of ζ. As such, $\mathscr{K}'''$ is a constant of motion, and $\delta\gamma$ may be calculated from

$$\delta\gamma = \pm(\mathscr{K}''' + b_0\cos\xi - a_1\xi)^{1/2}, \tag{8.35}$$

which is, of course, the same result already given by eqn (8.27), but with the constant $\mathscr{K}'''$ in the place of the constant I. Equation (8.35) spares us the explicit integration of one of the Hamilton equations. By integrating the remaining equation, namely

$$d\xi/d\zeta = \partial\mathscr{K}'''/\partial\delta\gamma = 2\delta\gamma \tag{8.36}$$

and using the known relationship between ζ and z, the expression (8.22) of $\xi(z)$ can also be recovered.

8.4 Adiabatic invariant

Up to now we have only dealt with the special situation in which the behaviour of the undulator and wave fields is such that the parameters a and b defined by eqns (8.19) have the particular form given by eqn (8.20). Under this condition, which is the ultrarelativistic correspondent of the case of constant acceleration considered in Section 8.1, the quantity I was found to be an invariant. For a more general form of the parameters a and b, i.e. of the undulator and wave fields, the equations of motion cannot be solved analytically. If, however, the variations of a and b are smooth, in the sense that over a synchrotron oscillation period a is an approximately linear function of ψ_u and b is approximately constant, one may expect that the solution will still be close to that expressed by eqn (8.22), but the parameter I will generally be drifting slowly together with a and b. This is indeed the case, and in fact the variation of I due to a smooth variation of b and $\mathrm{d}a/\mathrm{d}\psi_u$ can be calculated with the aid of a tool of mechanics known as adiabatic-invariant theory.

The adiabatic invariant is a quantity defined, for a general mechanical system provided with a time-independent Hamiltonian $\mathscr{H}(\xi, P_\xi)$ and undergoing periodic motion, as

$$J(\mathscr{E}) = \oint P_\xi(\mathscr{E}, \xi)\, \mathrm{d}\xi, \tag{8.37}$$

where $P_\xi(\mathscr{E}, \xi)$ denotes momentum calculated as a function of the, conserved, energy $\mathscr{E}$ and of position ξ by inverting the equation $\mathscr{H}(\xi, P_\xi) = \mathscr{E}$, and the integral is extended to a complete period of the motion. The adiabatic invariant is a function of energy alone, and is therefore itself a conserved quantity when the Hamiltonian is independent of time. The property by which the adiabatic invariant has deserved its name is that if the Hamiltonian of the system is slowly ('adiabatically') changed in time, that is if the Hamiltonian is not time-independent but its variation is slow on the scale of the period of oscillation, the adiabatic invariant is still, approximately, a constant of motion (for a proof see, e.g., Landau and Lifschitz 1976, sect. 49).

In order to find the adiabatic invariant associated with the motion of an electron in a tapered-undulator FEL, let us start out from the Hamiltonian $\mathscr{K}''$ of eqn (8.32) which, we recall, refers to the co-ordinate ξ and momentum $\delta\gamma$ considered as functions of the independent variable z. We may, just as was done in obtaining eqn (8.33), transform to a new Hamiltonian referring to the independent variable ζ such that $\mathrm{d}\zeta/\mathrm{d}z = k_u/\gamma_s$, namely $(\gamma_s/k_u)\mathscr{K}''$, and in addition define the deceleration parameter p, a function of z, as

$$p = -(\gamma_s/bk_u)\, \mathrm{d}\gamma_s/\mathrm{d}z. \tag{8.38}$$

The Hamiltonian $\mathscr{K}'''$ thus obtained is given by

$$\mathscr{K}''' = \delta\gamma^2 - b(\cos\xi + p\xi) \tag{8.39}$$

and is similar to that of eqn (8.34) except that b and p are now not assumed to be constant. Notice, incidentally, that eqn (8.39) coincides in form with the non-relativistic eqn (8.5).

Applying formula (8.37) to the Hamiltonian $\mathscr{H}'''$, and denoting as before the constant value which the Hamiltonian itself takes when b and p are constant by I, gives

$$J = \oint \delta\gamma(I, \xi)\,\mathrm{d}\xi = b^{1/2} \oint (I/b + \cos\xi + p\xi)^{1/2}\,\mathrm{d}\xi, \tag{8.40}$$

where the integral is extended over a complete closed orbit in phase space; that is, twice over the interval of the ξ-axis delimited by the two points where the integrand vanishes. Notice that the adiabatic invariant is defined only for closed orbits, and is an increasing function of the parameter I which varies from 0, for $I = -b(\cos\xi_s + p\xi_s)$ when the electron sits still at the synchronous phase $\xi_s = \arcsin p$, to

$$J_{\text{top}} = 2b^{1/2} \int_{\xi_1}^{\xi_2} (I_{\text{top}}/b + \cos\xi + p\xi)^{1/2}\,\mathrm{d}\xi = b^{1/2}F(p) \tag{8.41}$$

for $I = I_{\text{top}} = -b(\cos\xi_1 + p\xi_1)$ when the electron just scrapes the top of the quasi-potential well. (ξ_1 and ξ_2 in eqn (8.41) are the extreme edges of the potential well introduced in Section 8.1.) The function $F(p)$ defined by eqn (8.41) is plotted in Fig. 8.3.

When b and p are not constant but are slowly varying quantities, the adiabatic invariant J, still formally defined by the integral of eqn (8.40)

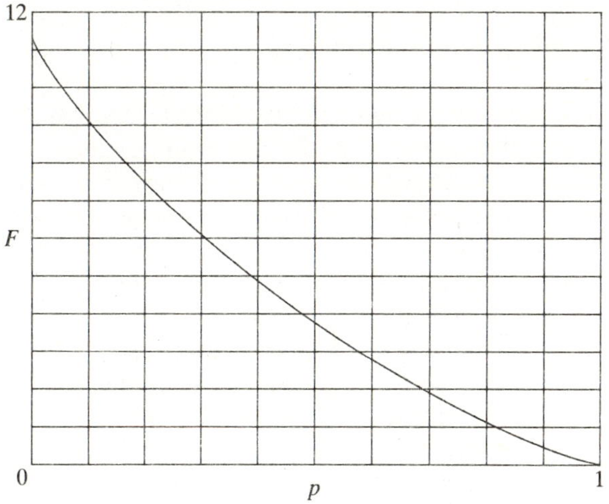

FIG. 8.3. Plot of the function $F(p)$ appearing in the expression (8.41) of the top adiabatic invariant J_{top}.

calculated for fixed values of b, p, and I, remains approximately constant while I changes. For this reason J is a more important quantity to watch than I. In particular, it may happen that with changing b and p a previously trapped electron becomes untrapped, and thus ceases to participate in the process of energy transfer. This drastic change in the character of the motion occurs when the value of J_{top}, which is given at any point along the undulator by eqn (8.41) and therefore changes together with b and p, drops below the value of J pertaining to the electron, which remains approximately constant and equal to the value determined at the moment of injection into the undulator.

In particular, we shall define a given tapering profile as 'detrapping' if the value of J_{top} decreases along the axis, so that those electrons which acquire a value of J close to J_{top} at the entrance are eventually detrapped and lost during their traversal, and 'overtrapping' if J_{top} increases along the axis, so that the electrons are pushed down towards the bottom of the potential well and the higher part of it remains unpopulated. It may be seen that both detrapping and overtrapping are harmful to the efficiency of energy conversion, the first because the detrapped electrons cease to participate in the process, and the second because choosing a lower value of J_{top} would allow a larger value of the deceleration parameter, and therefore, according to eqn (8.38), of $|d\gamma_s/dz|$. It follows that one of the objectives of the design of a tapered undulator should be to keep J_{top} constant, which means that the parameters b and p ought to satisfy the equation

$$b = \text{constant} \times F^{-2}(p). \tag{8.42}$$

8.5 Power balance in a tapered undulator

In the foregoing sections we have discussed how tapering the undulator structure may affect the motion of an electron beam in superposed undulator and wave fields, devoting particular attention to the energy of the electrons. In doing so we have implicitly assumed that the energy lost by the electrons will reappear in the travelling electromagnetic wave. Although acceptable under many practical conditions, this assumption is not always granted, and we now want to spend a few lines in its discussion.

Let us begin by recalling the observation made in Section 7.1, that energy conservation does imply that the energy lost by the electrons through electromagnetic interactions must in some way reappear as electromagnetic energy, but energy conservation alone does not at all imply that this energy will be coherently radiated into the same mode and at the same frequency which characterize the input wave. In fact, this is generally not true. Nevertheless, in the previous chapters (in particular, in Sections 5.3 and 7.1) we did find that most of the energy lost by the electrons turns up as power added to the input wave, when gain is small, owing to the reciprocity property of electromagnetic fields.

When gain is large and non-linear effects are present, energy may be channelled into unwanted frequencies and/or directions by various phenomena, and gain calculated according to energy conservation is only a maximum gain, which would be achieved if those effects were absent. It is expressed through the power increment P_{inc} as $G = 1 + P_{\text{inc}}/P_{\text{in}}$, where the maximum power increment is given by eqn (7.1):

$$P_{\text{inc}} = -m_e c^2 n_e \langle \gamma_f - \gamma_i \rangle. \tag{8.43}$$

Similarly, a maximum energy yield may be defined as $\eta = \langle \gamma_i - \gamma_f \rangle / (\langle \gamma_i \rangle - 1)$. Of course, the energy 'extraction', that is the relative amount of energy subtracted from the electron beam, is always exactly equal to this value, but it may be considered a 'yield' of useful energy only under ideal conditions.

Insofar as the energy lost by the electron beam may be assumed to reappear in the laser beam, and vice versa, a one-dimensional power balance equation may be written stating the constancy along the undulator of the sum of the e-beam power per unit transverse area, $\langle \gamma \rangle m_e c^2 n_e$, and the laser beam power per unit area, $\zeta (m_e c^2/e)^2 k^2 A_w^2/2$ (with A_w, as usual, dimensionless). With coefficients re-expressed in terms of the e-beam current density $j_b = en_e$ and of the Alfvén current $I_A \simeq 17\,\text{kA}$, this equation reads

$$A_w^2(z)/2 + (4\pi j_b/k^2 I_A)\langle \gamma(z) \rangle = \text{constant}. \tag{8.44}$$

To state more explicitly the conditions under which eqn (8.44) is valid, one may observe that if the gain length, though smaller than the total undulator length, is still larger than the typical scales of variation of the properties of the system with the axial co-ordinate, and in particular than the synchrotron oscillation period, as is required in any case for the adiabatic-invariant approximation, the undulator may approximately be seen as a series connection of a number of independent amplifiers, each having a small length compared to the gain length. Then, every partial amplifier is small-gain, even if the overall gain is large, and the reciprocity theorem tells us that gain is approximately equal to the maximum gain.

The above argument gives us a certain confidence that the actual energy yield of a tapered-undulator FEL will not be far from the maximum one, and in fact under most conditions it is reasonable to expect that it be so; this expectation may be defeated, however, by the presence of non-linear resonances, which may set in, quite in the same way as in every high-gain amplifier device, once gain grows beyond a certain threshold. When this happens small perturbations of the electron beam excited by non-linear effects, which would be negligible in a small-gain amplifier, become capable of being amplified by the undulator itself at a faster rate than the useful radiation and eventually disturbing operation of the device. They are known by the general name of 'FEL instabilities', and are one of the main topics of current research.

8.6 The one-dimensional equilibrium model of the tapered-undulator FEL

We are now ready to set up a basic one-dimensional model of a tapered-undulator FEL. We shall assume that electron energy is totally converted into laser energy (and vice versa), and that the undulator is much longer than the synchrotron oscillation period (i.e. that $A_w \gg M^2/2A_uN^2$ according to eqn (7.50)), so that, just as in Section 7.7, the average electron γ may be identified with $f_t\gamma_s + (1 - f_t)\gamma_i$. Supposing, in addition, that the undulator is not detrapping, so that f_t remains constant all the way through, eqn (8.44) may be rewritten as

$$A_w^2(z)/2 + (4\pi j_b/k^2 I_A) f_t \gamma_s(z) = \text{constant}. \tag{8.45}$$

The synchronous γ, γ_s, is determined as a function of the undulator parameters by eqns (8.28) and (8.19a), where $A_w^2/2$ can generally be neglected with respect to 1 for practical laser field amplitudes. These two equations together give

$$\gamma_s^2 = [1 + A_u^2(z)/2]k/2k_u(z) = M^2(z)k/2k_u(z) \tag{8.46}$$

which, incidentally, may be seen to coincide with the expression (7.38) of the synchronous energy in a uniform undulator, except for M^2 and k_u now being functions of z. The z-derivative of γ_s, on the other hand, is tied to the parameter $b = kA_uA_w/2k_u$ and to the deceleration parameter p which defines the shape of the quasi-potential pattern by eqn (8.38):

$$\gamma_s\, \mathrm{d}\gamma_s/\mathrm{d}z = -k_u bp = -kA_uA_w p/2. \tag{8.47}$$

b and p, in turn, allow the calculation of J_{top} from eqn (8.37) and thus the verification of the detrapping or overtrapping character of the undulator.

We can also replace eqn (8.45) by an equation directly giving the rate of increase of the laser power along the undulator; combining the derivative of eqn (8.45) with eqn (8.47) yields

$$\mathrm{d}A_w/\mathrm{d}z = 2\pi j_b A_u p f_t/k\gamma_s I_A. \tag{8.48}$$

Equations (8.41) and (8.45)–(8.48) offer a simplified description of the evolution of the mean properties of both the electron and radiation beams along the undulator for any given law of variation of the undulator parameters A_u and k_u, allowing the determination of the variation along the undulator of A_w, γ_s, and J_{top}. Conversely, considering A_u and k_u as unknowns, the same equations may be used in the design phase to obtain the profile of tapering which satisfies given constraints on the behaviour of the quasi-potential, e.g. eqn (8.42). They are valid under the two assumptions that the undulator is sufficiently longer than the synchrotron oscillation period for the electrons to reach an equilibrium distribution, so that the mean γ of the trapped portion of the electron beam may be identified with the synchronous γ, γ_s, and that the variation of the undulator parameters is sufficiently smooth (again, on the scale of length determined by the synchrotron oscillation period) that the adiabatic approximation applies. Because of these two assumptions we shall call these equations the 'equilibrium model' of the tapered FEL.

Although the approximations leading to the equilibrium model are somewhat heavy, in particular because, as was seen for the constant-parameter undulator in Section 7.8, the oscillations of the average electron γ around the equilibrium value are only weakly damped, this is practically the only model that is simple enough to afford a qualitative understanding of the significance of different choices of the undulator parameters. Once the general parameters have been grossly established, numerical simulations using more accurate models may be used to ascertain the importance of the imperfections of the theoretical one-dimensional model, including oscillations around the equilibrium values.

In particular, the equilibrium model may be used to calculate and design the profile of tapering. Let us examine two typical applications, a laser power amplifier and an electron accelerator.

8.7 The tapered undulator as a FEL power amplifier

As anticipated at the beginning of this chapter, one of the main assets of the tapered undulator as an amplifier is its capability to achieve an energy yield not restrained by the $1/N$ limit typical of uniform undulators (as given by eqn (7.66)), and therefore to produce a power output that is a non-negligible fraction of the input e-beam power. In fact, by designing the undulator so as to trap almost the whole of the electrons in potential wells and then gradually decrease the synchronous γ of eqn (8.46) while maintaining a high enough value of the top adiabatic invariant of eqn (8.41) for a sufficiently long distance, an energy yield close to unity can theoretically be achieved. In particular, let us now explore how the tapering profile should be designed and how long the undulator must be in order to achieve the desired energy conversion.

Two real functions can be arbitrarily assigned to define the undulator profile: the amplitude $A_u(z)$ and the wavenumber $k_u(z)$. Two constraints can therefore be satisfied. An almost mandatory constraint if operation is to be optimized within the one-dimensional equilibrium model is that the top adiabatic invariant of eqn (8.41) should remain constant along the undulator, so as to cause neither detrapping nor overtrapping. The second constraint can be assigned in various ways depending on functional or technological considerations. Two cases have been considered by Luchini (1985); that in which the greatest power output is sought to be obtained with the least number of poles in the undulator, corresponding[2] to A_u constant and equal to $\sqrt{2}$, and that in which, in order to make as compact an undulator as possible, k_u is fixed at the largest technologically feasible value and held constant while only A_u changes.

Whichever two constraints are chosen, together with eqns (8.45)–(8.48) they form a self-consistent system of equations which allows the determination of A_u, k_u, A_w, γ_s, and p (or, equivalently, the synchronous phase given by $\xi_s = \arcsin p$) along the undulator. The values of two constants need still be imposed, in addition to the initial γ of the electron beam: the initial value of the wave field at the input and the initial value of J_{top}. That the initial value of the wave field should be imposed is not surprising, since the whole process of amplification in a tapered undulator is non-linear, but is unusual for a designer of amplifiers, who is accustomed to working with linear devices. It means that operation of the tapered undulator can only be optimized for a predefined nominal input level. (In particular, the small-signal gain of a tapered undulator generally turns out to be lower than that achievable by a uniform undulator of comparable characteristics.) The initial value of J_{top}, on the other hand, determines the initial value of p and therefore, through eqn (8.4), the trapping range $\Delta\xi$ and the trapping fraction $f_t = \Delta\xi/2\pi$. The choice of J_{top} must result from a compromise: too low a value implies a high initial p and a low trapping fraction; too high a value implies that, since this value cannot be decreased further downstream if detrapping is to be avoided, the deceleration parameter p will remain low and energy will be transferred slowly.

If the solution of the system of equations considered above is followed, say numerically, in its evolution along the undulator, three regions will be recognized, fading smoothly one into another. The first is the bunching region, in which both the wave amplitude and the electron γ change relatively little while the deceleration parameter p shifts, as fast as the constant-J_{top} condition allows, from its initial value to a value near unity; the second is the amplification region, in which p is close to 1 and the wave amplitude grows, more or less linearly, according to eqn (8.48) while still not reaching a high enough value to affect the electron γ noticeably; the third is the energy-extraction region, in which the wave amplitude tops off at its final value while the electron γ suddenly decreases.

The maximum wave amplitude which can be attained if all the e-beam energy is converted into laser energy may easily be obtained from eqn (8.44) as

$$A_{w,\max}^2/2 = (4\pi j_b/k^2 I_A)\gamma_i, \tag{8.49}$$

where the input has been neglected with respect to the output laser field.

A simple estimate of the actual output level of a tapered undulator may be constructed by pretending that a linear-growth region characterized by $f_t = 1$, $p = 1$ and $\gamma_s \simeq \gamma_i$ extends over the whole undulator, and that the wave amplitude starts out from zero. Thus one obtains from eqn (8.48) a growth rate equal to

$$\mathrm{d}A_w/\mathrm{d}z = 2\pi j_b A_u/k\gamma_i I_A \tag{8.50}$$

which, by using eqns (8.46) and (8.49), may also be written as

$$\frac{\mathrm{d}A_w}{\mathrm{d}z} = \frac{A_{w,\max}^2}{2}\frac{A_u}{M^2}k_u. \tag{8.51}$$

Neglecting, for the scope of this estimate, the variability of the undulator parameters and replacing them by typical values, we may then write a nominal output amplitude simply by multiplying the growth rate of eqn (8.51) by the undulator length $L = Nl_u$ and adding the input amplitude $A_{w,i}$, as

$$A_{w,\mathrm{nom}} = A_{w,i} + \pi N A_{w,\max}^2 A_u/M^2. \tag{8.52}$$

Assuming $A_{w,\mathrm{nom}} \gg A_{w,i}$, one can then obtain the nominal energy yield as $A_{w,\mathrm{nom}}^2$ divided by $A_{w,\max}^2$, i.e.

$$\eta_{\mathrm{nom}} = \pi^2 N^2 A_{w,\max}^2 (A_u/M^2)^2. \tag{8.53}$$

Equation (8.53) enables us to state a first important result about the operation of a tapered undulator as a power amplifier: in order to offer an energy yield of order unity the tapered undulator must be designed for a maximum output field intensity such that

$$A_{w,\max} \simeq M^2/NA_u \tag{8.54}$$

(or $A_{w,\max} \simeq 1/N$ if the value of A_u is chosen which minimizes M^2/A_u, i.e. $A_u = \sqrt{2}$). A dimensionless vector potential satisfying condition (8.54) generally corresponds to a very high laser power density. Therefore the tapered undulator can only be used as a very-high-power laser device, and there is no way, apart from increasing the number of periods, N, beyond present practical limits, to scale it down to lower power levels. This characteristic is far from making it uninteresting, though, since the electron beam can indeed provide the energy density necessary for operation at these high power levels, whereas conventional lasers are limited by the maximum energy density that can be stored in their active medium. This difference makes the tapered-undulator laser amplifier a good candidate for such applications as, for instance, laser-driven nuclear fusion.

Dividing, instead, $A^2_{w,\text{nom}}$ by the square of the input vector potential $A^2_{w,\text{in}}$ gives the power gain,

$$G_{\text{nom}} = (1 + \pi N A^2_{w,\max} A_u / M^2 A_{w,\text{in}})^2, \tag{8.55}$$

which we have qualified, just as the other quantities above, by the subscript nom (nominal) to remind us that it refers to a situation in which a linear growth is imagined to take place over the whole undulator. A more accurate analysis may be found in the paper by Luchini (1985), where a parameterless laser-amplitude curve is described covering both the bunching and linear-growth regions. In the same paper the optimal value, within the equilibrium model, of the top adiabatic invariant and initial synchronous phase corresponding to various values of input field are also calculated, arriving at the plot presented in Fig. 8.4, of the ratio between the numerically obtained actual field amplitude and the nominal output level $A_{w,\text{nom}}$. As may be seen, this ratio eventually becomes of order unity, and therefore approximating A_w by $A_{w,\text{nom}}$ and G by G_{nom} in an order-of-magnitude estimate is justified.

Notice that the nominal amplitude gain $A_{w,\text{nom}}/A_{w,i}$ reported on the horizontal axis in Fig. 8.4 can be increased, the undulator parameters being fixed, just by decreasing $A_{w,i}$, and therefore it would seem that, amazingly enough, a larger output power is obtained with decreasing input power. Indeed, it would appear from the diagram that exactly the nominal output could be obtained simply by decreasing the input level to zero. This paradox is removed once it is remembered that Fig. 8.4 is calculated within the

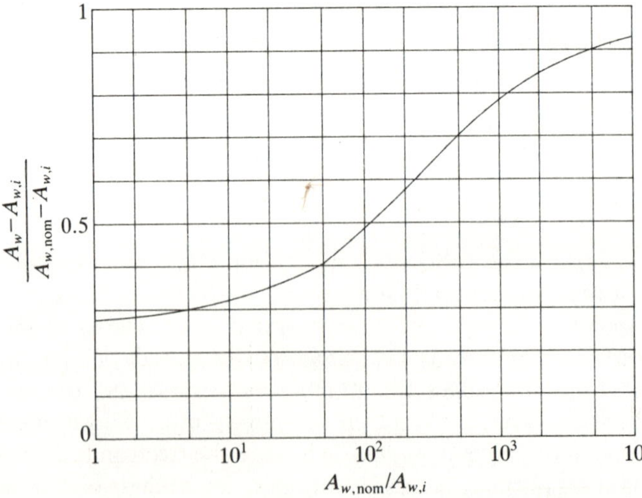

FIG. 8.4. Plot of the ratio $(A_w - A_{w,i})/(A_{w,\text{nom}} - A_{w,i})$ (actual amplitude increase over nominal amplitude increase) vs. nominal amplitude gain $A_{w,\text{nom}}/A_{w,i}$ as obtained from the numerical solution of eqns (8.45)–(8.48) for an optimized choice of the undulator profile (from Luchini 1985).

equilibrium model, and one of the hypotheses underlying the equilibrium model is that all the relevant quantities should vary slowly on the scale set by the synchrotron oscillation period. Imposing this condition on the wave amplitude at the undulator input gives

$$(\mathrm{d}A_w/\mathrm{d}z)L_{s,\min} \ll A_{w,\mathrm{in}}. \tag{8.56}$$

On taking $\mathrm{d}A_w/\mathrm{d}z$ from eqn (8.51) and $L_{s,\min}$ from eqn (7.48) (which, although derived for uniform undulators, may be assumed to give the correct order of magnitude for tapered undulators as well), eqn (8.56) gives a lower bound on input amplitude in the form

$$A_{w,\mathrm{in}}^{3/2} \gg (A_u/M^2)^{1/2}A_{w,\max}^2. \tag{8.57}$$

This condition, in turn, once put into eqn (8.55), yields an upper bound on gain:

$$G_{\mathrm{nom}} \ll N^2 A_{w,\max}^{4/3}(A_u/M^2)^{4/3}. \tag{8.58}$$

In particular, if $A_{w,\max}$ is chosen according to eqn (8.54), eqn (8.58) reduces to

$$G_{\mathrm{nom}} \ll N^{2/3}. \tag{8.59}$$

To end this section, let us mention an interesting modification of the tapered undulator, devised and experimented by a TRW–Stanford joint group (Neil *et al.* 1985), which may prove useful in future designs. Since a certain initial fraction of the undulator length is used for bunching the electron beam, and bunching is performed much better in an optical klystron than in an ordinary undulator, the idea is to use the first two sections of an optical klystron, a short undulator and a dispersion magnet, to prebunch the electron beam, which may then be injected into a tapered undulator characterized by a higher value of the deceleration parameter p. This design can afford a certain length reduction of the bunching section of the tapered undulator and, perhaps even more importantly, lower the input-field minimum threshold. (In addition, the TRW-Stanford undulator also had a variable tapering so as to make it possible to experiment with different tapering profiles.)

8.8 The inverse FEL: a laser-driven particle accelerator

Another possible application of the undulator is to couple energy in the direction opposite to that considered in the previous section, i.e. from the laser to the electron beam. The undulator thus becomes a laser-driven particle accelerator. Such a device works much in the same way as a microwave linac, but with an optical-wavelength accelerating field, and takes the name of inverse FEL. Its use has been proposed by many authors in order to overcome the technological energy-gradient limits of present linacs arriving, at most, at 10–20 MeV/m, and achieve energy gradients as large as 100–1000 MeV/m: Kolomenskii and Lebedev (1966), who actually proposed to use an axial magnetic field to put the electrons on spiralling trajectories; Palmer (1972), who proposed to use a helical magnetic undulator; an unpublished paper by M. Chodorow, R. Kompfner, and H. Motz, 'A laser electron accelerator', the results of which have been described by Motz (1979) and briefly in Section 1.7 of this book. Such a device has not yet been built, but as it is being considered by particle-accelerator designers we may now want to have a glance at it from the undulator side. Be warned, however, that this is the easiest part of the design: much tougher problems exist for generating a suitable laser beam (perhaps by means of an FEL!) and guiding the particle beam so that it is not destroyed by various beam instabilities, but those lie outside the scope of this book.

Since it is the very purpose of a particle accelerator to make γ change considerably, an inverse FEL must necessarily employ a tapered undulator. Contrarily to a laser amplifier, it will be designed with an input laser power exceeding the input e-beam power and with a negative value of the deceleration parameter p, so that energy is transferred from the laser to the electron beam. Apart from this difference, the previous equilibrium model of the tapered undulator applies equally well to the inverse FEL. In particular, the energy gradient achievable may be obtained from eqn (8.47), which must be combined, as before, with the synchronism condition (8.46) and two constraints on the undulator profile. If, for instance, it is chosen to design the undulator with constant values of the parameter p and of the undulator period k_u, and if the square of the undulator vector potential is large compared to unity so that eqn (8.46) may be approximated by $\gamma_s^2 = A_u^2 k/4k_u$, eliminating A_u between this equation and eqn (8.47) gives

$$d\gamma_s/dz = -pA_w(kk_u)^{1/2}. \tag{8.60}$$

Let us quote, for example, the configuration envisaged by Pellegrini (1982). A laser pulse of peak power of 2×10^{13} W at the wavelength of 1 μm, having a spot size of 0.25 cm, develops an electric field $E = 2.8 \times 10^{10}$ V/m, which corresponds to a dimensionless vector potential $A_w = eE/\omega mc = 8.7 \times 10^{-3}$. With an undulator period of the order of 10 cm and a deceleration parameter $p \simeq -1$, eqn (8.60) yields a γ-gradient of 170 m^{-1}, which corresponds to an energy gradient of 85 MeV/m. In fact, Pellegrini more precisely calculates that using either a constant undulator period of 10 cm and an undulator magnetic field intensity varying between 0.31 and 3.8 T, or a constant magnetic field of 1 T and a period varying between 3.8 and 23 cm, acceleration of the electron beam from 250 MeV to about 4 GeV can be accomplished in a length of 39 m, thus realizing a mean energy gradient of the order of 100 MeV/m.

Just as in the case of the previously considered laser amplifier, an inverse FEL undulator can only operate at the gradient indicated by the equilibrium model if the ratio between γ and its increment is such that the adiabaticity condition is satisfied, i.e. if

$$(\mathrm{d}\gamma_s/\mathrm{d}z)L_{s,\min} \ll \gamma_s. \tag{8.61}$$

In this case, however, it can be seen that eqn (8.61) imposes no additional restriction. On inserting the expressions (8.47) of $\mathrm{d}\gamma_s/\mathrm{d}z$ and (7.48) of $L_{s,\min}$, and eliminating l_u with the aid of the synchronism condition (8.46), eqn (8.61) becomes simply $A_u A_w/M^2 \ll 1$ which, it will be remembered from Section 7.5, is the condition that must be satisfied in any case in order that $L_{s,\min}$ is much greater than l_u.

8.9 Numerical simulations

The one-dimensional equilibrium model of the tapered-undulator FEL contains several approximations, the heaviest probably being the assumption of an equilibrium distribution of electrons in phase space because, as shown in Section 7.8, equilibrium is approached very slowly, and, though it allows a general understanding of the functioning of the device, is not sufficient for the design of a working piece of apparatus. Therefore all the laboratories and research centres interested in the construction of tapered undulators have developed computer simulation codes.

The basic equations that must be numerically solved in an FEL simulation code are the equations of motion of electrons, for instance in the form of the Hamilton equations (3.2) or (3.4), and the equations of electromagnetic wave propagation. Several choices are possible, the advantages and disadvantages of which we intend to outline below.

First, the simulation may involve a smaller or greater number of independent variables, with an obvious compromise between speed of calculation and accuracy. A complete simulation must account for three spatial dimensions, three independent components of electron momentum (or two transverse components plus energy), and an arbitrary space and time (or frequency) distribution of the electromagnetic field, working therefore with an electron distribution depending on six variables (plus time) and an electromagnetic field depending on three (plus time). Such a simulation is very expensive, and several researchers have chosen to adopt models of lower dimensionality, although some true three-dimensional simulations have also been made (see, e.g., McVey 1986).

In overviewing the techniques available and the difficulties involved in the numerical simulation of an FEL we shall, because of the largely different character of the electron dynamics and electromagnetic equations, consider each set of equations separately.

On the electromagnetic side, the basic equation is the inhomogeneous wave equation for the vector potential (4.2), which becomes the Helmholtz equation (4.10) if the analysis is restricted to single-frequency fields. The electromagnetic vector potential, however, is a rapidly oscillating function. An approximation that is almost mandatory to make the calculation feasible is to replace the Helmholtz equation by the paraxial wave equation (4.26), the unknown of which is the slowly varying quantity $\tilde{A} = A\mathrm{e}^{ikz}$. This approximation entails an additional advantage because the paraxial wave equation is a partial differential equation of the parabolic type, rather than elliptic like the Helmholtz equation, and may be solved numerically by marching in the increasing z direction. This is particularly convenient since the electron dynamics equations may also be solved by marching in the increasing z direction, and the values taken by the unknowns on successive planes perpendicular to the z-axis may be effectively uncoupled in the solution.

On each of these planes two alternative representations of the electromagnetic field are possible: a discretization in space, using the values of the vector potential at the vertices of a discretization grid, or a modal representation using a number of coefficients of an expansion of the field in a suitable set of functions, typically the eigenfunctions of the homogeneous paraxial wave equation, the Gauss–Laguerre or Gauss–Hermite modes. When space discretization is adopted, radiation boundary conditions must be explicitly imposed at the boundary of the solution domain, whereas with the modal representation they are implicit in the choice of modes. Examples of both representations may be found in the literature (for the first see, e.g., McVey 1986; and for the second, Tang and Sprangle 1984; Rosenbluth 1985), with a trend towards space discretization in the simulations running on the larger computers, where brute-force, extensive solutions with lots of detail are allowed, and towards modal representation on the smaller ones, where good global solutions with moderate resources are allowed.

An improvement on the numerical method based upon modal representations was introduced by Sprangle, Ting, and Tang (1987) under the name of 'source-dependent expansion'. It uses a set of Gauss–Laguerre modes having a waist varying with z in a way that depends on the source term in the wave equation, and is reported to give good approximations with only a very limited number of terms, so as to fall in a sort of middle ground between strictly numerical and approximate analytical methods.

If the simulation needs to take a multi-frequency field into account, like that which may be excited by non-linear effects even in the case of a single-frequency input, three possibilities exist. One is to adopt a discretization in frequency, by considering a discrete set of single-frequency components of the electromagnetic field coupled through the source term; another is to adopt a discretization in time, considering as unknowns the values taken by the electromagnetic field at successive instants within a suitable time window; a third possibility is to adopt a modal representation of the electromagnetic field in a suitable set of time functions.

The first method differs from the other two in its purpose: it is best suited for studying multiple-frequency generation phenomena, such as the sideband instability described in the next section, whereas the other two methods are better for studying finite-pulse effects, which are essentially due to the short electron pulses provided by most accelerating machines, and will be discussed in the next chapter. The first method also differs in requiring only marginal modifications to the single-frequency computer codes, whereas the other two require that the wave equation be solved in the time rather than in the frequency domain.

The equivalent in the time domain of the paraxial wave equation may be obtained by writing the electromagnetic vector potential $\boldsymbol{a}(t, \boldsymbol{x}, z)$ as $\boldsymbol{a} = \mathrm{Re}[\tilde{\boldsymbol{A}}(t, \boldsymbol{x}, z)\mathrm{e}^{\mathrm{i}(\omega t - \beta z)}]$, and assuming that $\tilde{\boldsymbol{A}}$ is a slowly varying function of t and z. Neglecting second derivatives of $\tilde{\boldsymbol{A}}$ with respect to t and z in the wave equation (4.2), much in the same way as in deriving the Leontovich paraxial wave equation, gives

$$\frac{\mathrm{i}\omega}{c^2}\frac{\partial \tilde{\boldsymbol{A}}}{\partial t} + \mathrm{i}\beta\frac{\partial \tilde{\boldsymbol{A}}}{\partial z} - \frac{\partial^2 \tilde{\boldsymbol{A}}}{\partial \boldsymbol{x}^2} = -\mu_0 \tilde{\boldsymbol{J}}. \tag{8.62}$$

The equations of electron dynamics may be formulated either in terms of the Boltzmann equation written for a discretized representation of the electron distribution function, or in terms of the equations of motion of a number of sample electrons having initial conditions specified in accordance with the given electron distribution at the entrance to the undulator.

In the first case, the electron current $\boldsymbol{j}$ which appears on the r.h.s. of the electromagnetic wave equation must be written according to eqns (6.5) and (6.6) as

$$-\mu_0 \boldsymbol{j}(t, \boldsymbol{x}, z) = \int \frac{\partial \mathscr{H}}{\partial \boldsymbol{P}} f(t, \boldsymbol{x}, \gamma, \boldsymbol{P}_\perp; z)\, \mathrm{d}^2\boldsymbol{P}_\perp\, \mathrm{d}\gamma, \tag{8.63}$$

or as the Fourier component of this integral at the working frequency ω if the approximation of considering a single frequency is taken.

In the second case, when calculating electron trajectories explicitly, the same current may be written as

$$-\mu_0 \boldsymbol{j}(t, \boldsymbol{x}, z) = \int \frac{\partial \mathscr{H}}{\partial \boldsymbol{P}} \left[\frac{\partial(t, \boldsymbol{x})}{\partial(t_i, \boldsymbol{x}_i)}\right]^{-1} f(t_i, \boldsymbol{x}_i, \gamma_i, \boldsymbol{P}_{\perp i}; 0)\, \mathrm{d}^2\boldsymbol{P}_{\perp i}\, \mathrm{d}\gamma_i \tag{8.64}$$

because, according to Liouville's theorem (see footnote 1 in Section 6.1),

$$f(t, \boldsymbol{x}, \gamma, \boldsymbol{P}_\perp; z)\, \mathrm{d}t\, \mathrm{d}^2\boldsymbol{x}\, \mathrm{d}\gamma\, \mathrm{d}^2\boldsymbol{P}_\perp = f(t_i, \boldsymbol{x}_i, \gamma_i, \boldsymbol{P}_{\perp i}; 0)\, \mathrm{d}t_i\, \mathrm{d}^2\boldsymbol{x}_i\, \mathrm{d}\gamma_i\, \mathrm{d}^2\boldsymbol{P}_{\perp i}, \tag{8.65}$$

t, $\boldsymbol{x}$, γ, and $\boldsymbol{P}_\perp$ being the time of transit, transverse position, energy, and transverse momentum at the abscissa z of an electron starting with initial conditions t_i, $\boldsymbol{x}_i$, γ_i, and $\boldsymbol{P}_{\perp i}$ at $z = 0$.

In all the numerical works known to the authors, preference has been given to the second route, because of the greater ease of solving ordinary differential equations (the electron Hamilton equations) rather than a partial differential equation (the Boltzmann equation) which, in addition, is written in six-dimensional space in a full simulation. The problem posed by the electron Hamilton equations is just of marginal difficulty, because the equations of different electrons are uncoupled and only initial conditions are given, and may be solved by any of the standard Runge–Kutta or predictor–corrector methods for ordinary differential equations. The only point deserving attention is that, unless polarization is circular, a choice exists between adopting the exact equations derived from the Hamiltonian (3.6) or the averaged equations derived from the Hamiltonian containing the effective mass (7.71). The difference is that in the first case the equations of motion are exact, but an integration step that is small compared to the undulator period must be chosen in order to track the oscillations of the electron around the guiding centre accurately, whereas in the second an approximation is involved but an integration step larger than the undulator period is allowed, because the oscillations around the guiding centre have been averaged out.

Once the electron trajectories have been calculated, one must evaluate the beam current which constitutes the source term in the electromagnetic wave equation. In evaluating the integral of eqn (8.64) a choice may again be operated.

In what has been called the 'Monte Carlo', or 'macroparticle', approach, integrals such as that of eqn (8.64) are not even considered. Random initial conditions are chosen with the aid of a random number generator, using a probability distribution corresponding to the given electron distribution at the undulator entrance, and a certain number N of electrons so chosen are followed in their motion through the undulator, calculating the current as if each electron carried one Nth of the total beam charge. This technique involves a statistical error the r.m.s. value of which may be estimated to be of the order of $N^{-1/2}$ when the processes of generating initial conditions for different electrons are assumed to be statistically independent.

The alternative approach is to adopt an approximate integration formula for the integral of eqn (8.64), using the values taken by the integrand function at suitable points forming, contrarily to the previous case, a regular pattern. Once these points have been chosen, evaluating the integrand again involves following the trajectories of a set of single electrons starting at these points, which may be done as before by solving the Hamilton equations of the electron with one of the standard methods for ordinary differential equations, and then applying the chosen integration formula. The Jacobian appearing in eqn (8.64) can very often be approximated by unity.

The choice of one or the other of the above methods must depend essentially on the number of dimensions of the simulation. In fact, as is always true of multi-dimensional integrals, the Monte Carlo method maintains an error of the order of $N^{-1/2}$, N being the number of electrons considered, independent of the number of dimensions, whereas the explicit integration formula has an error proportional to a given power n ($n - 1$ usually being called the order of the integration method) of the *distance* between integration points, which scales in general as $N^{-1/d}$, d being the number of dimensions of the integration domain. Therefore, in order for the explicit integration method to be superior to the Monte Carlo method, n must exceed $d/2$. This inequality is practically always realized for one- or two-dimensional integrals, so that in a one-dimensional FEL simulation involving only time and energy as variables the most convenient choice is the explicit integration method, but becomes much more difficult to achieve for six-dimensional integrals, so that in a full 3-D simulation the Monte Carlo method is preferable. For 2-D cases either method may be appropriate, and both have actually been used.

8.10 Instabilities

As anticipated in Section 8.5, one of the factors which may modify the ideal picture of a tapered undulator offered by the equilibrium model is instabilities. Under this name are collected all those phenomena which, driven by the non-linear interaction between the electron and laser beams, lead to amplification of perturbations of a wavelength different from the working wavelength, and thus to a waste of energy.

Free-electron laser instabilities are still a topic of continuing research, and it has not yet been clearly established whether or not they will be a major obstable to the development of very-high-gain FELs.

The common origin of most instabilities in non-linear systems is the birth of a resonance, once the device has gone sufficiently far from the linear regime to develop new kinds of oscillating motion characterized by their own frequencies, between these new frequencies and those of other parts of the system. In the case of the free-electron laser, the new oscillating motion which we have seen to characterize the non-linear regime is the 'pendulum', or synchrotron, oscillation of the electrons in the quasi-potential well.

The two main instability mechanisms which have been identified as operating in free-electron lasers are the synchrotron–betatron instability and the sideband instability.

The synchrotron–betatron instability is produced, just as in recirculating particle accelerators, when the frequencies of the synchrotron and betatron oscillations happen to be very close to each other or to small integral multiples of each other (in particular, in the case of the FEL, when the synchrotron frequency is about twice the betatron frequency). It may cause loss of electrons from the trapped bunch through resonant energy transfer from betatron to synchrotron oscillations. We shall not discuss the theory of this instability in detail because numerical simulations seem to indicate that it does not play a major role in the operation of FELs (Rosenbluth 1985;

Fawley, Prosnitz, and Scharlemann 1985; Tang and Sprangle 1985; Quimby 1986).

The sideband instability is produced when the synchrotron oscillation of frequency $\Omega = 2\pi c/L_s$, beating with the underlying fundamental electron beam perturbation at the input frequency ω, makes it possible for the undulator to amplify radiation components with frequency $\omega \pm n\Omega$, thus giving rise to sidebands in the amplification spectrum.

The linear stability analysis of the single-frequency FEL regime with respect to small perturbations with a different frequency may be performed in a manner very similar to the analysis of the linear large-gain FEL considered in Chapter 6, by starting with an unperturbed electron distribution in phase space f_0, and calculating with the aid of eqn (6.9) the evolution of a superposed small perturbation f_1. The only difference is that now the unperturbed electron distribution f_0 is the one corresponding to the working FEL regime rather than to the motion of the electron beam in the undulator alone. The resulting differential equations are, of course, much more complicated than those of Chapter 6 and, depending on the precise choice of the underlying 'unperturbed' FEL regime, may well defeat a complete analysis.

The simplest case, and the most studied one since the original exposition of the sideband instability given by Kroll, Morton, and Rosenbluth (1981), is obtained in a uniform undulator in which the electron distribution in the beam has attained equilibrium (which means, in particular, that enough synchrotron oscillations have been performed since the beginning of the undulation that all possible oscillation phases have become equally probable). Although this is not a very interesting regime in itself, because no further amplification of the laser beam takes place, the results carry over to the tapered undulator within the adiabatic-invariant approximation.

In order to set up the mathematical formulation of this problem we must reconsider the system formed by the Maxwell and linearized Boltzmann equations (eqns (6.8)–(6.11)), i.e. in one dimension and with the scalar and longitudinal potential neglected,

$$\frac{\partial f_0}{\partial z} + \frac{\partial \mathscr{H}_0}{\partial \gamma}\frac{\partial f_0}{\partial t} - \frac{\partial \mathscr{H}_0}{\partial t}\frac{\partial f_0}{\partial \gamma} = 0, \tag{8.66}$$

$$\frac{\partial f_1}{\partial z} + \frac{\partial \mathscr{H}_0}{\partial \gamma}\frac{\partial f_1}{\partial t} = \frac{\partial \mathscr{H}_1}{\partial t}\frac{\partial f_0}{\partial \gamma}, \tag{8.67}$$

$$\frac{\partial^2 \boldsymbol{a}_{1\perp}}{\partial z^2} - \frac{\partial^2 \boldsymbol{a}_{1\perp}}{\partial t^2} = \int \left(\frac{\partial \mathscr{H}_1}{\partial \boldsymbol{P}_\perp} f_0 + \frac{\partial \mathscr{H}_0}{\partial \boldsymbol{P}_\perp} f_1\right) \mathrm{d}^2 \boldsymbol{P}_\perp \,\mathrm{d}\gamma. \tag{8.68}$$

The unperturbed Hamiltonian $\mathscr{H}_0$ appropriate to the sideband-instability problem is the one corresponding to the electromagnetic vector potential (7.11); that is, according to eqns (7.14)–(7.16),

$$\mathscr{H}_0(\gamma, t, z) = \gamma - [M^2 - A_u A_w \cos \xi]/2\gamma \tag{8.69}$$

with

$$\xi = (k + k_u)z - kt. \tag{8.70}$$

The perturbation Hamiltonian $\mathscr{K}_1$ results from the addition of the perturbing field $\boldsymbol{a}_1(t, z)$ to the vector potential (7.11) and subsequent linearization of the Hamiltonian (7.14) with respect to $\boldsymbol{a}_1$.

The solution of eqn (8.66) assumes its simplest form when obtained in action-angle variables. The equation of motion corresponding to the Hamiltonian (8.69) is the pendulum equation. As is known from mechanics, the motion of a pendulum (or, for that matter, of any mechanical system undergoing periodic motion) may be described in terms of the canonical action-angle variables J, w, the first of which happens to coincide with the adiabatic invariant J of Section 8.4. The definition of the action variable J as a function of the Hamiltonian $\mathscr{H}$ is

$$J(\mathscr{H}) = \oint P_\xi(\xi, \mathscr{H})\, \mathrm{d}\xi \tag{8.71}$$

where $P_\xi(\xi, \mathscr{H})$ is the inverse function of the Hamiltonian $\mathscr{H}(\xi, P_\xi)$ and the integral is extended to a full period of oscillation. The angle variable, w, is the co-ordinate conjugate to J in a canonical transformation leading to J as the new momentum. It has the property that the physical co-ordinate and momentum are periodic functions of w of period 2π. After the canonical transformation to action-angle variables the Hamiltonian $\mathscr{K}_0$ becomes a function of J only and the Hamilton equations reduce to

$$\frac{\mathrm{d}J}{\mathrm{d}z} = -\frac{\partial \mathscr{K}_0(J)}{\partial w} = 0, \qquad \frac{\mathrm{d}w}{\mathrm{d}z} = \frac{\partial \mathscr{K}_0(J)}{\partial J} = k_p(J), \tag{8.72}$$

which have the trivial solution given by $J =$ constant and $w = w_i + k_p(J)z$, where k_p, the pendulum wavenumber, equals $2\pi/L_s$, with L_s given by eqn (7.35). At the same time the Boltzmann equation (8.66) becomes

$$\frac{\partial f_0}{\partial z} - \frac{\partial \mathscr{K}_0}{\partial w}\frac{\partial f_0}{\partial J} + \frac{\partial \mathscr{K}_0}{\partial J}\frac{\partial f_0}{\partial w} \tag{8.73}$$

and, since $\partial \mathscr{K}_0/\partial w = 0$, it has the trivial z-independent solution $f = f_0(J)$, with f_0 independent of w and an arbitrary function of J. This solution describes the equilibrium distribution of electrons in the quasi-potential generated by the superposition of the main wave field with the undulator, which is attained after a sufficient number of pendulum periods from the entrance of the undulator. The form of the function $f_0(J)$ depends on the input conditions and, in particular, but not only, on the energy distribution of the input beam.

Regarding the solution of eqns (8.67) and (8.68) it may be observed that, when f_0 is an equilibrium distribution, these—linear—equations are characterized by coefficients which depend only on the variables γ and ξ and are periodic functions of the latter. By an extension of Floquet's theorem, we may then expect that solutions exist of the form represented by

$$\boldsymbol{a}_1 = \mathrm{Re}[A_1(\xi)\mathrm{e}^{\mathrm{i}(\omega_s t - k_s z)}], \qquad f_1 = \mathrm{Re}[F_1(\gamma, \xi)\mathrm{e}^{\mathrm{i}(\omega_s t - k_s z)}], \quad (8.74\mathrm{a,b})$$

where A_1 and F_1 denote periodic functions of their argument ξ. These solutions are the modes of eqns (8.67) and (8.68). Just as the fundamental FEL modes studied in Chapter 6, these 'sideband' modes, so called because their peak gain occurs for a frequency slightly different from the fundamental, are characterized, for every given frequency ω_s, by one or more, possibly complex, discrete values of the propagation constant k_s. The imaginary part of k_s represents the amplification, or damping, rate of each mode.

Determining the modes of eqns (8.67) and (8.68) and the related propagation constants exactly is not easy, but some approximations are possible, exploiting again the analogy with the large-gain FEL model. In particular, it may be observed that, since $A_1(\xi)$ in eqn (8.74b) is a periodic function, $\boldsymbol{a}_1$ admits a discrete Fourier expansion as

$$\boldsymbol{a}_1 = \mathrm{Re} \sum_{n=-\infty}^{\infty} A_{1n}\mathrm{e}^{\mathrm{i}[(\omega_s + n\omega)t - (k_s + nk + nk_u)z]}. \quad (8.75)$$

As the velocity $\omega/(k + k_u)$ with which the phase ξ moves, that is, roughly, the velocity of the electrons, is different from the speed of light, c, only one of the components of this expansion can have a velocity very close to c. Just as in Section 6.9, it follows that this component, which may with no loss of generality be assumed to correspond to $n = 0$, will be amplified the most (or damped the least) by the wave equation and will prevail over the others under small-coupling (low amplification rate) conditions. It is therefore a good approximation, as long as the resulting amplification rate is small, to assume for $\boldsymbol{a}_1$ a simple exponential form.

Kroll, Morton, and Rosenbluth (1981) calculated the amplification rate of sideband modes within this approximation in the case of a δ-function unperturbed electron distribution $f_0(J)$ concentrated at the bottom of the potential well, as may be obtained at the end of a tapered undulator designed so as to produce strong overtrapping. They found that modes can exist in a discrete set of frequencies differing from the main wave frequency for, positive or negative, integral multiples of the small-oscillation pendulum frequency $\Omega = 2\pi c/L_{s,\min}$, the lower sideband being amplified and the upper one damped.

Riyopoulos and Tang (1987) considered a more general unperturbed electron distribution $f_0(J)$ and arrived at an analytical expression, too long to be reported here, for the growth rate $g_s = \mathrm{Im}(k_s)$ of the sideband wave which displays the following properties. Unlike the previous case, the frequency distribution of sideband modes is a continuum, due to the fact that electrons with different values of J assume all the possible oscillation frequencies contained between 0 and Ω. Upper and lower sidebands located symmetrically around the main signal frequency have approximately opposite growth rates, so that one of them is always unstable except in the trivial case of a uniform distribution such that $\mathrm{d}f_0/\mathrm{d}J = 0$. However, it cannot be said in general whether the upper or lower sideband is unstable, and several successive changes of sign of the growth rate occur in the cases considered by those authors.

The FEL sideband instability is still an argument of active research at the present time, and it remains to be understood how the picture given by the linear stability analysis changes when the amplitude of the sidebands becomes comparable to that of the main signal, and non-linear effects couple the sideband modes to each other and to the main wave. Numerical simulations seem to indicate that, at least under some circumstances, a multifrequency, possibly chaotic, regime may be entered with a negative influence on the coherence properties of the laser beam (Goldstein and Colson 1982; McVey 1986).

Notes

1. For the sake of simplicity we shall explicitly consider circular polarization only in this chapter. Nevertheless, the argument of Section 7.10 applies unchanged even when the undulator is tapered, and it remains true that the results obtained for circular polarization and the fundamental resonance can be extended to general polarization and resonance on higher harmonics simply by replacing A_u by the effective undulator potential $A_{ueff,n}$ of eqn (7.72) or (4.36).
2. The condition that must be fulfilled by the undulator parameters in order to achieve the maximum power output with the least number of poles may be found by maximizing the derivative of the wave amplitude A_w with respect to the undulator phase ψ_u, which is simply obtained by dividing both sides of eqn (8.48) by k_u. It may then be seen that the maximum of $\mathrm{d}A_w/\mathrm{d}\psi_u$ for given p, f_t, and γ_s is obtained by maximizing the ratio A_u/k_u under the constraint of eqn (8.46). This maximum is, in turn, achieved when $A_u = \sqrt{2}$ ($\langle a_u^2 \rangle = 1$).

References

Fawley, W. M., Prosnitz, D., and Scharlemann, E. T. (1984). Synchrotron–betatron resonances in free-electron lasers. *Phys. Rev. A* **30**, 2472–81.

Goldstein, J. C. and Colson, W. B. (1982). The control of optical pulse modulation due to the sideband instability in free electron lasers. *Proc. Int. Conf. on Lasers 1982*, New Orleans, 15–17 Dec. 1982.

Kolomenskii, A. A. and Lebedev, A. N. (1966). *Sov. Phys. JETP* **23**, 733.

Kroll, N. M., Morton, P. L., and Rosenbluth, M. N. (1981). Free-electron lasers with variable parameter wigglers. *IEEE J. Quantum Electron.* **17**, 1436–68.

Landau, L. D. and Lifschitz, E. M. (1976). *Mechanics*. Pergamon Press, Oxford, sect. 49.

Luchini, P. (1985). On the optimal profile of a tapered undulator. *FEL*84, 389–94.

Luchini, P. and Motz, H. (1985). An exact solution for the motion of an ultrarelativistic electron in a tapered undulator. *Appl. Phys.* **B37**, 47–54.

McVey, B. D. (1986). Three-dimensional simulations of free-electron laser physics. *FEL*85, 449–55.

Motz, H. (1979). Undulators and 'free-electron lasers'. *Contemp. Phys.* **20**, 547–68.

Neil, G. R., Edighoffer, J. A., Fornaca, S. W., Hess, C. E., Smith, T. I., and Schwettman, H. A. (1985). The TRW–Stanford tapered wiggler oscillator. *FEL*84, 199–202.

Palmer, R. B. (1972). Interaction of relativistic particles and free electromagnetic waves in the presence of a static helical magnet. *J. Appl. Phys.* **43**, 3014–23.

Pellegrini, C. (1982). Report of the working group on far field accelerators. *Proc. AIP Conf. on Laser Acceleration of Particles*, Los Alamos, 1982. AIP Conf. Proc. Vol. 91. American Institute of Physics, New York, pp. 138–49.

Quimby, D. C. (1986). Betatron–synchrotron resonances and misalignment in free electron laser oscillators with quadrupole focussing. *FEL*85, 456–63.

Riyopoulos, S. and Tang, C. M. (1987). The free-electron laser sideband instability reconsidered. *FEL*86, 226–34.

Rosenbluth, M. N. (1985). Two-dimensional effects in free-electron lasers. *IEEE J. Quantum Electron.* **21**, 966–9.

Slater, J. C. (1948). Design of linear accelerators. *Rev. Mod. Phys.* **20**, 473–518.

Sprangle, P., Ting, A., and Tang, C. M. (1987). Radiation focusing and guiding in the free electron laser. *FEL*86, 136, 142.

Tang, C. M. and Sprangle, P. (1985). Resonance between betatron and synchrotron oscillations in a free electron laser: a 3-D numerical study. *FEL*84, 61–8.

9

UNDULATOR OPERATION

This chapter deals with a collection of problems which do not pertain to the physics of the undulator proper, but rather to its interaction with the other parts that contribute to forming a working instrument, in particular the particle accelerator that produces the electron beam and the optical cavity that allows the undulator to be transformed into a free-electron laser.

Various modes of operation of undulators have been characterized in the previous chapters as (a) spontaneous emitter, (b) oscillator, (c) amplifier, and (d) accelerator modes. Radiation may be produced from any charged particle beam, and although we shall mostly be concerned with relativistic electron beams, highly relativistic proton and antiproton beams have been passed through undulators for the purpose of energy measurement.

The choice of the electron beam accelerator used to drive the undulator is governed by the purpose of the experimental arrangement and the scope of the laboratory. The electron accelerators available are (1) rf linacs, (2) induction linacs, (3) microtrons, (4) storage rings, and (5) electrostatic generators, each of which we shall discuss in turn.

9.1 Microwave-driven linac

There are two kinds of linac (LINear ACcelerator). Microwave-driven linear accelerators (rf linacs) use a slow-wave structure (iris-loaded waveguide) to achieve synchronism between the electron beam and a travelling electromagnetic wave, so that energy may be transferred from the latter to the former through the general mechanism described in Section 8.1. Actually, there is little difference of principle between the workings of a tapered undulator and of an rf linac. Just as we have seen in Chapter 8, only those electrons which happen to fall in a specified phase range become trapped in a quasi-potential well and exchange energy efficiently. Therefore the electron beam comes out of an rf linac in bunches (called 'micropulses'), separated by the period of the driving wave. The driving wave, in turn, is most conveniently furnished by klystron tubes which are themselves pulsed, producing a power of the order of tens of MW for a time ranging up to several microseconds. The general structure of the beam coming out of an rf linac is thus represented by bursts of micropulses (called 'macropulses') with a maximum duration of a few microseconds or tens of microseconds, and a repetition rate that goes from 1 up to, say, 100 Hz, depending on the cooling efficiency of the klystron tubes. If the linac is fed by a continuous electron beam, single micropulses have a repetition rate given by the frequency of the rf supply (generally, around 1 GHz) and a duration of a fraction of the rf period. The linac, however, is often preceded by a specially designed accelerating rf cavity, called the prebuncher, which 'time-focuses', so to speak, the electron beam, compressing the electrons into short bunches having the right phase to be accelerated by the travelling-wave structure. In addition, the electron-emitting cathode itself is almost always pulsed, rather than continuously powered. In this case the micropulse repetition rate corresponds to the prebuncher and cathode frequency, which is generally a subharmonic of the linac's, and the micropulse duration, depending mainly on the quality of the prebuncher, may range down to a few picoseconds.

The energy of the beam produced by a linac is virtually unlimited, since an unlimited number of accelerating sections (each yielding several MeV or tens of MeV) may be strung in series. In practical terms it is limited by cost; in the largest existing linacs tens of GeV are achieved.

Beam current is limited by the maximum electrical load that the rf supplies, of both the prebuncher and the accelerating sections, can withstand. A peak current of 100 A in a 30-ps micropulse has been achieved in the Los Alamos FEL experiment (see Chapter 11).

In addition to the time structure, energy, and current of the electron beam, fundamental parameters for undulator operation are the beam emittance and energy spread. The concept of energy-spread is self-explanatory. Emittance, for those who are unfamiliar with it, is a parameter defined roughly, in each of two perpendicular directions, as the product of the beam's diameter and full-width angular spread.[1] Its importance derives from a fundamental theorem of paraxial electron optics (analogous to the Langrange–Smith–Helmholtz theorem of Gaussian optics of ordinary light), which states that any values of diameter and angular spread corresponding to a given emittance may be achieved by a suitable lens system (e.g. a magnetic field) but the emittance itself cannot be changed. In other words, emittance is the conserved beam brightness in a linear system. In fact, an even stronger result holds, that the emittance (the normalized emittance in a relativistic beam) cannot be reduced (but *can* be increased), even by active devices such as the accelerating structures and the prebunchers.[2] It follows that emittance in a linac is determined mainly by the emission properties of the cathode and by diffusion (space charge) effects in the accelerating chain. However, since the cathode can usually provide more current than the linac can accelerate, beam emittance may be improved just after the cathode by trimming the beam prior to being introduced into the prebuncher.

9.2 Induction linac

In an induction linac the beam current limitation imposed on rf linacs by the limited peak operating power of their rf supply is overcome by generating the accelerating field from a pulse-forming network rather than from an rf oscillator. This higher current is, however, obtained at the expense of a much larger beam width, or more precisely emittance, which is necessary in order to contain space-charge effects within tolerable limits. The electron beam is initially furnished by a diode capable of producing a beam current greater than 10 kA at around 1 MeV. Subsequent acceleration is accomplished by a travelling magnetic field produced by current from a capacitor bank discharged into an array of induction coils, suitably switched to give rise to a pulse travelling with the beam. The resulting electron pulse consists of a single bunch several tens of nanoseconds long, which can be repeated at a rate determined by the time taken by the dc power supply to recharge the capacitor bank (typically around 1 Hz).

Figure 9.1 shows a schematic diagram of an induction accelerator module. This module can be viewed as a 1 : 1 current transformer, the primary of which is the power feed and the secondary of which is the beam. The toroidal ferrite core of the transformer surrounds the accelerating column and provides an accelerating electromotive force equal to the voltage that is applied at the primary, e.g. 250 kV in the ETA and ATA accelerators built at Livermore (Orzechowski *et al.* 1985). The primary circuit must be supplied with a current of 10 kA at this voltage, and for this purpose special pulse-forming devices (water blumleins) are adopted.

Just as with the rf linac, the beam energy achievable with an induction linac is limited only by the cost of building additional modules, but the energy gain

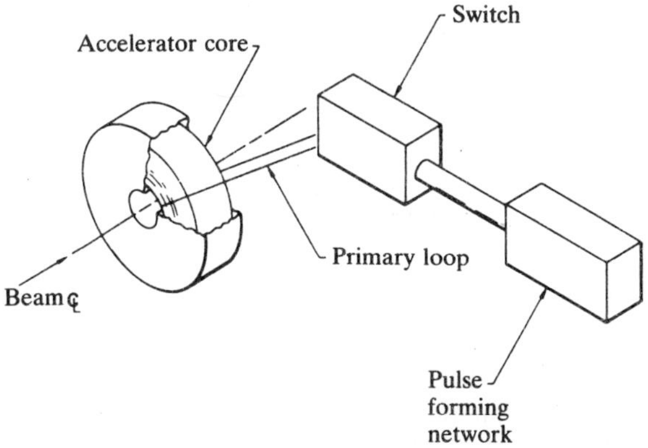

FIG. 9.1. Working principle of an induction accelerator module (from Orzechowski *et al.* 1985; © 1985 IEEE).

per module is much lower. Energies range from a few MeV to 50 MeV (the design energy of the ATA accelerator).

The high beam current that is characteristic of induction linacs puts very severe requirements on the cathode, which determines, for the reasons explained earlier, the emittance and energy spread of the beam. It is difficult to obtain from this type of machine a beam sufficiently monoenergetic and collimated to be useful for an undulator working in the optical region, but it has delivered a large power output in experiments performed in the centimetre- and millimetre-wave region, and experiments in the infrared are in progress. Cathode technology is improving at a sustained rate, though.

9.3 Microtron

The microtron is a compact and relatively inexpensive accelerating device suitable for energies of a few tens of MeV. It uses a single microwave cavity for repeated acceleration and a single flat-pole magnet for confining the electrons in circular trajectories. During each transit an energy increment is

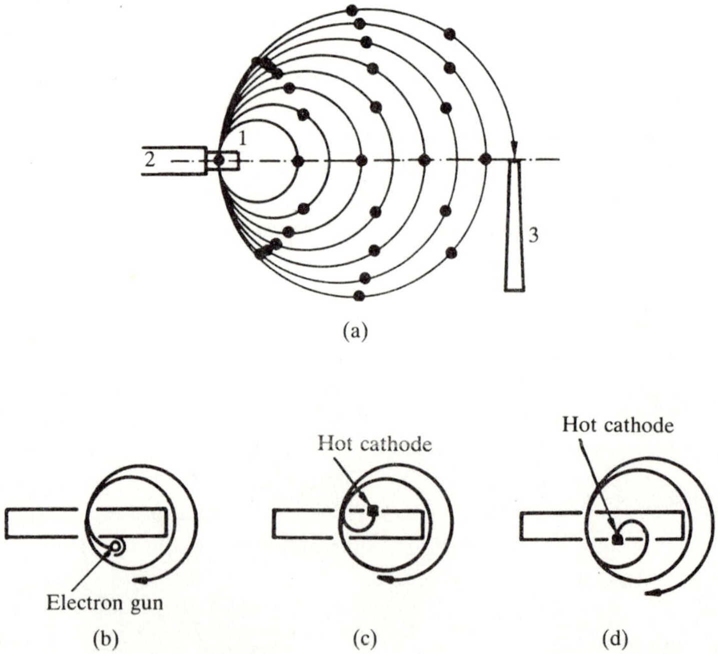

FIG. 9.2. Schematic diagram of a microtron (a) and particulars of the accelerating cavity and of the electron injection trajectory: 1, accelerating cavity; 2, rf feed waveguide; 3, shielded magnetic channel for electron beam extraction. (b) Wernholm injection. (c) Kapitza injection. (d) Melekin injection. (From Dattoli *et al.* 1983.)

added so that the electron bunch describes a larger circular orbit which is tangential to the previous one at the location of the cavity. The particular synchronism condition of the microtron, which requires that an electron may acquire just enough energy in each turn to increase its transit time by one whole period of the microwave, gives lateral and longitudinal focusing properties to the device which ensure proper confinement of the bunch. After a predetermined number of turns, usually between 20 and 30, the bunch is extracted by a magnetic screen intercepting the outermost orbit.

The arrangement is illustrated in the schematic Fig. 9.2, which also shows the location of the electron source in the accelerating rf cavity according to Wernholm, Kapitza, and Melekin.

Like the rf linacs, microtrons are usually fed rf power by pulsed klystrons. This gives their beams a similar time structure, with micropulses lasting a few tens of picoseconds and repeating at a subharmonic of the frequency of the rf supply, and macropulses lasting several microseconds and repeating at a maximum rate determined by the cooling characteristics of the klystron, ranging from 1 Hz up to, say, 100 Hz. The maximum beam current is determined by loading of the rf supply and by the emission properties of the cathode material, and may attain a peak of a few amperes in the micropulse. Emittance is comparable with that of rf linacs, while energy spread and beam current are both smaller.

9.4 Storage ring

In various laboratories around the world, storage rings have been dedicated to function as radiation sources for user communities. They are designed to keep a single or a few electron bunches, initially injected by a lower-energy accelerator, turning around for a long time, typically many hours, gradually accelerating them up to energies of GeVs. In typical high-energy experiments, for which storage rings were originally designed, the electrons are then extracted tangentially from the ring and made available at experimental stations suitably collocated in the proximity. A storage ring, however, also provides the unique possibility, precious for the generation of synchrotron radiation, of making the electrons interact with suitable 'insertion devices', e.g. undulators, placed along the ring itself, where the same electron bunch can perform repeated passages with high repetition rate.

The radiation from the beam, bent in a magnetic field—the synchrotron radiation—is used for a great variety of experiments in physics, engineering, biology, and medicine. Undulators are a natural way of enhancing synchrotron emission, and they have been routinely adopted for this purpose for many years now, at practically all storage-ring sites in operation. This

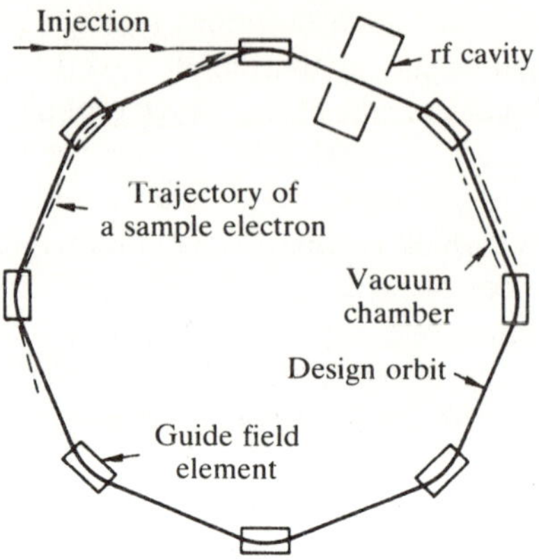

FIG. 9.3. General scheme of the construction of a storage ring (from Sands 1970).

application only exploits the spontaneous emission of the undulator. The next step, storage-ring free-electron lasers, is today in its experimental stage. The first operation has been achieved in 1983 (see Chapter 11).

The dynamics of electrons in a storage ring is a complicated subject which has been lucidly explained by Sands (1970). The designer of a ring has to be careful to avoid instabilities that result from oscillations of the electrons in their orbits. He is helped by the damping effect of the radiation emitted in the machine, and he uses extra magnets for such purposes, which may be arranged to form very short undulators, or wigglers, and which, in addition, concentrate the radiation in a manner useful for some experimenters.

The general scheme of the construction is illustrated in Fig. 9.3, that shows bending magnets and an rf cavity which makes up for the energy losses by acceleration. In Fig. 9.4 the so-called 'lattice' (i.e. the succession of magnetic fields encountered by the electrons around the ring) is shown in more detail; it includes a succession of focusing and defocusing magnets, and it is possible to include straight sections where, for instance, the beam may be steered into an undulator.

We cannot treat all the problems arising out of the operation of a storage ring because the physics of this device is beyond the scope of this book. We shall confine ourselves to some basic concepts. Particles in the ring describe orbits which may or may not be closed, depending on their energy and initial position and momentum. For every energy there will be a closed orbit, corresponding to suitable initial conditions, and the motion of all the particles is described with reference to the closed orbit of a particle having the nominal design energy $\mathscr{E}_0$. The independent co-ordinate around the machine is the distance s along this orbit. Lateral deviations are described by co-ordinates x parallel and y perpendicular to the plane of the orbit; we shall generically call either of them u.

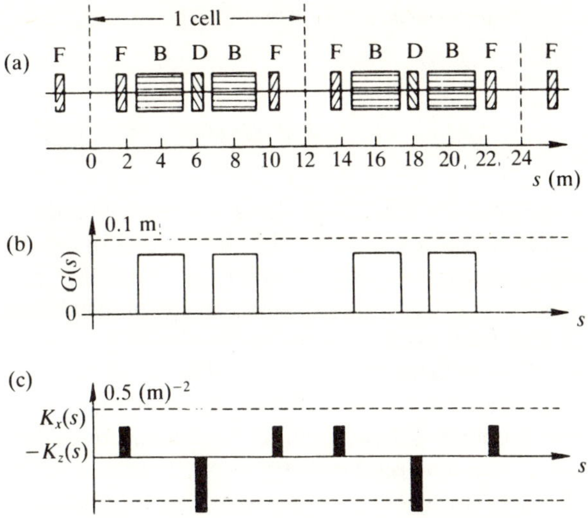

FIG. 9.4. A storage-ring 'lattice' (from Sands 1970).

The closed orbit corresponding to an energy different from $\mathscr{E}_0$ will be different from the design orbit. Its deviation $u_0(s)$ may be expressed, to first order in the energy deviation $\Delta\mathscr{E} = \mathscr{E} - \mathscr{E}_0$, as

$$u_0(s) = (\Delta\mathscr{E}/\mathscr{E}_0)\eta(s). \tag{9.1}$$

The characteristic function $\eta(s)$, generally called the 'dispersion function' may be computed from the magnetic fields for any given configuration of bending and focusing magnets (a given 'lattice'). By definition it is always periodic, satisfying

$$\eta(s) = \eta(s + C),$$

where C is the circumference of the ring.

For any given energy, only a particle starting with a particular set of initial conditions will perform a closed orbit. Particles with different initial positions carry out transverse oscillations, which are called betatron oscillations, about the equilibrium orbit.

The betatron oscillations are very important. If they are allowed to build up, particles are lost from the beam. The deviations from the design orbit may be obtained as $u = u_0 + u_\beta$, where the non-periodic function $u_\beta(s)$ is described by

$$u_\beta(s) = a\beta^{1/2}(s)\cos[\varphi(s) + \varphi_0], \tag{9.2}$$

where a is an arbitrary constant and $\beta(s)$ is the so-called betatron function, which may be computed from the strength of the focusing (quadrupole) magnets and is again periodic with period C. The phase $\varphi(s)$ is the betatron phase, defined by

$$\varphi(s) = \int_0^s \beta^{-1}(s)\,\mathrm{d}s. \tag{9.3}$$

The quantity $\nu = \varphi(C)/2\pi$ is called the tune or operating point of the storage ring. (The definition of some writers differs by a factor of 2π). Its value is the number of lateral betatron oscillations per turn. The choice of the operating point influences the interaction between betatron oscillations and ring imperfections, and must not fall on an integral or half-integral number.

For an ensemble of particles the amplitude, a, of betatron oscillations appearing in eqn (9.6) will have a statistical distribution, which the interplay of random excitation and damping makes Gaussian. Let us denote by σ_a the r.m.s. value of a characterizing this distribution. The transverse displacement u, according to eqn (6.9), also has a Gaussian distribution, and its r.m.s. value is $\Delta u = \beta^{1/2}\sigma_a/\sqrt{2}$. By the same token, the derivative $\mathrm{d}u/\mathrm{d}s$, that is the angle at which the electron crosses the design orbit, has a Gaussian distribution and an r.m.s. value approximately equal to $\sigma_a/(2\beta)^{1/2}$. The product of twice the r.m.s. values of the orbit displacement and angle is the beam emittance

$$\varepsilon = 2\sigma_a^2, \tag{9.4}$$

and is constant around the ring as required by the Langrange–Smith–Helmholtz theorem.

When the emittances in the two transverse directions and the betatron functions are given, the r.m.s. beam cross-section is determined anywhere along the machine. The emittance is most important from the point of view of undulator operation. It determines how small the beam cross-section and the opening angle of the beam at the undulator entrance can be made. The requirements will be discussed below.

So far we have discussed only the lateral focusing of the electron beam by the ring magnets. The longitudinal position of the electrons within the circulating bunch and their energy are stabilized by the counteracting effects of synchrotron radiation, which continuously subtracts energy from the electrons, and the radio frequency power system, which compensates for the energy losses at every passage through the accelerating cavity. Because of the restoring forces created by these two actions, electrons that do not have exactly the design values of energy and longitudinal position (i.e. phase relative to the rf accelerating field) perform oscillations, called energy or 'synchrotron' oscillations, around these values. In a first approximation these oscillations may be considered to be harmonic and, when an ensemble of electrons is considered, the amplitude of oscillation may be assumed to have a Gaussian distribution. There will turn out to be a Gaussian energy spread, measured by the r.m.s. value, $\sigma_{\mathscr{E}}$, of energy deviations, and a phase spread associated with the r.m.s. bunch length.

The origin of synchrotron oscillations may be qualitatively described as follows. An electron on the design orbit must perform one full turn of the ring in an integral multiple of the period of the rf field, so as always to enter the accelerating cavity with the same phase. An electron with a slightly different energy will run on a different closed orbit according to eqn (9.5), the length of which we may write, always in a linear approximation, as

$$L = (1 + \alpha\Delta\mathscr{E}/\mathscr{E}_0)L_0, \tag{9.5}$$

L_0 being the length of the design orbit, and α a constant named the dilation factor. (The effect of betatron oscillations on the length of the orbit is of second order and may be neglected.) Since electrons travel around the ring with a velocity practically equal to the speed of light c, the off-energy electron will accumulate a time delay of $(\alpha L_0/c)(\Delta\mathscr{E}/\mathscr{E}_0)$ per turn and a corresponding phase lag (or lead) of $\alpha k L_0 \Delta\mathscr{E}/\mathscr{E}_0$, k being the wavenumber of the accelerating field. At the same time, the energy of the electron will decrease by a small amount in each turn because of synchrotron emission, which is independent of phase, and re-increase by the potential 'kick' of the rf cavity, which is a more or less sinusoidal function of phase. On the average over many turns, the electron will behave as a dynamical system, in which phase is the 'co-ordinate', energy deviation is the 'momentum', and $1/\alpha$ is the 'mass', acted upon by a sinusoidal potential (a pendulum!). As long as phase deviations are small, the longitudinal position and energy deviation of the electron will perform almost harmonic oscillations. The value of energy deviation for which the phase difference becomes so high as to let the pendulum turn over, and the electron leave the bunch, defines the energy acceptance of the storage ring.

A peculiar characteristic of the storage ring with respect to the accelerating devices considered in the previous sections is that emittance and energy spread are not determined by the properties of the gun from which the electrons are initially emitted, but rather by the dynamic balance between excitation and damping of betatron and synchrotron oscillations in the ring itself.

Damping is provided by synchrotron emission which, when its dependence on electron energy is taken into account, may be seen to generate an additional force proportional to energy deviation, which acts as a braking term in the equations of motion of both synchrotron and betatron oscillations. The damping time turns out to possess a very simple approximate expression, being just the ratio of the electron energy to the rate of energy emission, i.e. the time that the electron would take to radiate away all of its energy, for synchrotron oscillations, and twice as much for betatron oscillations.

The main source of excitation of energy oscillations, at least at low beam densities, is the discrete quantum character of the emission of radiation. This may be modelled, in a semi-classical approach, as a statistical process with a Poisson distribution. If the mean number of quanta of energy u emitted per unit time is denoted by N, and the damping time by τ, the r.m.s. energy spread is given by

$$\sigma_{\mathscr{E}}^2 = \tau N u^2. \tag{9.6}$$

Using reasonable estimates of τ, N, and u, Sands obtains

$$\sigma_{\mathscr{E}}^2 = \mathscr{E}_0 u_c, \tag{9.7}$$

where $\mathscr{E}_0$ is the design energy of the ring and u_c is the mean energy of the synchrotron radiation quanta, which is not far from hc/λ_c, with $\lambda_c = R/\gamma^3$, the cut-off wavelength of synchrotron radiation mentioned in Chapter 2 (eqn (2.4)).

Strictly coupled to the energy oscillations are the longitudinal oscillations of the electrons, which determine the bunch length. In a typical storage ring the bunch length is designed to be a fraction of the order of 10 per cent of the distance between successive bunches.

Owing to the random direction of the emitted quanta, the discrete radiation process will also excite lateral betatron oscillations and thus determine the emittance of the electron beam. From Sands the estimate

$$\varepsilon \simeq \frac{2}{\nu^2}\left(\frac{\sigma_{\mathscr{E}}}{\mathscr{E}_0}\right)^2 \tag{9.8}$$

may be derived.

The foregoing estimates of energy spread and emittance descend from a single-electron analysis, but actual values may be higher because of collective interactions.

Collective interactions are also responsible for the limitation of beam current. Particularly significant in this respect is the Touschek effect. When two particles performing betatron oscillations collide, some of their transverse momentum is transformed into longitudinal momentum and their energy changes by equal and opposite amounts. When the energy change is larger than the energy acceptance of the storage ring both particles are lost, and the lifetime of particles in the ring is correspondingly reduced if this happens often. This is the Touschek effect. Since the probability of a two-particle collision increases with the square of electron density, there results a limitation of instantaneous bunch current.

9.5 Undulator operation in a storage ring

Originally built for experiments in high-energy physics, storage rings have been successively dedicated as generators of uv light and X-rays for various applications, being capable of providing a much more intense beam than alternative sources. Initially, just the synchrotron radiation from bending magnets was used, but as soon as the generation of synchrotron light became the main purpose of several large storage rings it was realized that an undulator inserted in the ring could provide a much better output. Today, most storage rings include a short undulator (wiggler) of, say, five periods, and in some cases a long undulator has been installed just to exploit its spontaneous emission.

The installation of an undulator in a storage ring must cope with the fact that the undulator magnet becomes part of the lattice, and must perturb the operating parameters, and in particular the shape of the electron orbit of the ring, as little as possible. Notice, on the other hand, that it is just because electron bunches are recirculated and may be exploited more than once that operation of an undulator as a spontaneous emitter is convenient in a storage ring, whereas before the birth of free-electron lasers nobody would have thought to put an undulator in a linear accelerator.

In order not to alter the operation of the storage ring, the undulator must be equivalent, from the electron-optical point of view, to the straight section it replaces. In zeroth order, when the undulator field is assumed to vary with z only, this means that the undulator must not displace or bend the electron beam. This translates into two integral constraints on the magnetic field which will be discussed in the next chapter on undulator construction. In the next order, the focusing effect of the unavoidable lateral variation of the undulator field, which was discussed in Section 3.8, comes into play, and must be kept small by limiting the undulator length or the magnetic field strength. When this limitation conflicts with performance of the undulator it may become necessary to redesign the whole optics of the machine to take the presence of the new element into account. However, in the storage ring FEL experiments designed to date, which use rather short undulators, the alteration of betatron oscillations is generally quite small.

A beneficial side-effect of putting an undulator in a storage ring is that the resulting increase in total synchrotron radiation emitted increases the damping of both synchrotron and betatron oscillations, with a consequent improvement in performance. Sometimes a short wiggler may be designed in just for this purpose.

If the purpose is to provide a powerful, narrow-band light source, the undulator may be transformed into a free-electron laser by placing an optical resonator cavity around it, as will be discussed in greater detail in Section 9.9. By properly adjusting the length of the cavity, the round-trip time of the radiation in the resonator is made equal to the time interval between electron bunches. In this way a light pulse emitted by one bunch comes back into the undulator just in time to be amplified by the next bunch and, starting with spontaneous emission, power is built up in the cavity in the form of a recirculating light pulse, much in the same way as in a mode-locked conventional laser.

A storage-ring free-electron laser differs from FELs using other kinds of accelerators in that the electron beam is not lost after the passage in the undulator, and therefore efficiency may be high even if only a very small part of the electrons' energy is radiated out in each turn. However, the very recirculation of the electrons through the ring creates a peculiar regime of operation for storage-ring FELs. In fact, contrarily to simple spontaneous emission which has a damping effect on synchrotron oscillations, laser action adds to the beam energy spread and, if the increment cannot be compensated for by damping in other parts of the ring, energy spread continues to increase until the bunch is destroyed or until the undulator gain is depressed enough to achieve a balance.

To understand the origin of this effect we have to remember from Chapter 5 that the first-order effect of a laser wave interacting with an electron in an undulator is an energy change given, in dimensionless form, by

$$\gamma_1 = \int_0^L \frac{\boldsymbol{a}_0}{\gamma_0} \cdot \frac{\partial \boldsymbol{a}_1}{\partial t}\, \mathrm{d}z, \tag{9.9}$$

where $\boldsymbol{a}_0 = \mathrm{Re}(\boldsymbol{A}_u e^{ik_u z})$ is the dimensionless undulator vector potential and $\boldsymbol{a}_1$ is the dimensionless vector potential of the laser wave. When $\boldsymbol{a}_1$ represents a pure sinusoidal plane wave, eqn (9.9) reduces to eqn (5.13). The energy change given by eqn (9.9) depends on the initial phase of the electron and, once averaged over the whole beam, reduces to zero, so that the leading contribution to the mean energy variation is found in second order, as was seen in Section 5.4.

As far as the electron energy distribution is concerned, however, a first-order effect *is* present. Electrons with different initial phases are given different energy variations. Since the laser wavelength is small (in particular, much smaller than the electron bunch length), laser phase changes rapidly, and for the scope of dynamics of the ring may be considered a uniformly distributed random variable. It follows that the mean square energy spread of the beam $\sigma_{\mathscr{E}}^2$ increases by the mean square value of $\mathscr{E}_1 = mc^2\gamma_1$ at every passage through the undulator.

The calculations giving this mean square value have already been performed in Chapter 5. In particular, eqn (5.56) reads

$$\langle \mathscr{E}_1^2 \rangle = \frac{e^2 L^2}{8\gamma_0^2} \sum_{n=0}^{\infty} |\boldsymbol{A}_{ueff,n} \cdot \boldsymbol{E}_w^*|^2 \frac{\sin^2 \nu_n}{\nu_n^2}, \tag{9.10}$$

where $\boldsymbol{E}_w$ is the (dimensional) electric field of the wave. Assuming that the polarization of $\boldsymbol{E}_w$ is the correct one for maximum interaction (i.e. the same as that of $\boldsymbol{A}_{ueff,n}$) and keeping only the single term in the sum that is resonant (i.e. has $\nu_n \simeq 0$) we may recast eqn (9.10) as

$$\langle \mathscr{E}_1^2 \rangle = \frac{A_{ueff,n}^2}{8\gamma_0^2} \frac{\sin^2 \nu_n}{\nu_n^2} e^2 L^2 E_w^2. \tag{9.11}$$

It may be seen that a laser field interacting with the undulator produces a random excitation of energy deviations in the electron beam with a mean square value proportional to E_w^2; that is, to the power of the laser field itself. When the beam is continuously recirculated in a storage ring this additional excitation affects the regime of synchrotron oscillations. The precise way in which the electron distribution evolves must be studied by adding an appropriate term (sometimes called the 'FEL noise' term) in the storage-ring equations, which are beyond the scope of this book. We can, however, give the following order-of-magnitude considerations.

The damping rate, D, of synchrotron oscillations in a storage ring was recalled in the previous section to be approximately given by

$$D \simeq P_s/\mathscr{E}_0, \tag{9.12}$$

where P_s is the total synchrotron radiation power emitted by each electron in the entire ring and $\mathscr{E}_0$ is the mean electron energy. If synchrotron oscillations are excited by the FEL at a rate of $\langle \mathscr{E}_1^2 \rangle$ per turn, and simultaneously lose a fraction D of their amplitude $\sigma_{\mathscr{E}}^2$ per unit time, a steady-state balance is reached when the rate of production and damping balance each other, i.e. when

$$\langle \mathscr{E}_1^2 \rangle = DT\sigma_{\mathscr{E}}^2, \tag{9.13}$$

T being the revolution period around the ring.

In a laser oscillator starting up from zero, $\langle \mathscr{E}_1^2 \rangle$ will grow together with the intracavity field until $\sigma_{\mathscr{E}}^2$ is pushed beyond the energy acceptance of either the undulator or the ring itself. If the energy acceptance of the ring is exceeded, the beam will be lost. If, on the other hand, the energy acceptance of the undulator is exceeded first, undulator gain will be depressed, and the intracavity field will stabilize at about that level. The energy acceptance of an undulator is roughly given by $\sigma_{\mathscr{E}} \simeq \mathscr{E}_0/2N$ (as in eqn (5.83)), where N is the number of periods. (In order to consider operation on the nth harmonic rather than on the fundamental frequency, it is sufficient to replace N by nN in this and all the following formulae.) Thus the amplitude of the intracavity field will stabilize on a value such that

$$\langle \mathscr{E}_1^2 \rangle \simeq DT(\mathscr{E}_0/2N)^2 = TP_s\mathscr{E}_0/4N^2. \tag{9.14}$$

A very simple and useful relation is obtained by expressing $\langle \mathscr{E}_1^2 \rangle$ in terms of the output power of the laser. According to Madey's theorem, eqn (5.49), the average energy $\langle W_{we} \rangle$ released by each electron to the electromagnetic field in every passage through the undulator is given by half the derivative of $\langle \mathscr{E}_1^2 \rangle$ with respect to $\mathscr{E}_0$. It may be easily calculated from eqn (9.11) that in the neighbourhood of the point of maximum gain $\langle W_{we} \rangle \sim 2N\langle \mathscr{E}_1^2 \rangle/\mathscr{E}_0$ (simply because $\mathrm{d}\langle \mathscr{E}_1^2 \rangle/\mathrm{d}\mathscr{E}_0$ is the derivative of a function which is different from zero over a typical range of $\mathscr{E}_0/2N$). Equation (9.14) then gives $\langle W_{we} \rangle \sim TP_S/2N$;

that is, after multiplying both sides by the total number of electrons in the ring and dividing by T,

$$P_{out} \sim P_{stot}/2N, \tag{9.15}$$

where P_{out} is the total power released by the electron beam to the electromagnetic field, which coincides with the output laser power but for mirror absorption in the optical resonant cavity, and P_{stot} is the total synchrotron radiation power emitted by the beam in the ring.

Equation (9.15) is known as the Renieri limit (Renieri 1979*a*,*b*; Dattoli and Renieri 1980) and shows that the maximum average coherent power that can be generated by a storage-ring FEL is proportional to the total incoherent synchrotron radiation power emitted in the ring itself divided by the number of periods of the undulator. This is a very strong limitation under the typical operating conditions of the storage-ring FELs designed and operated up to now, and constrains the output power to a level much lower than the saturation mechanism considered in Chapter 7 would allow (but still anything but low in practical terms). Of course, eqn (9.15) does not mean that the laser power is less useful than the synchrotron emission because the former is concentrated in a narrow beam and in a thin spectral line, whereas the latter is spread all over the ring's circumference and in a wide spectrum. Notice also that eqn (9.15), although valid in a pulsed as well as in a continuous regime, refers to the average output power, the average being taken over a time of the order of the damping time of the ring (typically, several milliseconds). A much higher peak power may be extracted in pulses that are short compared to the damping time, provided that a large enough gap is allowed for the electron beam to 'cool down' between pulses.

9.6 Electrostatic machines

An electrostatic accelerator, just as its name implies, accelerates electrons using an electrostatic field. An insulated electrode is charged to a high negative voltage by an electrostatic (e.g. van de Graaff) generator and then an electron beam emitted by a cathode connected to the high-voltage terminal is driven through an accelerating column in which the potential, by means of a succession of capacitively coupled discs, is gradually brought to ground. Beam energy of an electrostatic accelerator is limited by the maximum electrostatic potential that can be practically generated, of the order of 10 MV, but beam uniformity may be very good.

In the application of electrostatic accelerators to FELs a very important role is played by the capacity, C, of the high-voltage electrode. The van de Graaff generator is only capable of providing a d.c. charging current, I_C, typically of less than 1 mA, which is not sufficient for undulator operation. When the electron beam extracts a greater current, I_B, the intensity of which can be directly controlled by a grid placed after the cathode, the voltage, V, will decrease according to

$$\mathrm{d}V/\mathrm{d}t = -(I_B - I_C)/C. \tag{9.16}$$

In order for the beam to interact efficiently with an undulator having N periods, the relative beam energy variation should be kept smaller than the energy width of the undulator gain curve, namely $1/2N$, so that

$$(\Delta V/V)_{\max} = 1/2N. \tag{9.17}$$

According to eqn (9.16), this variation takes place in a time

$$\Delta t_{\max} = CV/2N(I_B - I_C) \tag{9.18}$$

and this is therefore the maximum pulse length that can be provided to the undulator. Before another pulse can be initiated the condenser must be charged up again. The charging rate is $\mathrm{d}V/\mathrm{d}t = I_C/C$, and the initial potential will be recovered after a time

$$\Delta t_C = C\Delta V/I_C = (I_B/I_C - 1)\Delta t_{\max} \tag{9.19}$$

whence the maximum pulse repetition rate is determined.

The pulse duration of the FEL can be much improved if the spent electron beam is recovered and accelerated back to the high-voltage electrode, thus lowering its rate of discharge. As a side-effect the efficiency of the system also increases, since only a small fraction of the beam's energy must be replenished. This has been done in the experiment carried out at the University of California at Santa Barbara (see Section 11.5) by sending the spent beam backwards through the accelerating column, letting it be decelerated by the adverse field until it is dumped in a specially designed collector held at a potential a few tens of kV below that of the main emitting electrode. They have achieved higher than 90% recovery, i.e. a tenfold increase in pulse length, thus extracting 50 μs pulses at a current of 1.2 A and an energy of 3 MeV with a drift of 0.6 keV/μs (Amir *et al.* 1986).

9.7 Parameter values required for undulator operation

We shall now discuss various machine parameters, beam current, emittance, and relative energy spread required for undulator operation in their various roles.

9.7.1 *Spontaneous radiation (synchrotron radiation) emitter*

Using an undulator just for its spontaneous radiation is, of course, the less demanding application as to machine parameters, but the low energy yield attainable per pass makes recovery of the electron beam a necessity, so that the storage ring is the only practically sensible choice. Indeed, several storage rings around the world have been dedicated as synchrotron radiation sources, and many of them derive their 'synchrotron' radiation from undulators rather than just from bending magnets.

In this context the name 'wiggler' or 'multipole wiggler' is usually given to an undulator composed of just a few (say, five or less) periods, which displays a broad and complicated frequency and angle emission distribution with an intensity a few times higher than that obtainable from a single bending magnet; the name 'undulator' is reserved for a device which has a larger number of periods and displays the characteristic emission distribution of Fig. 4.2, with an *fwhm* linewidth of $0.89\lambda/nN$ in wavelength and $0.94M/\gamma(nN)^{1/2}$ in angle around the resonant wavenumber $k_r = 2k_u/(M^2/\gamma^2 + \vartheta_{we}^2)$ and its odd harmonics (as in eqns 4.42–4.44).

The peak spectral brightness (power per unit frequency, beam area, and solid angle) of spontaneous undulator radiation is given, in mks units, by eqn (4.46) evaluated for $\vartheta_{we} = 0$ and $v_n = 0$, i.e.

$$S_\omega = k^2 L^2 A_{ueff,n}^2 \zeta e j_b / 32\pi^2\gamma^2, \tag{9.20}$$

and is of the order of N^2 times that obtainable from a single bending magnet of the same bending radius (i.e. peak magnetic field).

The minimum wavelength that can be achieved is limited only by the energy of the electron beam. Intense u.v. and X-ray radiation is routinely generated in this way for a variety of applications from 1–2 GeV storage rings.

9.7.2 *FEL oscillator*

Placing an undulator with sufficient gain inside an optical resonator, the details of which will be dealt with in the following sections, allows radiation fed back by the resonator to be amplified again and again in the undulator and yields an FEL oscillator.

In an oscillator, the emphasis will be on narrow bandwidth. Energy spread of the electrons due to all sources, cathode temperature, etc., should be kept as small as possible. If it is to work in the pulsed regime the length of the micropulse should not be too short, so that the beam which travels more slowly than the light does not slip behind the light pulse; and the length of the macropulse should also not be too short, so that sufficient passes of the light pulse through the undulator may take place for the radiation amplitude to build up in the optical resonator. As will be seen in greater detail later, the transit time of the light between mirrors must be closely matched to the interval between pulses, which it must be very closely, but not quite, an integral multiple of.

In this section we want to recall the requirements put on the driving accelerating machine by an undulator suitable for an FEL oscillator. In order to achieve a gain per pass of, say, 10 per cent, e-beam current density must at least attain the value which can be derived from eqn (5.24),

$$j_b \gtrsim 0.19\gamma^3 N^{-3} A_u^{-2} I_A/\pi^2 l_u^2 \tag{9.21}$$

and, if the best possible choice of all the other parameters is assumed, e-beam current must at least attain the value which can be derived from eqn (5.82),

$$I_b \gtrsim 0.003(M^2/A_{ueff,n}^2)\gamma I_A/N^2. \tag{9.22}$$

As might be expected, these values decrease with increasing number of periods N. However, the largest N that can be adopted, is limited, further than by technical problems in making the undulator, by the constraint of eqn (5.83) on electron energy spread,

$$\delta\gamma/\gamma \ll 1/2N, \tag{9.23}$$

in excess of which eqn (9.21) and (9.22) cease to be valid. In addition, assuming that suitable magnetic optics are used to achieve the best balance between electron beam diameter and angular spread, the constraint of eqn (5.84),

$$\varepsilon \ll \lambda \tag{9.24}$$

on beam emittance, where λ is the output wavelength of the laser, must be respected.

The minimum operating wavelength of a free-electron laser oscillator is determined mainly by the limitations (9.23) and (9.24) on emittance and energy spread, and by the lack of suitable high-reflectivity mirrors for making the optical resonator at very small wavelengths. No present experiment has operated below visible wavelengths, even if a large effort is being put into obtaining operation in the ultraviolet and positive results may be expected soon; several experimenters having at their disposal electron beams with high current but not very high energy and not very good emittance and energy-spread characteristics have chosen larger operating wavelengths in the infrared or even in the millimetre- and submillimetre-wave region.

If energy spread and emittance are low but space is limited, as might happen, for instance, in a storage ring not specifically designed for undulator operation, so that it is not possible to achieve the largest number of periods, N, allowed by eqn (9.23), a better choice is offered by the optical klystron, which has a higher gain for a given e-beam current than a simple undulator of the same length but limits energy spread and emittance more severely.

9.7.3 *FEL amplifier*

An FEL amplifier, useful for amplifying the laser beam produced by a separate, FEL or conventional, oscillator to a high final power level, will necessarily be a large-gain device. Correspondingly, its small-signal gain and its e-beam requirements will not be expressed by the formulae of Chapter 5, but by those of Chapter 6. In particular, the relevant energy-spread limitation is given by eqn (6.89):

$$\delta\gamma/\gamma \ll l_u/4\pi L_g. \tag{9.25}$$

In order to offer a good energy yield, this amplifer will employ a tapered undulator, possibly preceded by an optical-klystron injection section, as described at the end of Section 7.7. To produce an energy yield of order unity (i.e. not decreasing with increasing N as in an untapered undulator) it will necessarily have to be designed for an output field amplitude such that

$$A_{w,\max} \simeq 1/N \tag{9.26}$$

(as in eqn (8.54)). Therefore, the driving particle accelerator will have to supply a beam current density, according to eqn (8.49), such that

$$4\pi j_b/k^2 I_A \simeq 1/(2N^2\gamma), \tag{9.27}$$

i.e. on replacing k by $2\gamma^2 k_u/M^2$,

$$j_b \simeq \gamma^3 N^{-2} M^{-4} 2\pi I_A/l_u^2. \tag{9.28}$$

A different application, which might also be classified as a non-linear kind of amplifier, is that of the optical klystron for frequency multiplication in the way described in Section 7.11. Such a device has already been built experimentally, as reported in that section, and is the only kind of undulator which has until now produced u.v. radiation in other than the spontaneous-emission mode of operation.

9.7.4 *Electron accelerator*

The fourth possible mode of operation of an undulator is as an electron accelerator, as described in Section 8.8. This mode does not put particular requirements on the electron beam but rather on the laser, since it is the latter that supplies the power. The necessary laser power is within the reach of the large lasers built for such applications as laser-driven nuclear fusion, but accelerator physicists do not seem to have decided yet whether or not laser acceleration is more cost-effective than other accelerating techniques.

When free-electron lasers leave the experimental phase and begin to be built routinely they themselves may become the preferential power supply of laser accelerators. A device formed by a free-electron laser and an electron-accelerating undulator, tightly coupled, would bear some resemblance to the two-beam accelerators which are now being built using pairs of coupled rf cavities, one of which extracts power from a relatively low-energy beam so that the other can use this power to accelerate a higher-energy one.

9.8 Optical resonator cavities

An optical resonator is an arrangement formed by two mutually facing mirrors, which is capable of selecting electromagnetic modes of a particular direction and frequency in the optical range. Combined with an amplifier of optical waves, a medium in which a population inversion has been created, or an undulator, it constitutes a laser oscillator, much in the same way as an L–C resonator combined with an amplifier constitutes an electronic oscillator circuit. The theory of optical resonators is dealt with at length in any book on lasers, and in several books on general optics, and does not change in any way when it is to be applied to free-electron lasers, so that a thorough treatment in this book is unnecessary. For the reader's convenience, in this section we shall recall the basic elements of optical-resonator theory which will be needed in the next section to illustrate the differences between optical resonators designed for conventional and free-electron lasers.

The frequency selective properties of optical resonators arise because a wavefront undergoing a stationary motion back and forth between the mirrors must arrive in phase with itself at the end of a complete round trip. Therefore, in a first-approximation one-dimensional model, stationary modes are found to exist only at those frequencies such that twice the distance between the mirrors equals an integral multiple of the wavelength.

Thus the optical resonator acts as a comb filter, allowing the existence of a set of modes spaced in frequency by an equal amount $\Delta\omega$ which, if the distance between the mirrors is denoted by D, is given by

$$\Delta\omega = 2\pi c/2D. \tag{9.29}$$

If the bandwidth of the amplifier coupled to the optical resonator (i.e. inserted between the mirrors) is of the same order as this $\Delta\omega$ a single mode will oscillate and a very monochromatic laser beam will be obtained. In the opposite case, on the other hand, the laser output will generally be comprised of several modes. Note that mode separation decreases with increasing cavity length. Good selectivity may be obtained in a long cavity by interposing an additional filter, typically a diffraction grating or a Fabry–Perot interferometer which is itself a small optical resonator, but at the cost of increased radiation losses.

In the absence of additional optical components such as windows used to confine the active medium or the above-mentioned selective devices, radiation losses in an optical resonator are due to imperfect mirror reflectivity and to radiation spillover at the mirror edges. Either of these effects can also be used to couple out the laser beam, by using a partially transmitting mirror at one end of the resonator or by collecting radiation scraped off around a suitably sized mirror. Optical resonator losses may be characterized by the rate of power reduction per turn, β, which in the case of reflection losses is simply the product of the two mirror reflectivities, or alternatively by the ring-down time

$$\tau = \frac{2D}{-c \log \beta} \simeq \frac{2D}{c(1-\beta)}, \tag{9.30}$$

i.e. the time necessary for the energy of a stored mode to decrease by a factor 1/e, or again by the Q factor, defined in analogy with the Q factor of a microwave cavity or circuit as $Q = \omega\tau$.

Resonator losses provide a non-zero bandwidth, of absolute value $1/\tau$ and relative value $1/Q$, to each resonance line, and at the same time determine the minimum gain required of the amplifier to start up oscillation, through the obvious condition $\beta G > 1$.

For reasons that will soon become evident, an optical resonator is generally constructed with curved mirrors. The shape of the mirror surfaces may be characterized by their curvature radii R_1 and R_2, assumed positive if the surfaces are outward-convex and negative if inward-convex. The wavefronts of the forward (going from the left to the right mirror) and backward (vice versa) waves may likewise be characterized by their curvature radii R_f and R_b which, according to the laws of geometrical optics, vary linearly with the axial co-ordinate z as

$$R_{f,b} = z - z^{(s)}_{f,b}, \tag{9.31}$$

$z^{(s)}_f$ and $z^{(s)}_b$ being the co-ordinates of the centre of curvature, or virtual source, of either wave. A wavefront with curvature $C = 1/R$, when reflected from a surface with curvature $C_1 = 1/R_1$, acquires a curvature $C' = 2C_1 - C$. Therefore, taking the origin of the z axis at the left mirror, so that the right mirror is located at $z = D$, we can write the equations imposing that the forward and backward wavefronts transform into each other upon reflection as

$$\frac{1}{z^{(s)}_f} + \frac{1}{z^{(s)}_b} = \frac{2}{R_1}, \qquad \frac{1}{D - z^{(S)}_f} + \frac{1}{D - z^{(s)}_b} = \frac{2}{R_2}. \tag{9.32a,b}$$

These may be reduced to the single quadratic equation

$$(D - R_1/2 - R_2/2)[z^{(s)}]^2 - D(D - R_2)z^{(s)} + D(D - R_2)R_1/2 = 0, \tag{9.33}$$

the two roots of which yield $z^{(s)}_f$ and $z^{(s)}_b$. The solution of eqn (9.33) is

$$z^{(s)}_{f,b} = \frac{D(D - R_2) \pm [D(D - R_2)(D - R_1)(D - R_1 - R_2)]^{1/2}}{2D - R_1 - R_2}. \tag{9.34}$$

As is seen, $z^{(s)}_f$ and $z^{(s)}_b$ may be either real or complex, depending on the sign of the discriminant $\Delta = D(D - R_2)(D - R_1)(D - R_1 - R_2)$, which may be

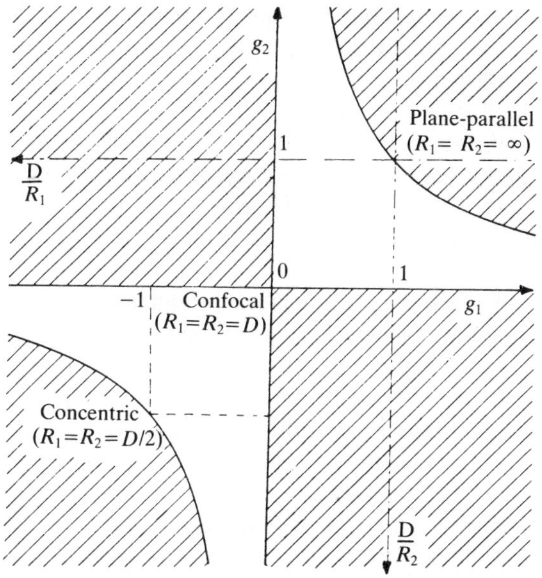

FIG. 9.5. Kogelnik–Boyd stability plot for optical resonators (from Solimeno and Torre 1985). The shaded area is the unstable region.

verified to be negative when $0 < g_1 g_2 < 1$, g_1 and g_2 being defined as $g_{1,2} = 1 - D/R_{1,2}$, and positive otherwise. Optical resonators characterized by a positive value of Δ are termed unstable, because it may be shown that the spot size of a light beam going back and forth between the mirrors (assumed to be of infinite extension) grows up indefinitely in time: optical resonators with a negative value of Δ, on the contrary, support field configurations of finite extension, and are named stable. The distribution of stable and unstable resonators in the plane of the parameters g_1, g_2 is illustrated by the Kogelnik–Boyd stability plot of Fig. 9.5. Noteworthy limiting cases, separating the stable from the unstable region, are the plane resonator ($R_1 = R_2 = \infty$), the concentric resonator ($R_1 = R_2 = D/2$), and the confocal resonator ($R_1 = R_2 = D$).

Of course, strictly speaking, wavefronts originating from a complex source point have no meaning in geometrical optics: but, as was already observed in Section 5.11, a wave which formally has a complex source point is nevertheless a solution of the paraxial wave equation, namely the Gaussian beam. In fact, if the problem of resonator modes is restated, in the next approximation to geometrical optics, using the Fresnel diffraction integral (4.23) or, equivalently, the paraxial wave equation (4.26), it will be found that eqn (9.34) retains its validity, and that when Δ is negative the resonator mode is composed of two Gaussian beams originating from the complex virtual source points $z_f^{(s)}$ and $z_b^{(s)}$.

Since a Gaussian beam falls off exponentially in amplitude in the radial direction, it will also represent sufficiently well the mode of a stable resonator with finite but sufficiently large mirrors, and a stable resonator will have exponentially small spillover losses.

On the contrary, an unstable resonator constantly loses some power in the radial direction and therefore has non-zero losses even in the geometrical-optics approximation. The actual determination of its modes cannot be carried out without explicitly taking diffraction at the mirror edges into account.

As is true of all mode-supporting electromagnetic devices, once we leave the one-dimensional approximation and begin to consider transverse structure we must expect to find not one but a sequence of modes, so that each of the previously considered 'longitudinal' resonance lines divides up into a multitude of lines associated with different transverse modes. In the case of stable optical resonators with infinite (i.e. sufficiently large) mirrors, the modes are represented by parabolic cylinder functions. Higher transverse modes have an increasingly higher cross-section and, just as longitudinal mode selectivity is determined by the overlap between the spectral distribution of modes and the spectral response of the amplifier, transverse mode selectivity is determined by the overlap between the modes' and the amplifier's cross-section. In particular, if the active cross-section of the amplifier is comparable with that of the fundamental Gaussian mode, only this mode will oscillate, whereas if the active cross-section of the amplifier is much larger several modes will be contemporaneously present, with a consequent degradation in transverse spatial coherence.

We now discuss the fundamental mode further. For a given design of an optical resonator, one wishes to know the minimum spot size and its location, and the spot size at the mirrors.

The first question is easily answered by combining eqn (9.34) with the formulae for Gaussian beam size given in Section 5.11. In particular, the minimum spot size is given by $w_0 = (z_R/k)^{1/2}$, z_R being the imaginary part of the complex number given by eqn (9.34), and the abscissa z_w of the location where it occurs (waist of the Gaussian beam) is the real part of the same number. More generally, as shown in Section 5.11, the cross-section of the Gaussian beam at any position z is given by

$$\{[(z - z_w)^2 + z_R^2]/kz_R\}^{1/2}.$$

At the positions of the mirrors, $z = 0$ and $z = D$, this formula yields

$$w_1 = [(z_w^2 + z_R^2)/kz_R]^{1/2}, \qquad w_2 = \{[(D - z_w)^2 + z_R^2]/kz_R\}^{1/2}. \quad \text{(9.35a,b)}$$

In particular, as may be expected from symmetry considerations, if $R_2 = R_1$ the waist occurs at $z_w = D/2$. It is all too easy to see that, in this case, $z_R = [(D/2)(R - D/2)]^{1/2}$ and thus

$$w_0 = (\lambda/2\pi)^{1/2}(D/2)^{1/4}(R - D/2)^{1/4} \quad \text{(9.36)}$$

and

$$w_{1,2} = (\lambda D/2\pi)^{1/2}\{R^2/[2D(R - D/2)]\}^{1/4}. \qquad (9.37)$$

Note that spot size at the mirrors tends to infinity as R tends to $D/2$ (concentric resonator) or to infinity (plane resonator). In these, in fact, as well as in unstable cases, the radiation beam fills up all the space available, and its shape is essentially affected by diffraction at the mirror edges. Therefore concentric, plane, and unstable resonators always have non-negligible spill-over losses.

In a stable resonator, on the other hand, the ratio of mirror radius to spot size, the square of which is usually indicated as the Fresnel number, determines the field amplitude at the mirror edge and therefore the extent of spillover losses. With a radius larger than about three times the spot size at the mirror, spillover power losses are reduced below 10^{-4} and become negligible with respect to the reflection losses of even the best mirrors.

Even easier is the inverse problem of determining the radii of curvature of the mirrors so as to obtain a given waist position and size (or equivalently, Rayleigh range) of the Gaussian beam. The answer to this problem, in fact, is given directly by eqn (9.32), where one has to substitute the complex numbers $z_w \pm iz_R$ for $z_1^{(s)}$ and $z_2^{(s)}$.

9.9 Matching an optical cavity to the undulator

There is an essential difference between the requisites that a resonator must have for a free-electron laser and for a conventional laser.

In a conventional laser, power density is limited by the properties of the active medium, which can only store energy from the pump and release it to the laser beam with a limited spatial density. For this reason, resonators for high-power conventional lasers must provide a mode cross-section as wide and uniform as possible, and this requirement is usually satisfied with unstable resonators, in spite of their less favourable output beam quality. Stable resonators are reserved for comparatively low-power lasers in which energy density is small and beam quality can be privileged.

On the contrary, as was seen in Sections 5.11 and 8.7, efficient operation of a free-electron laser *requires* a very high intracavity power density, so that there is all the interest in making the mode as thin and intense as possible inside the undulator without, however, having power densities at the mirrors so high as to burn them. In addition, gain is often not very high, thus calling for a low-loss configuration. This combination of circumstances directs the choice towards nearly concentric stable resonators.

In particular, in Section 5.11 we have seen that an optimal condition for the amplification of a Gaussian beam is attained, if the electron beam is sufficiently thin and has sufficiently low momentum and energy spread, when the Rayleigh range of the Gaussian beam is roughly $\frac{1}{4}$ of the undulator length. The elements of resonator theory recalled in the previous section now enable us to discuss the implications that this criterion has on the design of the resonator.

Because of the need to leave room for the magnets used to drive the e-beam in and out of the undulator, and because of problems related to transit time that we shall discuss later, the resonator must usually be significantly longer than the undulator. Let us assume for example, that it is three times longer. The above criterion then leads us to design a resonator with a Rayleigh range $z_R = D/12$. Such a small value of z_R/D involves a working point rather close to the stability limit and in such a situation one should expect problems. We shall, indeed, find that the resonator becomes very sensitive to misalignment, and in particular to mirror displacement and tilt.

If, for instance, we consider a symmetric resonator and put $z^{(s)}_{f,b} = D/2 \pm iD/12$ into eqns (9.32), we find $R_1 = R_2 = 37D/72$, very close to the value $D/2$ corresponding to a concentric resonator. In an exactly concentric resonator, having $R_1 = R_2 = D/2$, the angular position of the optical axis is undetermined, as may be easily visualized by imagining the two mirrors to be part of a single spherical shell; and therefore, in a nearly concentric resonator, this position becomes the more sensitive to perturbations of all kinds the smaller the difference $R_1 + R_2 - D$ is.

Let us quantitatively determine the sensitivity to mirror tilt of the position of the optical axis in a nearly concentric resonator. With reference to Fig. 9.6, suppose that mirror 1 is rotated by an angle α in the plane of the figure, to position 1′. The optical axis is, by definition, a line perpendicular to both mirrors. Since we only want to consider small displacements, we may look for the new position of the optical axis by using the local representations of the mirror profiles through their tangents and curvatures, i.e.

$$z_1 = \alpha x_1 + x_1^2/2R_1 \tag{9.38}$$

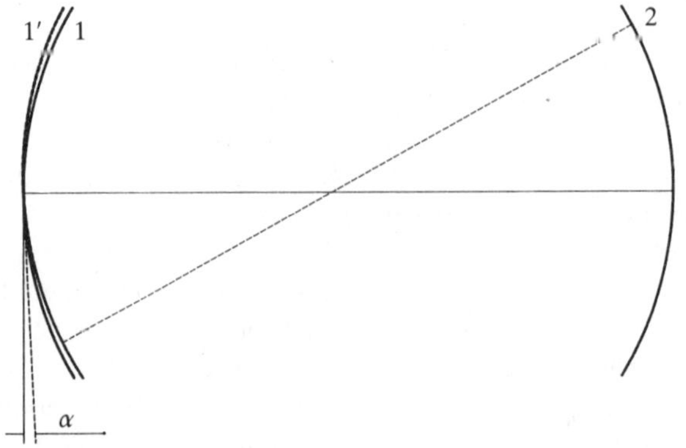

FIG. 9.6. Effects of mirror tilt on the position of the optical axis in a resonator.

for mirror 1′, and

$$z_2 = D - x_2^2/2R_2 \tag{9.39}$$

for mirror 2. By differentiating eqn (9.38) it may be seen that a line perpendicular to mirror 1′ at position x_1 has the angular coefficient $-(\alpha + x_1/R_1)$, and therefore must intersect mirror 2, after a distance approximately equal to D, at the position x_2 given by

$$x_2 = x_1 - (\alpha + x_1/R_1)D. \tag{9.40}$$

Imposing the condition that this line be also perpendicular to mirror 2 at x_2 gives the additional equation

$$-\alpha - x_1/R_1 = x_2/R_2 \tag{9.41}$$

which, together with eqn (9.40), allows the determination of x_1 and x_2. The result is expressed by

$$x_2 = \frac{x_1}{g_2} = \frac{R_1 R_2}{R_2 + R_1 - D}\alpha, \tag{9.42}$$

where, as before, $g_2 = 1 - D/R_2$. In a concentric resonator, where $R_1 + R_2 = D$, the denominator in eqn (9.42) vanishes and the sensitivity to mirror tilt becomes infinite. In a nearly concentric resonator it is finite but very high. For instance, in the case considered above, of $R_1 = R_2 = 37D/72$, eqn (9.42) gives $-x_1 \simeq x_2 = 9.5\alpha D$. It goes without saying that under these conditions the mirror curvatures themselves must be realized with low tolerances in order not to pass the stability limit.

Acceptable mirror tolerances must be carefully considered in designing an optical resonator for a free-electron laser; it is often necessary to partially trade off the optimal fit of the resonator mode to the undulator in favour of a greater stability of the mirror configuration.

9.10 Short-pulse operation of an FEL oscillator: pulse slippage and laser lethargy

The designer of a free-electron laser oscillator must choose the beam energy and magnet spacing so as to produce the desired frequency. The available beam current will determine how long the undulator must be in order to produce a gain greater than the mirror losses, and greater enough to build up the laser power in the cavity in a time shorter than the macropulse duration of the electron-feeding accelerator. Since the gain bandwidth of the undulator is comparatively large, many longitudinal modes will simultaneously oscillate and the precise length of the optical resonator will have little, if any, bearing on frequency tuning. Instead, whenever the electron beam has a micropulse structure, as happens in all microwave-driven accelerators, resonator length is critically tied to micropulse spacing.

As, in an ultrarelativistic beam, the electrons move at a speed very close to the speed of light, when the electrons arrive in short bunches of a few centimetres in length it is possible for the laser light itself to take the form of short pulses moving more or less in synchronism with the electron bunches. In this way, for the whole travel in the undulator, each laser pulse 'sees' the peak micropulse current, and is correspondingly amplified much more than if it were uniformly distributed in time and thus its amplification were determined by the average current over the macropulse. The micropulse and macropulse currents may easily differ by a factor of 100. In order to exploit this favourable regime, however, the designer must make sure that the length of the optical resonator be just such that a laser pulse travelling along with a particular electron bunch may catch the next electron bunch on the next turn around. Approximately, this requires that twice the resonator length be equal to the distance between micropulses, which is in turn close to an integral multiple of the wavelength of the rf field driving the electron accelerator. To obtain a more precise picture of the pulsed regime of the FEL one has to take several effects into account.

According to eqn (3.25b), the axial velocity inside the undulator of an electron of nominal energy γ_0 is given by

$$v_z = c/[1 + M^2/2\gamma_0^2] \simeq c[1 - M^2/2\gamma_0^2], \tag{9.43}$$

where $M^2 = 1 + A_u^2/2$. We assume that the transverse momentum and energy spread of the e-beam are not so high as to lead to sensible bunch lengthening over the length of the undulator. Thus eqn (9.43) may be taken to give the velocity of an electron bunch. The electron bunch will gradually loose ground with respect to a co-propagating light pulse travelling with velocity c, and at the end of the undulator of length L it will have slipped behind by a distance

$$s = (M^2/2\gamma_0^2)L, \tag{9.44}$$

which is called the slippage distance. Since $L = Nl_u$ and $(M^2/2\gamma_0^2)l_u = 2\pi/k_r = \lambda_r$, the central wavelength of emitted radiation, another expression of the slippage distance is $s = N\lambda_r$, which could also have been obtained from the synchronism condition which states that the electron passes over one radiation period for each undulator period. It is not unusual for the slippage distance to be of the same order of magnitude as the electron bunch length, as was the case, for instance, in the 1977 Stanford FEL.

In an oscillator configuration in which the light pulse is recirculated over and over in the undulator, a self-consistent pulse shape will arise (dubbed a 'supermode' by Dattoli and Renieri 1979), which reproduces itself when it is amplified by the electron bunch that simultaneously drifts by the slippage distance. This self-consistent shape may be calculated numerically, as has been done by several authors (see, e.g., *FEL*81).

When slippage is taken into account one realizes that the optimal distance between the cavity mirrors necessary to maintain synchronism between the light and electron pulses does not exactly equal half the distance l_b between successive electron bunches. It would still seem to equal $\pi c/\omega_b$, where ω_b is the repetition frequency of electron bunches, but even this result is not completely correct, as it relies on the assumption that the light pulse travels with speed c. In the presence of amplification the group velocity of the light pulse will be slightly lower because of a phenomenon known as laser lethargy.

Laser lethargy was first noted by Hopf in connection with mode-locked conventional lasers. It also occurs in an undulator in the pulsed mode of operation. The phenomenon has been analysed by Al-Abawi *et al.* (1979) and Moore and Scully (1980), and they show that the light pulse is effectively slowed down because more is radiated into the hind part of the light pulse than into the front part. In the small-gain regime, which characterizes most FEL oscillator configurations, the phenomenon may be explained as follows. According to the description of the small-gain regime adopted in Chapter 5, the electromagnetic wave acting on the electrons may be approximated by the input field alone, and the fields radiated by the electrons summed up with each other and the input field only at the exit section of the undulator. The output is therefore the sum of the unperturbed input field, which propagates through the undulator with the speed of light, c, and the fields radiated by the electrons at various positions, which carry along a delay increasing linearly with the position of the source along the undulator. The greatest part of the radiated field is emitted near the end of the undulator, where the energy bunching has been finally converted into position bunching, and lags behind the unperturbed field by roughly the slippage distance s. The self-consistent field is obtained on imposing that the sum of the unperturbed and radiated fields should reproduce the shape of the pulse presenting itself at the entrance of the undulator but for an axial translation. This translation, intermediate between that of the unperturbed field, i.e. 0, and that of the radiated field, almost s, produces the effect of a light pulse travelling with a speed lower than c, which is the lethargy effect.

The exact form of the self-consistent field, as obtained from the numerical calculations, is complicated. It turns out that for every positive lag, δs, defined as

$$\delta s = 2\pi c/\omega_b - 2D \tag{9.45}$$

one can find a self-consistent mode characterized by a certain amplification rate. A typical plot of the amplification rate versus δs may be seen in Fig. 9.7, which reports a comparison between measurements taken on the Stanford FEL and the results of numerical calculations performed by Colson (1982). It may be noticed that the amplification rate increases rapidly from zero for $\delta s = 0$ to a maximum, and then falls off at a much lower rate. The position of the maximum may be estimated from the foregoing description of the self-

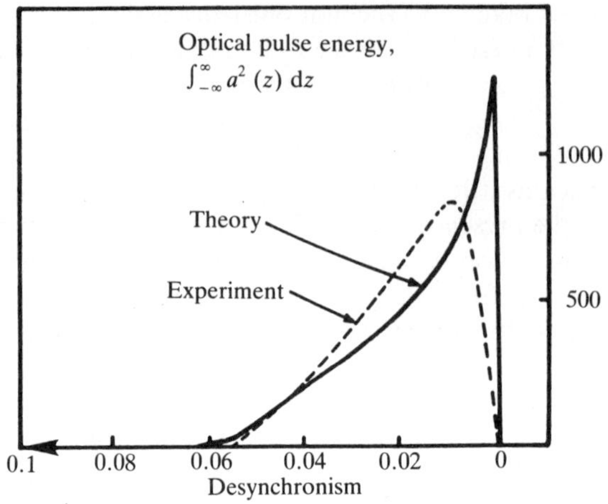

FIG. 9.7. Calculated and experimental curves of the amplitude (total energy content) of the self-consistent mode of a pulsed FEL oscillator vs. cavity length desynchronism δs, in arbitrary units (from Colson 1982; © 1982 Addison-Wesley Publishing Co.).

consistent field as arising from the superposition of the unperturbed field, which has zero lag, and the radiated field, which has a lag of approximately s, to be at

$$\delta s \simeq gs, \tag{9.46}$$

g being the single-pass differential gain calculated in Chapter 5, that is, the relative amplitude of the radiated with respect to the unperturbed field.

Perhaps the most important observation coming out of the analysis of the pulsed mode of operation of FEL oscillators is that operation in this regime can only be obtained by tuning and stabilizing the optical-cavity length to within a very narrow range, a fraction of the product gs, which often means a fraction of a wavelength. In all the experiments performed to date great care has been devoted to the necessary stabilization, sometimes secured through a closed-loop active control system using interferometric telemeters and piezoelectric or voice-coil actuators.

Notes

1. The precise definition of emittance, which we give for the sake of completeness but which is not needed for the present discussion, is, in each of the two directions of the transverse plane in which betatron oscillations take place independently,

$$\varepsilon_{1,2} = 4[\langle \delta x_{1,2}^2 \rangle \langle (\mathrm{d}\delta x_{1,2}/\mathrm{d}z)^2 \rangle - \langle \delta x_{1,2}\, \mathrm{d}\delta x_{1,2}/\mathrm{d}z \rangle^2]^{1/2},$$

where angular brackets denote an ensemble average over the statistical electron distribution and $\delta x = x - \langle x \rangle$. When position ($x_{1,2}$) and direction of motion ($\mathrm{d}x_{1,2}/\mathrm{d}z$) are non-correlated, it reduces to the product of twice the r.m.s. position deviation, i.e. the full-width position spread, and twice the r.m.s. angular deviation, i.e. the full-width angular spread, as stated in the text. Some authors define emittance as π times the above value.

2. According to the Liouville theorem of Hamiltonian mechanics, the motion of a set of instances of a deterministic system that follows Hamilton's equations, starting with all possible initial positions falling inside a given region of phase space, leads successively through regions of the same area as the initial one. Supposing that the electrons start off in a rounded-out region well characterized by global parameters such as its r.m.s. extension along the co-ordinate and momentum axes, the product of these parameters can only remain constant if the contour of the region of phase space occupied by the electrons is subjected to a linear transformation (this is the case considered by paraxial Gaussian optics), or increase if the presence of non-linear interactions makes the contour increasingly tangled, although always of the same area. When the effect of diffusive processes (electron collisions) is non-negligible, the area itself increases. The product of the r.m.s. extensions along the co-ordinate and momentum axes is proportional to the above-defined emittance, which must therefore either remain constant or increase, if the electron γ does not change considerably. In an accelerating column where γ does change it is the normalized emittance $\gamma\varepsilon$ which is proportional to phase-space area and must either remain constant or increase.

The strict analogy with statistical mechanics and the foundations of thermodynamics cannot be escaped. The foregoing result parallels the second principle of thermodynamics, in the form stating that free evolution or work alone without heat transfer (i.e. interaction with a deterministic system such as an rf cavity) can only keep constant or increase the entropy content of a body. The only way to decrease the phase area occupied by the electron beam, other than by selectively scraping away part of the electrons, is to exploit an interaction with a suitable non-deterministic system (a low-temperature 'heat sink'). Such a system is, for instance, the incoherent synchrotron radiation field, the cooling action of which plays an important role in storage rings. However, no similar mechanism is usually significantly active in linacs.

References

Al-Abawi, H., Hopf, F. A., Moore, G. T. and Scully, M. O. (1979). Coherent transients in the free-electron laser: laser lethargy and coherence brightening. *Optics Communs* **30**, 235–8.

Amir, A., Elias, L. R., Gregoire, D. J., Kotthaus, J., Ramian, G. J. and Stern, A. (1986). Spectral characteristics of the UCSB FEL 400 μm experiment, *FEL*85 35–8.

Colson, W. B. (1982). Optical pulse evolution in the Stanford free-electron laser and in a tapered undulator. *FEL*81, 457–88.

Dattoli, G. and Renieri, A. (1979). Classical multimode theory of the free-electron laser. *Lett. Nuovo Cimento* **24**, 121–8.

Dattoli, G. and Renieri, A. (1980). Storage ring operation of the free-electron laser: the oscillator. *Nuovo Cimento* **59B**, 1–39.

Dattoli, G., Marino, A. and Renieri, A. (1983). The FEL—microtron activity at the CNEN Frascati center. In *Free electron lasers* (ed. S. Martellucci and A. N. Chester), pp. 295–309. Plenum Press, New York.

Moore, G.T. and Scully, M. O. (1980). Coherent dynamics of a free-electron laser with arbitrary magnet geometry. I. General formalism. *Phys. Rev. A* **21**, 2000–08.

Orzechowski, T. J., Scharlemann, E. T., Anderson, B., Neil, V. K., Fawley, W. M., Prosnitz, D., Yarema, S. M., Hopkins, D. B., Paul, A. C., Sessler, A. M. and Wurtele, J. S. (1985). High-gain free-electron lasers using induction linear accelerators. *IEEE J. Quantum Electron.* **21**, 831–44.

Renieri, A. (1979*a*). Storage-ring operation of the free-electron laser: the amplifier. *Nuovo Cimento* **53B**, 160–78.

Renieri, A. (1979*b*). Free-electron laser amplifier operation in storage-ring. *IEEE Trans. Nucl. Sci.* **26**, 3827–9.

Sands, M. (1979). The physics of electron storage rings: an introduction. SLAC Report 121, Stanford Linear Accelerator Center, Palo Alto, California, U.S.A.

Solimeno, S. and Torre, A. (1985). Optical resonators for free-electron lasers. *FEL*84, 404–27.

10

MAGNET DESIGN

In this chapter consideration will be given to magnet design. The principle has already been described in Chapter 1 in connection with the original undulator, which had intertwined soft-iron pole pieces of north and south polarity attached to soft-iron bars of the respective polarity, and kept at constant magnetic potential by permanent magnets (magnetron magnets). In recent designs both linear and helical undulators (twisters) have been realized, using either permanent magnets or electromagnets, with or without iron, or a combination of both. We shall examine the various options in turn.

10.1 Objectives of magnet design: field errors

In all of the previous chapters we have assumed the existence of a magnetic undulator field which, at least on the axis, has an exactly sinusoidal dependence on z. In practice, of course, this field can only be realized with finite precision. One of the objectives of magnet design is thus to keep field errors under control. In addition, a high field intensity (at least one such that $A_u = O(1)$ but possibly higher) is desirable because radiation emission and amplification increase with it, and as short a period as possible is desirable because it allows operation at small wavelengths: furthermore, this high field and small period must be achieved while allowing a sufficient gap between the magnet poles to pass the electron and laser beam in. These requirements push present magnet technology to its limits. Finally, consideration must be given to the problem of injecting the electron beam into the undulator and extracting it at the other end, and to the techniques for making continuously variable magnets, particularly useful if the undulator is a tapered one.

Let us first address the issue of field errors. The most important types of field errors are, in order of increasing harmfulness: harmonicity errors, i.e. systematic deviations from a sinusoidal profile; phase errors, longitudinal positioning imperfections which may induce loss of coherence; and random amplitude errors, which result in unwanted sideways beam drift ('walk-off').

Deviation of the field from a sinusoidal profile (harmonic distortion) involves, as long as the pattern is still perfectly periodic with zero mean, only the presence of higher harmonics of the undulator wavenumber, and in the spontaneous emission and small-signal regimes it is harmful only insofar as the effective amplitude of the useful first harmonic is reduced. Since the higher harmonics of the undulator field act as superposed additional undulators of shorter period, however, a too high harmonic content may cause the production of an excessive amount of u.v. or soft-X-ray radiation, which is liable to damage the mirrors in a visible FEL's optical resonator.

Phase errors must not be allowed to build up to the point of affecting the synchronism condition of the electron with the beat wave created by the undulator and laser fields. Avoiding phase errors is a purely mechanical issue, simply implying that a tolerance of a small fraction of the undulator period must be maintained on the absolute distance of any part of the undulator from a fixed reference point. However, in a very long undulator comprised of several hundred periods it may require non-trivial construction techniques.

Beam walk-off caused by random field amplitude errors is the most severe problem, and reducing it to an acceptable level generally requires some form of verification and correction of the finished assembly. At the origin of this problem is that magnetic field errors generate a random force which drives the electron in a sort of doubly Brownian[1] random walk, and may eventually lead it far away from the nominal straight path. When this happens, interaction with the laser wave, which travels in a straight line independently of the undulator field, becomes impossible.

In order to assess the effect of field errors quantitatively we have to reconsider the basic equations of electron motion laid out in Chapter 3. Since those equations are written in terms of the electromagnetic vector potential, we need an expression for the vector potential of the undulator field. On assuming, for the sake of simplicity, that the magnetic field vector $\boldsymbol{b}$ generated by the undulator is a function of the axial co-ordinate only, but an arbitrary one, it can be seen that a vector potential satisfying the equation curl $\boldsymbol{a} = \boldsymbol{b}$ may always be obtained as

$$\boldsymbol{a}(z) = \int_0^z \boldsymbol{b}(z') \wedge \hat{z}\, \mathrm{d}z', \tag{10.1}$$

and is itself a function of z only. Under these conditions the Hamiltonian is independent of transverse co-ordinates and, according to the Hamilton equations of motion (3.4), transverse momentum $\boldsymbol{P}_\perp$ is conserved. The transverse position of the electron is given by the first of eqns (3.22); using the ultrarelativistic expression (3.15) of the Hamiltonian and assuming $\boldsymbol{P}_\perp = 0$ we obtain

$$\boldsymbol{x}(z) = -\int_0^z \frac{\boldsymbol{a}(z'')}{\gamma}\, \mathrm{d}z'' = -\int_0^z \gamma^{-1} \int_0^{z''} \boldsymbol{b}(z') \wedge \hat{z}\, \mathrm{d}z'\, \mathrm{d}z''. \tag{10.2}$$

When γ is constant and $\boldsymbol{b}(z)$ is a sinusoidal function, of course, eqn (10.2) gives nothing other than the by now familiar expression (3.24a) of the electron motion; when $\boldsymbol{b}(z)$ is not perfectly periodic, though, the double integration present in eqn (10.2) produces an accumulation of errors which, after a number of periods, makes the electron beam wander off the design trajectory.

Let us now calculate the r.m.s. expected value of the electron beam displacement after N periods of an undulator. When γ is constant or of negligible variation, eqn (10.2) may be transformed by a simple integration by parts into

$$\boldsymbol{x}(z) = \gamma^{-1} \int_0^z (z - z')\hat{\boldsymbol{z}} \wedge \boldsymbol{b}(z')\,\mathrm{d}z'. \tag{10.3}$$

We shall assume that the undulator field is generated by $2N$ magnets placed with alternate polarity (more elaborate configurations which will be considered in the next sections may be adopted in practice, but they do not change the order of magnitude of the result). Let $(-1)^n \alpha_n \boldsymbol{B}_u g(z - z_n)$ be the field, of nominal maximum intensity $\boldsymbol{B}_u$, generated by the nth magnet located at $z_n = nl_u/2$. We shall assume the field pattern $g(z)$ of different magnets to be identical, but multiplied by a random coefficient $\alpha_n = 1 + \varepsilon_n$, where the fluctuation ε_n has zero mean and r.m.s. value σ, the fluctuations of different magnets being statistically independent. This situation characterizes, for instance, an undulator made out of identical electromagnets energized by different power supplies, and is also an acceptable qualitative representation of permanent-magnet undulators. Under these assumptions the total magnetic field will be given by $\boldsymbol{b} = \boldsymbol{b}_0 + \delta\boldsymbol{b}$, where $\boldsymbol{b}_0$ is the nominal undulator field and

$$\delta\boldsymbol{b} = \sum_{n=1}^{2N} (-1)^n \varepsilon_n \boldsymbol{B}_u g(z - z_n) \tag{10.4}$$

is the random field fluctuation. Inserting this expression into eqn (10.3), where the magnetic field appears linearly and therefore the nominal field and the fluctuation simply superpose, gives a position fluctuation

$$\delta\boldsymbol{x} = \gamma^{-1} \sum_{n=1}^{2N} (-1)^n \varepsilon_n \hat{\boldsymbol{z}} \wedge \boldsymbol{B}_u \int_0^z (z - z')g(z' - z_n)\mathrm{d}z'. \tag{10.5}$$

Assuming, as is reasonable, that the function $g(z)$ has a relatively short range of the order of a half-period, we may approximate the integral on the r.h.s. of eqn (10.5) as

$$\int_0^z (z - z')g(z' - z_n)\mathrm{d}z' = \int_{-z_n}^{z - z_n} (z - z_n - z')g(z')\mathrm{d}z' \simeq (z - z_n)\int_{-\infty}^{\infty} g(z')\mathrm{d}z'. \tag{10.6}$$

In order to realize an almost sinusoidal pattern, the function $g(z)$ ought to resemble a half-period of a sinusoid. Although, in practice, the pattern will be slightly different, with tails extending out to infinity, we may take as value of the integral of g from $-\infty$ to ∞ that corresponding to a half-period of a sinusoid, i.e. l_u/π. Equation (10.5) then becomes

$$\delta \boldsymbol{x} = \frac{l_u}{\pi\gamma} \hat{\boldsymbol{z}} \wedge \boldsymbol{B}_u \sum_{n=1}^{2N} (-1)^n \varepsilon_n (z - z_n) \tag{10.7}$$

and its mean square value, calculated according to the assumptions that $\langle \varepsilon_m \varepsilon_n \rangle = \sigma^2 \delta_{mn}$ and $z_n = nl_u/2$, is

$$\langle \delta x^2 \rangle = \left(\frac{l_u}{\pi\gamma} \right)^2 B_u^2 \sigma^2 \sum_{n=1}^{2N} (Nl_u - nl_u/2)^2, \tag{10.8}$$

i.e. after approximating the sum by $(l_u/2)^2(2N)^3/3$ (a result which may be obtained by approximately replacing the sum by an integral or, equivalently, keeping only the leading term in its exact expression as a polynomial in N),

$$\langle \delta x^2 \rangle = 2N^3 l_u^4 B_u^2 \sigma^2 / 3\pi^2 \gamma^2. \tag{10.9}$$

For the purpose of assessing the importance of the lateral electron drift caused by field errors, we may compare the r.m.s. value of $\boldsymbol{\delta x}$ with the amplitude of the wiggling motion of the electron (3.29):

$$\Delta x_u = A_u/\gamma k_u = B_u/\gamma k_u^2. \tag{10.10}$$

Hence one obtains the ratio

$$\langle \delta x^2 \rangle^{1/2} / \Delta x_u = 10.3 N^{3/2} \sigma. \tag{10.11}$$

As can be seen, the electron drift increases rapidly with the number of periods N, and even for $\sigma = 1$ per cent, which is about the minimum value that permanent-magnet manufacturers can presently guarantee, its expected amplitude already becomes greater than that of the wiggling electron motion for $N \geqslant 5$. It is thus evident that for any practical number of periods the electron drift caused by statistically independent field errors is intolerable, and must be reduced by means of correction magnets, or by measuring the exact field generated by each of the undulator magnets and preselecting them so as to make their errors compensate each other.

10.2 End poles

Somewhat related to the just discussed issue of beam walk-off due to field errors is the problem of correctly injecting the beam into and extracting it from the undulator. If it is desired, as is necessary, for instance, in a storage ring, that the beam may come out of the undulator on the same line as where it went in, it is necessary to ensure that both the lateral displacement $\boldsymbol{x}(z)$ of eqn (10.3) and its derivative become zero again at the end of the undulator. This requirement translates into the two integral constraints:

$$\int_0^L b(z)\mathrm{d}z = 0 \qquad \int_0^L b(z)z\,\mathrm{d}z = 0. \tag{10.12a,b}$$

A simple sinusoidal field pattern truncated abruptly at both ends can only satisfy eqn (10.12) if both injection and extraction occur on the side of the orbit, at the points shown in Fig. 10.1(a) where the electrons' direction of motion is parallel to the axis. The necessary magnetic field pattern is the one shown in the same figure, with the field falling suddenly to zero at the centre of each of the end poles.

Of course, such a discontinuous field is not practically realizable; it remains true in general, however, that in order to satisfy eqns (10.12) the first and the last poles of the undulator must generate a field such that its integral along the z-axis equals half that of the field of other poles. In other words, the end poles must be 'half-poles'. A practical field profile and electron trajectory might be those represented in Fig. 10.1(b).

Notice that, in general, the electron oscillations caused by the undulator field are not symmetrical about the unperturbed trajectory that the electron would follow in the absence of the undulator, so that it is important that the end poles be of the same polarity (i.e. if the undulator begins with a half north-up pole, it must end with a half north-pole), otherwise the resulting

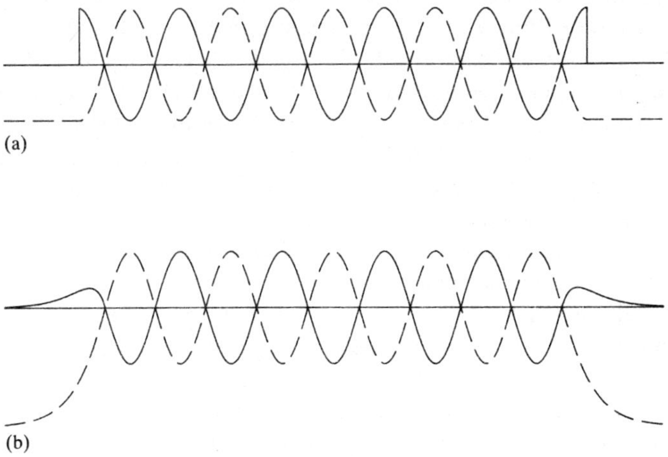

FIG. 10.1. Undulator field profile and corresponding electron trajectory, with the end poles shown by considering an undulator with only five periods. The magnetic field intensity (solid line) and electron trajectory (broken line) are drawn on the same plot despite the fact that the vectors $\boldsymbol{b}$ and $\boldsymbol{x}$ are orthogonal in space. (a) Abrupt field termination; (b) gradual field termination.

injection and extraction trajectories will not be aligned, although they will be parallel. Oscillations that are symmetric about the unperturbed line of motion can only be produced if not only the field profiles of the end half-poles, but also those of the poles next to them, are specially designed.

If very precise alignment is required, either the gap or the field intensity of the end half-poles must be variable, so that fine tuning can be operated with the undulator in place.

10.3 Permanent-magnet linear undulator

The original undulator of 1953 was a linear undulator formed by soft-iron pole pieces of north and south polarity attached to soft-iron bars of the respective polarity. They were kept at constant magnetic potential by permanent magnets (magnetron magnets), disposed as seen on Fig. 1.6. A modification of the structure, necessary to lead the beam in and out, was operated at both ends as described in the previous section.

An electromagnetic (em) linear undulator may be obtained as a straightforward modification of this design by replacing the permanent magnets by suitable coils.

During the set-up of the first free-electron laser experiment in the 1970s, the interest of scientists shifted momentarily towards electromagnetic circular undulators, which we shall examine in a later section; but, soon after, attention returned to linear undulators, because of their simpler construction and because they do not encircle the beam line on all sides, and so allow an easier placement of test equipment. The aim was also, and still is today, to use permanent magnets, because they are more compact (thus allowing a smaller period), yield as high or higher a field and, of course, draw no power.

A simple way to achieve the sinusoidal pattern required of the undulator field using permanent-magnet blocks was indicated by Halbach (1981), and has since been known as the Halbach design. Halbach indicates rare-earth cobalt (REC)—in particular, samarium cobalt—as a suitable magnet alloy. It has a very high remanent field, of the order of one tesla.

Since REC has a magnetic permeability not very different (about 4 per cent) from that of a vacuum, the fields of different magnets may be linearly superposed on each other (the magnets are 'transparent' to each other's fields), and therefore a sinusoidal pattern may be simply realized by a sequence of identical magnet blocks having their easy axis of magnetization rotated by increments of $2\pi/M$ rads, M being the number of magnet blocks per period. The larger M is, the better is the approximation to a sinusoidal pattern. A particularly convenient value from the vantage point of the construction is $M = 4$, which can be realized by putting square magnet blocks side-by-side as in Fig. 10.2. The on-axis field distribution produced is

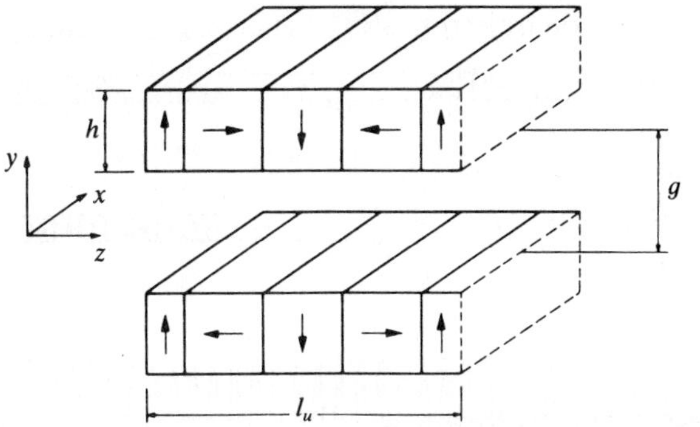

FIG. 10.2. Halbach-design permanent magnet undulator with $N = 4$. (from Walker 1985).

very nearly sinusoidal, with an amplitude given (Winick and Spencer 1980) by

$$B_u = 2B_r \frac{\sin(\pi/M)}{\pi/M}(1 - e^{-2\pi h/l_u})e^{-\pi g/l_u}, \tag{10.13}$$

where B_r is the remanent field of the magnetic material (of the order of 1 T for REC), and g and h are the dimensions shown in Fig. 10.2. Typically, using $M = 4$ and square magnet blocks with $h = l_u/4$ and $B_r = 0.9$ T gives $B_u = 1.284e^{-\pi g/l_u}$ T.

Many existing undulators have been constructed according to Halbach's design. As an example we may quote Hu and Elias's (1986) undulator, a five-period sample of which appears in Fig. 10.3. It may be noticed that end bars have been rotated in order to tune up the half-poles for correct steering in and out of the e-beam. The magnetic field, theoretical electron trajectory, and configuration of a 32-period undulator are shown in Fig. 10.4. Figure 10.5 shows the actual trajectory measured after constructing the undulator, where

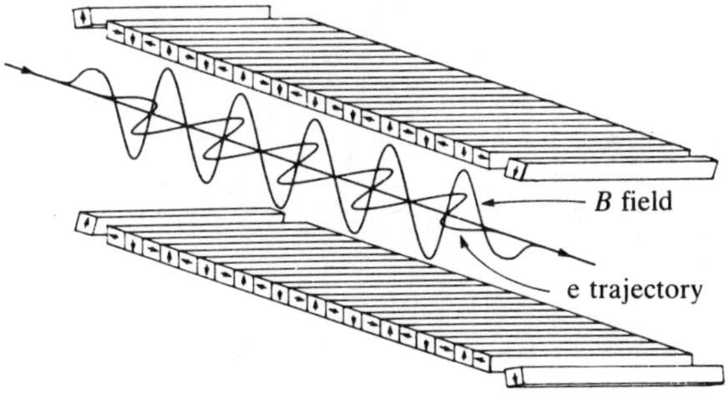

FIG. 10.3. A five-period sample of the Halbach-design UCSB undulator (from Hu and Elias 1986).

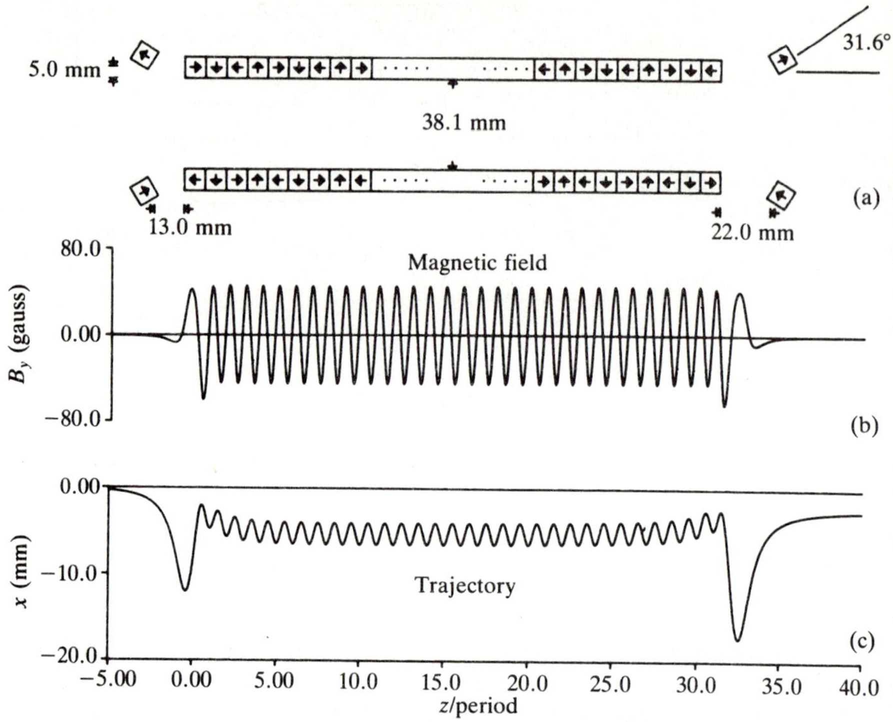

FIG. 10.4. Illustration of (a) a 32-period sample of the UCSB undulator having four rotated field-clamping magnets, (b) the corresponding ideal magnetic field on axis, and (c) the theoretical electron trajectory (from Hu and Elias 1986).

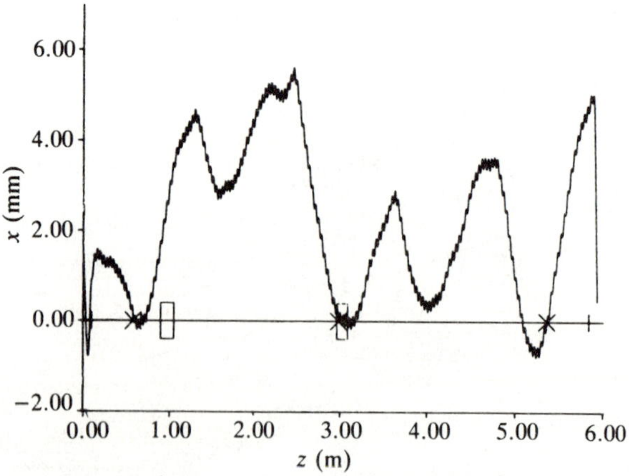

FIG. 10.5. Measured electron trajectory in the 160-period UCSB undulator, illustrating beam walk-off (from Hu and Elias 1986).

the effect of random field errors is clearly seen. Hu and Elias reckon that FEL gain was reduced by as much as a factor of 2 in their experiment because of beam walk-off.

10.4 Hybrid magnet systems

Although convenient from the practical point of view, an undulator made out of permanent magnets only does not yield the highest possible magnetic field. A higher field may be generated by iron-assisted permanent magnets; that is, just as was done in the original undulator, by conveying the magnetic flux of the magnets to the interaction region through suitable iron pole pieces. An example of a recent design following this scheme, depicted in Fig. 10.6, is the Boeing - Spectra Technology undulator (Robinson *et al.* 1987 and references therein). An undulator of this kind has a higher harmonic distortion than the Halbach design but, as was said in Section 10.1, harmonic distortion does not really harm undulator performance. A particular contrivance adopted in the design of the BAC–STI undulator is the use of canted poles to create a uniform radial field gradient in the undulator and symmetrize the focusing action it has on the beam. (It will be remembered from Section 3.8 that the unavoidable transverse gradient of the undulator magnetic field has a long-term focusing effect on the electrons which gives rise to betatron oscillations. In a linear undulator, however, this effect takes place non-symmetrically in the plane of the magnets and in the plane orthogonal to it, unless some action is taken, as in the present case.)

The delicate point about hybrid undulators is that saturation in the iron creates non-linear effects which make it difficult to predict the field reliably enough to keep random beam walk-off under control. It is therefore

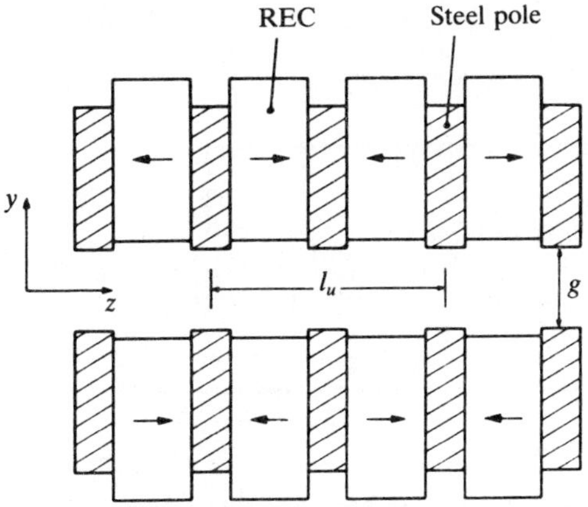

FIG. 10.6. Hybrid REC-steel undulator (from Walker 1985).

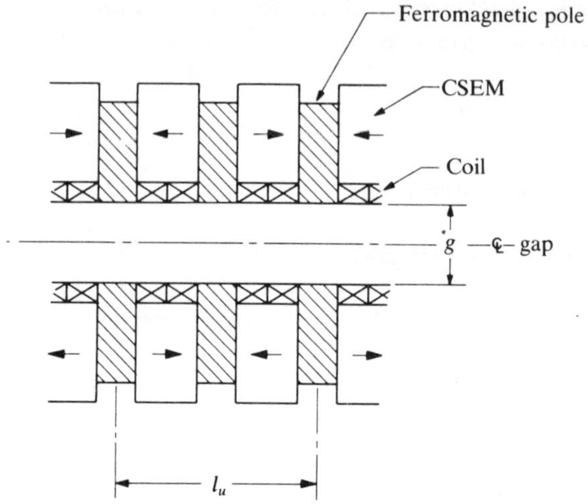

FIG. 10.7. Hybrid REC-steel undulator with tuning electromagnets located on the inside (CSEM refers to 'charge-sheet equivalent materials', e.g. REC) (from Halbach 1986*a*).

necessary, after the construction of a hybrid undulator, to perform a verification and correction of the field profile by mechanically trimming the individual poles. Alternatively, as remarked by Halbach (1986*a*), electromagnetic coils may be added to a permanent magnet structure to make sections individually tunable, and to increase the field level where it might otherwise be limited by saturation in the iron; they may also be useful to achieve the field profile needed for a tapered structure.

A cross-section of the typical configuration that is obtained by adding correction electromagnets to the scheme of Fig. 10.6 is shown in Fig. 10.7. In order to maximize the field and ease the handling of the electromagnets, it

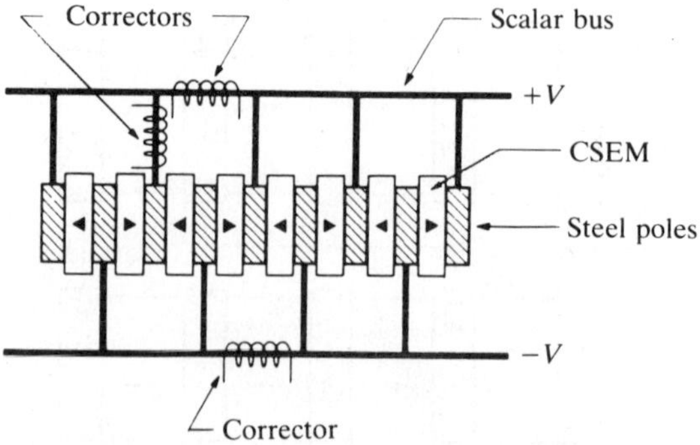

FIG. 10.8. Hybrid REC-steel undulator with external tuning electromagnets (from Halbach 1986*a*).

may be better to locate them on the outside of the assembly. This consideration led Halbach to propose the further refinement depicted in Fig. 10.8, which shows a hybrid magnet assembly projected on the plane of the electron orbit. The poles are supposed to be at constant scalar magnetic potential and they are connected to corresponding constant-potential bus-bars.

10.5 Electromagnetic undulator

Up to now we have considered permanent magnets as the main source of the undulator field. If a very large tuning range is desired in a tapered undulator, however, one may be willing to use electromagnets to generate the main field. With pure electromagnets, saturation in the iron still creates a major problem, and Halbach (1986*a*) shows that the highest field level compatible with reasonable power dissipation in d.c. electromagnet coils is lower than that achievable by permanent magnets (although this limit does not apply to pulsed, possibly ironless, electromagnets). He proposes to correct this situation using again a hybrid design, in which, however, electromagnets generate the main field and permanent magnets are put near the regions of strongest iron saturation in such a manner as to redistribute the magnetic field lines more uniformly within the iron and push the saturation threshold further. Notice that, when used in this way, the permanent magnets do not affect the magnetic field in the undulator gap but only in the iron, and therefore their tolerances do not contribute to field errors, which remain determined by the characteristics of the electromagnets. A simple permanent-magnet-assisted em undulator is shown in Fig. 10.9. To obtain an even stronger undulator, Halbach (1986*a*) proposes the structure illustrated in Fig. 10.10, which he calls a *strong electromagnetic wiggler*. At a field level of 1.8 T in the iron, with Sm–Co alloy used in the places indicated in the figure, he calculates that a field of 0.6 T in the gap can be reached, slightly higher than that of a pure

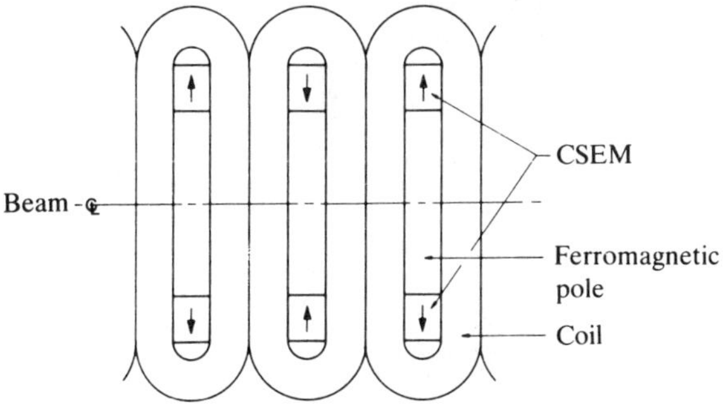

FIG. 10.9. Permanent-magnet-assisted electromagnetic undulator, in which permanent magnets have the function of avoiding saturation in the iron poles (from Halbach 1986*a*).

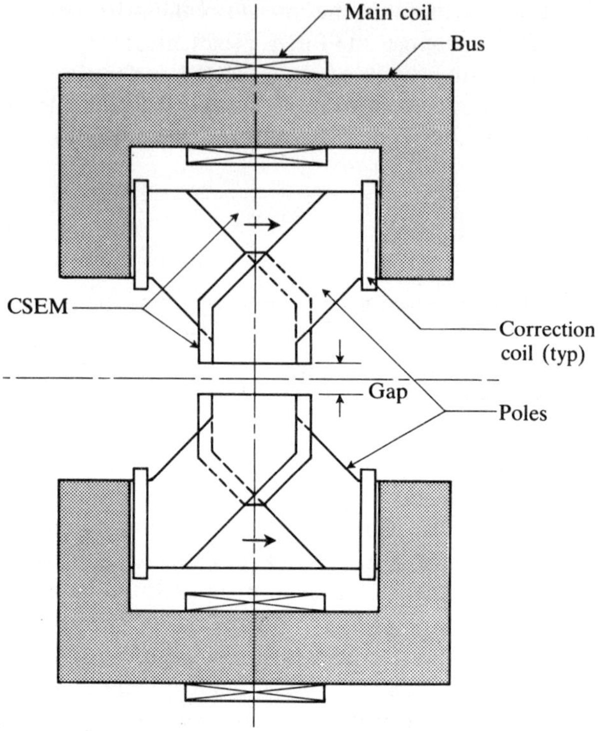

FIG. 10.10. Strong electromagnetic wiggler (from Halbach 1986*a*).

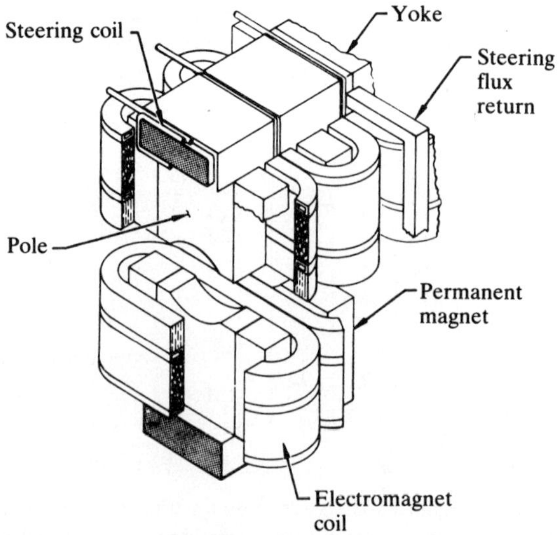

FIG. 10.11. Schematic section of the Paladin REC-assisted electromagnetic undulator (from Deis *et al.* 1987).

permanent-magnet hybrid undulator but with better tolerances and tunable over a reasonably wide range. A peculiar characteristic of this design is that, as opposed to other electromagnetic undulators, it can never be turned off completely, because doing so would require exceeding, with a reverse field, the saturation limit of the iron. The range in which it can be tuned without inducing saturation goes from about half maximum to maximum field.

The above design principles have been implemented in the Paladin undulator constructed at the Lawrence Livermore National Laboratory. This is a 25.6 m long 640-pole undulator, capable of generating a peak field of 0.32 T. A schematic drawing of a small section of the Paladin undulator is shown in Fig. 10.11. As may be seen, it follows the basic permanent-magnet-assisted em design of Fig. 10.9. Notice also that curved pole tips have been adopted to provide symmetric focusing of betatron oscillations in the two transverse directions. Measured field errors in this undulator were at 0.14 per cent r.m.s. (Deis *et al.* 1987).

The same working group has also built a prototype of the strong electromagnetic wiggler, the schematic design of which is shown in Fig. 10.12.

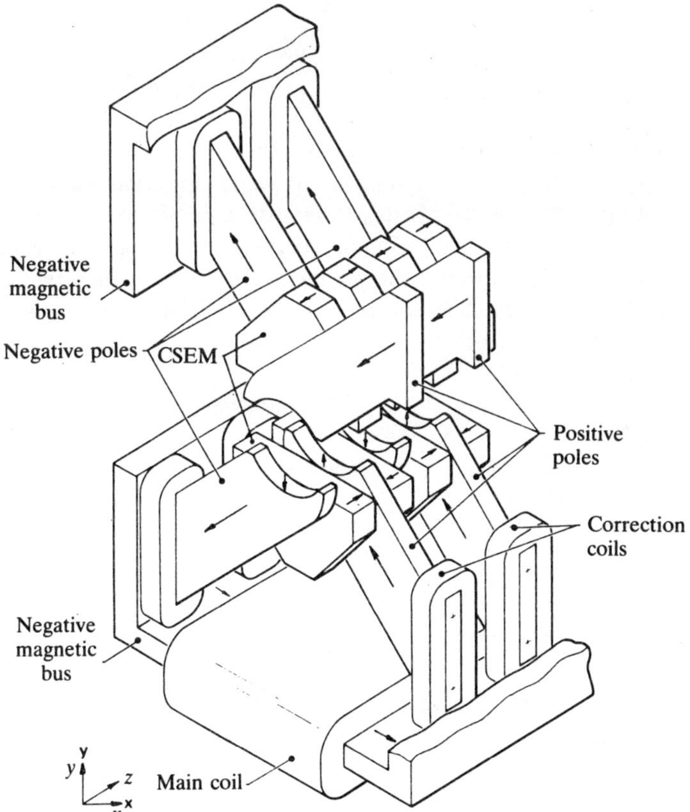

FIG. 10.12. Schematic of the strong electromagnetic wiggler prototype (from Deis *et al.* 1987).

10.6 Desirable excitation patterns for tapered undulators

One possible way to construct a tapered undulator is to make a constant-period electromagnetic undulator with independently fed coils which can be excited at different magnetic-field levels. The problem of designing such an electromagnetic undulator so that beam walk-off may be kept small under changing excitation conditions has been dealt with by Halbach (1986*b*).

We are now treating a configuration similar to that of Fig. 10.9, but with successive coils separately energized. We want to determine an excitation pattern for successive coils which guarantees a perfectly periodic orbit. Halbach's idea is to divide up each pole's coil into independently fed sections, connecting together several sections to each power supply in such a way that they compensate each other's beam displacement independent of the other power supplies.

The transverse co-ordinate of the orbit, x, is given by eqn (10.2) which, if the variation of γ is negligible, becomes eqn (10.3). The condition for periodicity of the orbit is

$$\int_0^{nl_u} (z - z')\boldsymbol{b}(z')\mathrm{d}z' = 0. \tag{10.14}$$

The field $\boldsymbol{b}$ is the sum of the fields due to the coils. The simplest magnet pattern satisfying eqn (10.14) is composed of a single pole surrounded by two half-poles of opposite polarity. If the undulator is composed of several such arrangements, eqn (10.14) will still be satisfied by the combination field, owing to linear superposition of the magnetic fields, even when the coil sets are independently fed at different current levels. One then arrives at the excitation pattern shown below:

Power supply number	Pole number 1	2	3	4	5	6	7	...
1	1	−2	1					...
2			1	−2	1			...
3					1	−2	1	...
...	....	....	....	....	....	....	....	...

where numbers in the double-entry table indicate the relative strength (number of turns), v_{mn}, of coils located on a given pole n and connected to a given power supply m.

If γ is not constant, as in a high-yield tapered FEL amplifier or in a laser accelerator, but γ^{-1} can be represented by a power series in z as $\gamma^{-1} =$

$g_0 + g_1 z + g_2 z^2 + \ldots$, eqn (10.2) may be rewritten, after integration by parts, as

$$x = \int_0^z \left[g_0(z - z') + g_1 \frac{(z - z')^2}{2} + g_2 \frac{(z - z')^3}{3} + \ldots \right] \hat{z} \wedge b(z')\mathrm{d}z', \qquad (10.15)$$

and in this case sideways drift of the orbit can, according to Halbach, be avoided in Nth order by subjecting the v_{mn} to equations

$$\sum_{n=0}^{N+2} v_{mn} n^j = 0, \qquad j = 0, 1, 2, \ldots, N + 1, \qquad (10.16)$$

the solution of which is given by a binomial-coefficient excitation pattern:

$$v_{mn} = (-1)^n \binom{N + 2}{n}. \qquad (10.17)$$

An excitation pattern for $N = 2$ is shown below:

Power supply number	Pole number											
	1	2	3	4	5	6	7	8	9	10	11	...
1	1	−4	6	−4	1							...
2			1	−4	6	−4	1					...
3					1	−4	6	−4	1			...
4							1	−4	6	−4	1	...
...	....											...

This choice makes the orbit periodic, because the influence of the coefficients g_n which represent the non-uniformity of γ has been eliminated by equating moments to zero.

10.7 Helical undulators

We start with a structure which can be implemented in a very simple manner: a pulsed helical undulator consisting of an ironless copper coil energized by a short current pulse of, say 1000 A has been used in the ENEA–Frascati experiment (see Chapter 11).

The desired spiralling field pattern can be obtained by means of a double helix like that of Fig. 1.9, which may be visualized by assuming that an open helix is wound around a tube and then a second helix is wound in the spaces between the turns of the first helix. When currents of opposite polarity are passed through the two coils the axial magnetic field is cancelled and the spiralling transverse field pattern remains.

The helical undulator used in the 1977 Stanford FEL was also of similar construction, except that, being included in the structure of the superconducting linac used to generate the electron beam, the undulator coil itself was made of superconducting material and therefore could be powered continuously rather than in short pulses.

According to Walker (1985), the on-axis magnetic field amplitude due to two identical coaxial solenoids separated by half the pitch l_u and carrying current I in opposite directions, so as to cancel the axial field, is given by

$$B_u = 2\mu_0 \frac{I}{l_u} [\xi \mathrm{K}_0(\xi) + \mathrm{K}_1(\xi)], \tag{10.18}$$

where $\xi = \pi g/l_u$, g is the internal diameter, and K_0 and K_1 are modified Bessel functions. Using an approximation of the Bessel functions that is good to 10 per cent for $\xi > 1$, one can write the dimensionless vector potential approximately as

$$A_u \simeq \frac{I}{1.8\ \mathrm{kA}} \mathrm{e}^{-0.9\xi}. \tag{10.19}$$

10.8 Permanent magnets for helical fields

Twister structures for circularly polarized undulator fields may also be constructed from permanent magnet blocks (Halbach 1981). Short, segmented dipoles may be made from magnetic material and arranged along an axis, rotating each with respect to the previous one so that the desired period along this axis is obtained. Let there be N such dipoles. The field on the axis will then be given by:

$$B_u = B_r \frac{\sin(\pi/N)}{\pi/N} \frac{\sin(2\pi/M)}{2\pi/M} [T(k_u r_1) - T(k_u r_2)], \tag{10.20}$$

where

$$T(x) = \mathrm{K}_0(x) + \frac{x}{2} \mathrm{K}_1(x)$$

and M is the number of geometrically identical (nearly trapezoidal) blocks used for the construction of the dipole slices. These blocks fit between two concentric circles with radii r_1 and r_2, as shown on Fig. 10.13 for the case of a dipole made of eight blocks. If M trapezoidal blocks per slice are used to

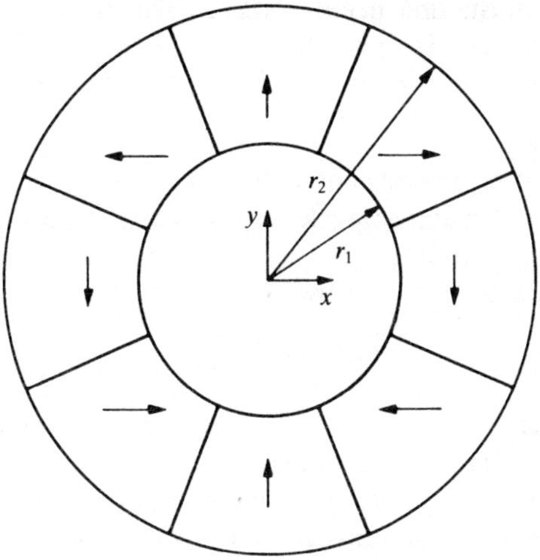

FIG. 10.13. Geometry of REC helical magnet (from Walker 1985).

construct a dipole magnet, their easy axes of magnetization should make consecutive angles $4\pi/M$ with each other. The larger M is, the better will be the approximation to a dipole field.

By the same technique, higher multipole magnets may also be constructed, arranging the trapezoidal blocks so that their easy axis of magnetization makes consecutive angles $2(n+1)\pi/M$ with each other in the case of a $2n$-pole magnet. For instance, quadrupoles for alternating-gradient focusing of the beam can be built according to this principle with an angle of rotation of the blocks of $3\pi/8$ ($M = 16$).

10.9 The two-stage concept

We have only discussed magnetostatic undulators but, as was explained in Chapter 3, an undulator can also be constituted by a travelling electromagnetic wave. Despite the fact that travelling electromagnetic waves cannot provide as high a dimensionless vector potential as a magnetostatic undulator can, they do provide as small an undulator period as wanted, and therefore offer an opportunity for building a small-wavelength free-electron laser using not-very-high energy electron beams.

Exploiting this idea, Elias (1979) proposed to build a two-stage FEL, in which submillimetre radiation gathered inside the optical cavity of a standard magnetostatic FEL oscillator is used as an undulator for a second FEL, producing, from either the same or a second electron beam, visible radiation.

The principle is illustrated in Fig. 10.14. The beam interacts with the undulator to produce radiation at the approximate wavelength

$$\lambda_1 = l_u/2\gamma^2 \tag{10.21}$$

in the wider optical resonator shown. With good mirrors the power density built up at this intermediate wavelength between them may be made high enough to make the standing wave act like an undulator for coherent wave production by the same beam at the much shorter wavelength

$$\lambda_2 = \lambda_1/4\gamma^2 = l_u/8\gamma^4 \tag{10.22}$$

in the thinner optical resonator. In the arrangement shown in the figure, holes are provided in the larger mirrors of such a size that the shorter-wavelength radiation can reach through to the smaller mirrors but only negligible loss of the longer-wavelength radiation is caused. Another possible arrangement in which the longer-wavelength radiation flows in a three-mirror ring resonator and interacts with the electron beam in two of its sides has also been considered (Kimel *et al.* 1986) and is shown in Fig. 10.15.

The main disadvantage of the two-stage FEL is that the relatively low dimensionless vector potential of the intermediate wave makes a large number of periods necessary in order to achieve a high enough gain of the travelling-wave undulator for the second-stage FEL to oscillate. While a large number of periods *can* easily be realized, simply by providing a long enough interaction region (for instance, at an intermediate wavelength of 700 μm an interaction length of 1.4 m corresponds to 2000 periods of the travelling wave, which are equivalent to 4000 periods of a magnetostatic undulator), it carries along a tight limitation on electron-beam energy spread, according to eqn (9.23), and on spectral purity (monochromaticity) of the intermediate wave. The practical workability of the two-stage concept has not been, until today, verified. An experiment aiming at producing 1 μm radiation from an electrostatic accelerator by this technique was planned at

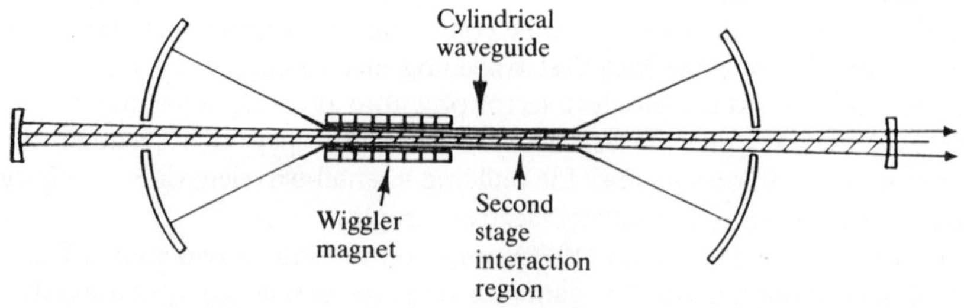

FIG. 10.14. Conceptual design for the collinear resonator cavities of a two-stage FEL (from Segall *et al.* 1984).

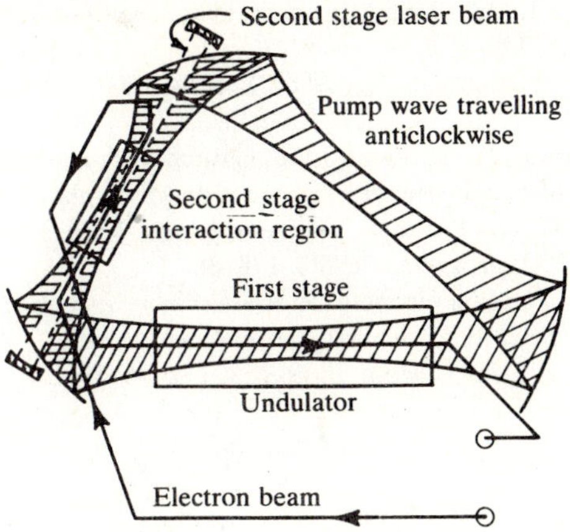

FIG. 10.15. Ring-laser design of a two-stage FEL (from Kimel *et al.* 1986).

the University of California at Santa Barbara (Kimel, Elias and Ramian 1986), but has never yet been undertaken.

Notes

1. The reader will remember that a Brownian random walk is obtained from the integral of a stochastic process having vanishing correlation between the values taken at different times (white noise). Since, as will be seen below, the lateral position of an electron in an undulator results from integrating twice a field error which, unless particular action is taken, may be assumed to occur independently in different magnets, one may call it 'doubly Brownian'. It follows that, as will appear from eqn (10.11) below, the r.m.s. lateral deviation increases as the power $\frac{3}{2}$ of distance along the undulator (or number of periods, which is the same) rather than as the power $\frac{1}{2}$ characteristic of Brownian motion.

References

Deis, G. A., Burns, M. J., Christensen, T. C., Coffield, F. E., Kulke, B., Prosnitz, D., Scharlemann, E. T. and Halbach, K. (1987). Electromagnetic wiggler technology development at the Lawrence Livermore National Laboratory. *Proc. Tenth International Conference on Magnet Technology*, Boston, MA, U.S.A., 23–6, September, 1987.

Elias, L. R. (1979). High-power, CW, efficient, tunable (UV through IR) free-electron laser using low-energy electron beams. *Phys. Rev. Lett.* **42**, 977.

Halbach, K. (1981). Physical and optical properties of rare earth cobalt magnets. *Nucl. Instr. Meth. Phys. Res.* **187**, 109–17.

Halbach, K. (1986*a*). Some concepts to improve the performance of DC electromagnetic wigglers. *FEL*85, 115–19.

Halbach, K. (1986*b*). Desirable excitation patterns for tapered wigglers. *FEL*85, 95–9.

Hu, R. C. J. and Elias, L. R. (1986). Gain depression due to undulator field errors in UCSB FEL. *FEL*85, 120–4.

Kimel, I., Elias, L. R. and Ramian, G. (1986). The UCSB two-stage FEL experiment. *FEL*85, 320–7.

Robinson, K. E., Quimby, D. C., Slater, J. M., Churchill, T. L. and Valla, A. S. (1987). Field certification of a high-strength tapered hybrid undulator. *FEL*86, 62–71.

Segall, S. B., Takeda, H., von Laven, S., Diament, P. and Ward, J. F. (1984). The KMS Fusion two-stage FEL program. *FEL*83, 178–88.

Walker, R. P. (1985). Periodic magnets for free-electron laser. *FEL*84, 366–80.

Winick , H. and Spencer, J. E. (1980). Wiggler magnets at SRRL—Present experience and future plans. *Nucl. Instr. Meth. Phys. Res.* **172**, 45–53.

11

EXPERIMENTS AND APPLICATIONS

While the undulator used for its spontaneous emission is a mature device, the undulator used as a free-electron laser is, today, still at an experimental (although advanced) stage. We conclude with a survey of some of the first free-electron laser experiments performed, selected for the purpose of exemplification amongst the many which we do not have enough space to describe, and of their proposed applications.

11.1 The 1977 Stanford FEL

We have already mentioned in Chapter 1 the Stanford FEL experiment of 1977, the most significant parameters of which are collected in Table 11.1. In this experiment a superconducting helical undulator was mounted on a superconducting rf linac. Owing to the high Q-factor ($\sim 10^8$) of its superconducting rf cavities, the Stanford SuperConducting Accelerator (SCA) has a much higher macropulse duration, of the order of many milliseconds, than conventional rf linacs. This is a direct consequence of the long decay time of the rf field stored in the accelerating cavities, which is associated with such a high Q.

In a second series of measurements taken in 1981, accurate data were accumulated on the electron and laser pulse shape (Eckstein *et al.* 1982). Owing to improvements in the electron pulse shape obtained by modifying the prebunching system of the linac, an average power of 5 W and a peak power of 130 kW could be obtained in this second run.

11.2 The Los Alamos FEL

The free-electron laser experiment performed at Los Alamos in 1981–4 provides another example of an FEL driven by an rf linac. The Los Alamos FEL was the fourth successful FEL experiment, according to Warren *et al.* (1985), and produced infrared light at around 10 μm wavelength with a peak micropulse power of 10 MW (over 40 ps micropulses) and a mean macropulse power of 6 kW (over 100 μs macropulses).

Table 11.1 Parameters of the Stanford FEL

Undulator parameters
Type of undulator: helical, superconducting magnet
Period, $l_u = 3.3$ cm
Length, $L = 5.28$ m
Number of periods, $N = 160$
Magnetic field intensity, $b_u = 0.23$ T
Dimensionless vector potential, $A_u = 1$

Accelerator parameters
Type of accelerator: superconducting rf linac
Electron energy, $\mathscr{E} = 43.5$ MeV
Energy spread, 0.05%
Emittance, 0.06 mm mrad
Micropulse duration, 3 ps
Micropulse repetition rate, 11.8 MHz
Micropulse peak current, 2.6 A
Macropulse duration, 1.5–20 ms (18 000–250 000 micropulses)
Macropulse repetition rate, 10–20 Hz
Macropulse mean current, 130 μA

Optical resonator parameters
Mirror distance, $D = 12.7$ m
Mirror radii, $R_{1,2} = 7.5$ m
Rayleigh range, $z_R = 2.7$ m
Ouput coupling, 1.5%
Round-trip losses (including output), 2.8%

Output characteristics
Wavelength, $\lambda = 3.4$ μm
Peak power during micropulse, 7 kW
Mean power during macropulse, 0.36 W

Given the relatively high energy spread and emittance of the linac, the undulator was designed with a smaller number of poles and a greater output wavelength than the Stanford undulator. Despite the smaller number of poles, the high peak current provided by the Los Alamos linac allowed a small-signal gain of as much as 40 per cent. The general layout of the experiment may be seen in Fig. 11.1. Its main operating parameters are reported in Table 11.2.

A thorough series of measurements was performed on the output laser beam from the Los Alamos experiment. It showed a very good spatial coherence indeed, the beam being only about 5 per cent wider than a diffraction-limited one, but it also displayed a frequency spectrum several orders of magnitude wider than a conventional laser's (about 1 per cent width), accompanied by a complicated frequency and time structure. This

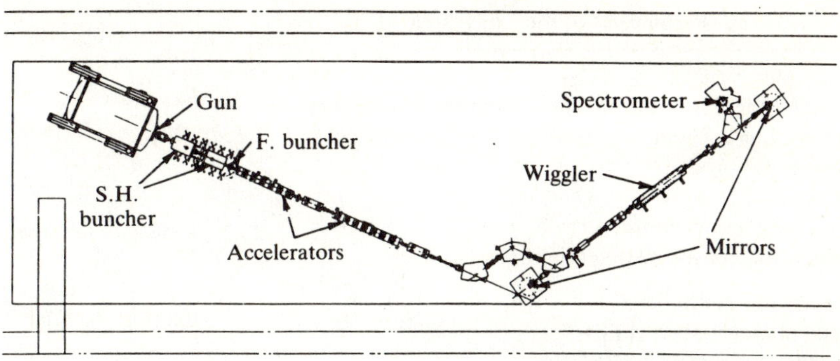

FIG. 11.1. Layout of the Los Alamos FEL set-up (from Stein *et al.* 1985; © 1985 IEEE).

effect was imputed to various instabilities related to the sideband instability[1] discussed in Section 7.8. No similar structure had appeared in the amplification experiment. (In a previous amplification experiment, a high-power (1 GW) CO_2 laser beam had been fed into a tapered undulator, obtaining extraction of up to 4 per cent of the electron beam energy.) Hence it was concluded that feedback from the cavity mirrors plays a decisive role in the onset of the instability.

The role of the sideband and other similar instabilities has been explored by the Los Alamos staff through numerical simulations performed both before and after the experiments (Goldstein *et al.* 1986; Warren *et al.* 1986; and references therein), a good agreement being found between a one-dimensional time-discretized model of the evolution of the optical pulse and the experiment. Warren *et al.* (1986) give the following description and physical interpretation of the experimental findings.

In the experiment, three different regimes appear to characterize operation of the laser at different power levels. At low optical power, obtained either by low beam current or by desynchronizing the cavity length so that a single light pulse is not amplified by all the electron micropulses, a single narrow spectral line is observed, with features in agreement with the single-frequency theoretical calculations. Above a threshold of about 25 MW intracavity power, a single sideband appears at a wavelength longer than the fundamental, with amplitude and width changing with the power level. Above a second threshold of about 100 MW the sideband divides up into several components of varying amplitude and separation, giving rise to a complicated, spiked spectrum.

In a numerical simulation which follows the optical pulse as it is successively amplified by 2000 electron micropulses, corresponding regimes are encountered as pulse energy grows. For the first part of the process, growth is smooth and the pulse shows no modulation. At the power level of the second

Table 11.2 Parameters of the Los Alamos FEL

Undulator parameters
Type of undulator: linear, Sm–Co permanent magnet
Period, $l_u = 2.73$ cm
Length, $L = 1$ m
Number of periods, $N = 37$
Peak magnetic field intensity, $b_u = 0.3$ T
Dimensionless vector potential, $A_u = 0.75$
Accelerator parameters
Type of accelerator: rf linac
Electron energy, $\mathscr{E} = 21$ MeV
Energy spread, 0.5%
Emittance, 3π mm mrad
Micropulse duration, 40 ps
Micropulse repetition rate, 21.66 MHz
Micropulse peak current, 50 A
Macropulse duration, 100 μs ($\sim$ 2000 micropulses)
Macropulse repetition rate, 1 Hz
Macropulse mean current, 25 mA
Optical resonator parameters
Mirror distance, $D = 6.92$ m
Mirror curvature radii, $R_1 = 3.80$ m, $R_2 = 3.35$ m
Rayleigh range, $z_R = 0.63$ m
Output coupling, 7%
Round-trip losses (including output), 7.5%
Output characteristics
Wavelength, $\lambda = 10$ μm
Micropulse peak power, 10 MW
Macropulse mean power, 6 kW
Energy extraction, 1%

regime the optical pulse begins to show a weak irregular amplitude modulation. Further down, the pulse appears strongly modulated and actually spiked. Although the time structure of the pulse could not be deduced from the spectral measurements taken during the experiments, because of a lack of phase information, the spiked simulated pulse would have a spectrum like that measured in the third regime. The authors conclude that the spiking model is a good candidate for explaining the observed behaviour. Noticing that spike separation tends to be close to the slippage distance, s, (defined by eqn (9.44) they propose, as a possible qualitative mechanism, that the spikes are due to a non-linear self-narrowing of the optical pulse, which takes place when the transit time of the electrons in the undulator approaches half a synchrotron period. The slippage distance would then be the appropriate separation between pulses in order that they interact with different electrons and do not influence each other.

The Los Alamos FEL has since been operated with either a uniform or a tapered undulator on a wide wavelength range (9–35 μm with a single undulator). After an upgrade of the accelerator structure, a power output of 40 MW peak has been achieved with an energy extraction of 2 per cent from a tapered undulator. Present experimenters seek a further improvement of energy efficiency by an energy recovery technique: the spent electron beam at the exit from the undulator is sent through an rf decelerating structure, i.e. a set of rf cavities identical to those used for acceleration but entered by the electron bunches with the opposite phase, which returns power to the accelerating structure. Apart from the obvious increase in overall efficiency, which may not be the main aim in an otherwise very costly machine, energy recovery has additional advantages. One is an increase in electron beam current, important in those systems such as the electrostatic accelerator and the rf linac where the current-limiting factor is power supply available for acceleration. Another is a decrease of electron energy at the beam dump, with a consequent decrease in the amount and harmfulness of the ionizing radiations produced. At the same time, the automatic reduction in the electron beam current which occurs when energy recovery is interrupted diminishes the danger of damaging the vacuum-pipe walls should a misaligned beam hit them.

11.3 The ENEA-Frascati FEL

In the experiment run at the ENEA-Frascati Centre, a 20 MeV microtron and a pulsed electromagnetic undulator were used to generate infrared light at around 10 μm. The undulator was initially built with circular polarization, simply by winding a five-turn coil around a Plexiglass vacuum pipe in the double-helix pattern of Fig. 1.9. Connected to a capacitor bank and a resistor so as to form a critically damped resonant circuit with a decay time of 59 μs, the coil was pulsed at a voltage of up to 10 kV, achieving a maximum magnetic field of 0.22 T, corresponding to a dimensionless vector potential $A_u = 0.7$. The critically damped resonant circuit allowed a 10 μs time window during which the magnetic field was constant to within 0.15 per cent, a tolerance small enough not to affect the undulator resonance condition negatively. With the coil immersed in oil for insulation and cooling, a pulse repetition rate of 0.2 Hz was possible (Bizzarri *et al.* 1985).

In the first set of experiments run with the above undulator the beam current provided by the microtron was insufficient to let the FEL oscillate. The microtron was then upgraded to attain the beam parameters shown in Table 11.3 and, at the same time, the helical undulator was replaced by a linear one, formed by a similar winding supported by a square Plexiglass rod

Table 11.3 Parameters of the ENEA FEL

Undulator parameters
Type of undulator: linear, pulsed electromagnet
Period, $l_u = 2.4$ cm
Length, $L = 1.25$ m
Number of periods, $N = 50$
Dimensionless vector potential, $A_u = 1.4$
Maximum repetition rate, 0.2 Hz
Accelerator parameters
Type of accelerator: microtron
Electron energy, $\mathscr{E} = 20$ MeV
Energy spread, 0.12%
Emittance:
6π mm mrad (parallel to the undulator)
2π mm mrad (perpendicular to the undulator)
Micropulse duration, 20 ps
Micropulse peak current, 2.4 A
Macropulse duration, 5 μs (150 micropulses)
Macropulse mean current, 120 mA
Optical resonator parameters
Mirror distance, $D = 4.937$ m
Mirror radii, $R_1 = 2.5$ m, $R_2 = 3$ m
Rayleigh range, $z_R = 0.8$ m
Output coupling, 0.2%
Round-trip losses (including output), 1.4%
Output characteristics
Wavelength, $\lambda = 10.6$ μm

passed through by a 0.1 mm stainless steel vacuum pipe. In this way a higher dimensionless vector potential ($A_u = 1.4$) was achieved. The number of passes in the resonator during the period of the electron macropulse (5 μs) allowed by the driving klystron tube was insufficient to let the laser really operate, but the beginning of an oscillation was nevertheless observed in 1985, as a slight increase of the optical pulse intensity when the cavity was aligned (Bizzarri *et al.* 1986).

11.4 The Orsay FEL

The free-electron laser installed on the ACO storage ring at Orsay was the first storage-ring FEL to operate, in 1983, at the visible wavelength of 650 nm (Billardon *et al.* 1983). In order to exploit the high quality of the e-beam produced by the storage ring, and to compensate somewhat for its low

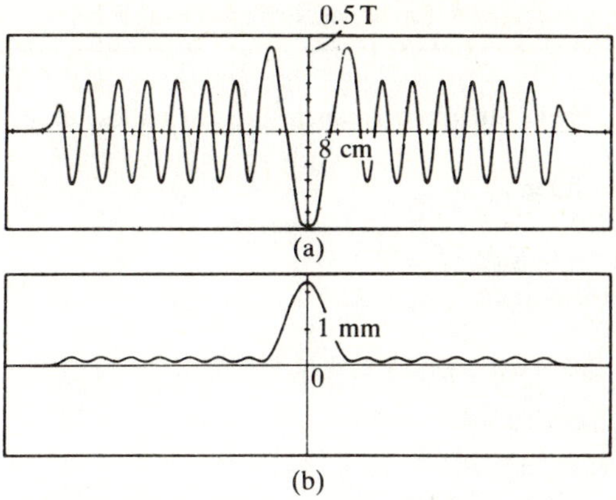

FIG. 11.2. Magnetic field (a) and electron trajectory (b) in the Orsay optical klystron (from Billardon *et al.* 1985; © 1985 IEEE).

current, this FEL's undulator was built in an optical-klystron configuration, the magnetic-field profile and calculated electron trajectory of which are reproduced in Fig. 11.2. Nevertheless, small-signal gain was as low as 10^{-3}, and oscillation of the optical cavity could only be obtained with the use of special multilayer mirrors having a reflectivity better than 99.95 per cent (see Table 11.4).

Initially, the FEL would only oscillate in a narrow spectral region around the reflectivity peak of the mirrors (645–655 nm). After a redesign of the storage ring which led to a different operating point (set of tune numbers) and a more favourable value of the β-functions in the undulator section, tunability over the range 575–644 nm with one set of mirrors and 463–486 nm with a second set of mirrors was achieved, the limits being set primarily by the reflectivity curve of the available mirror coatings (Billardon *et al.* 1987).

Power output from the ACO FEL was low, a few mW on average. One reason for this was the low extraction efficiency of the high-reflectivity mirrors, which let through only about 1/30 of the power they did not reflect. Another reason was that the total synchrotron radiation power emitted in ACO, which determines the maximum average laser power because of the Renieri limit (9.15), was only a few watts at the energy of 200 MeV, much lower than that which the storage ring was originally built for. The Orsay group demonstrated the possibility of obtaining a higher output power by Q-switching the optical resonator, i.e. disaligning the optical cavity and then bringing it periodically back into alignment, say, by deflecting the electron beam trajectory slightly or by desynchronizing the electron and light pulses through a minute change (3×10^{-6} is sufficient) of the rf power frequency, so that energy is extracted from the laser in strong intermittent pulses rather than continuously. They achieved a peak and a mean power during the Q-switched pulse that was 300 times higher than in continuous operation

Table 11.4 Parameters of the Orsay FEL

Undulator parameters
Type of undulator: linear Sm–Co permanent-magnet optical klystron
Period, $l_u = 7.8$ cm
Overall length, $L = 1.3$ m
Number of periods, $N_{1,2} = 7$ each of two sections
Peak magnetic field intensity, $b_u = 0.31$ T
Dimensionless vector potential, $A_u = 2.3$
Length of dispersive section, 24 cm
Maximum field in dispersive section, 0.58 T

Accelerator parameters
Type of accelerator: storage ring
Electron energy, $\mathscr{E} = 160$–245 MeV
Energy spread, 0.09–0.13%
Pulse duration, 0.5–1 ns
Pulse repetition rate, 27.2361 MHz
Pulse peak current, 1.3 A
Beam lifetime, 1–4 h
Mean current, 16–100 mA

Optical resonator parameters
Mirror distance, $D = 5.5$ m
Mirror radii, $R_{1,2} = 3$ m
Rayleigh range, $z_R = 1$ m
Output coupling, 3×10^{-5}
Round-trip losses, 7×10^{-4}–17×10^{-4}

Output characteristics
Wavelength, $\lambda = 463$–486 nm, 575–644 nm, 645–655 nm (with different mirror sets)
Peak power, 0.15 W
Mean power, 3 mW

(Billardon *et al.* 1986), and expect an even higher output power, of the order of 1 MW in Q-switching mode, from their new FEL facility on the Super-ACO storage ring now being built.

The most difficult problem faced during the Orsay experiment was probably posed by mirror technology. A multilayer coating technique using about 20 alternating layers of two substances with different indices of refraction is capable of providing the required reflectivity, but these mirrors are easily damaged by stray radiation, in particular by ultraviolet radiation which is copiously present in the higher harmonics of the spontaneous-emission spectrum of the undulator. A screening of mirror materials had to be conducted (Billardon *et al.* 1985; Elleaume *et al.* 1985*a*,*b*), after which the experimenters settled for a configuration with alternating SiO_2 and TiO_2 layers, using SiO_2 for the outermost layer. According to Velghe and Elleaume

(1987), neither degassing from various components of the vacuum system nor γ-ray irradiation during the injection phase of the storage ring had any identifiable effect on the mirrors, whereas rapid degradation was produced by ultraviolet radiation. Power loss, initially of the order of 10^{-4}, grew rapidly to 4–5 $\times$ 10^{-4} when the mirrors were exposed to undulator radiation from the ACO FEL and then, more slowly, in a time of the order of hours, up to about 10^{-3}. At the same time, the bandwidth of the mirrors was also degraded, losses off the centre frequency of the reflectivity curve increasing more than at the centre. The first, rapid, degradation of mirror reflectivity seems to be due, according to Velghe and Elleaume, to bulk damage from photons of around 30 eV energy. The second, slower phase, during which the bandwidth narrowing takes place, appears to be caused by surface pollution from organic substances, which are decomposed by u.v. rays and react with the surface. The bandwidth narrowing is explained by the electric field's having a node, i.e. a zero, in the surface layer at the centre design wavelength of the mirror but not at offset wavelengths, so that reflectivity at the centre wavelength is less sensitive to surface effects.

11.5 The UCSB FEL

Elias *et al.* (1985) have constructed a free-electron laser using an electrostatic (van de Graaff) accelerator to produce the electron beam. In such a machine, electrons sprayed onto an insulating belt to a condenser charge it to a high voltage and provide the undulator beam from a gun, followed by an accelerating column of diaphragms with gradually decreasing potential. The arrangement is shown in Fig. 11.3.

As explained in Section 9.6, in a pulsed electrostatic accelerator the maximum current and pulse duration are interrelated through the capacity of the high-voltage electrode. In the UCSB experiment energy recovery was used to increase the peak beam current up to about 1 A in 50 μs long pulses. The output wavelength was in the submillimetre wave range, 400 μm for an electron energy of 3 MeV. The other main parameters are shown in Table 11.5.

The most innovative part of the UCSB design was the beam recovery system necessary to obtain a sufficient beam current and duration from the electrostatic accelerator. One of the most delicate components, that had to be purposely designed in order to achieve a recovery rate of 90–95 per cent, was probably the beam transport system used to drive the electron beam from the electrostatic column to the undulator and back into the electrostatic column for deceleration. It was designed using computer simulation codes and

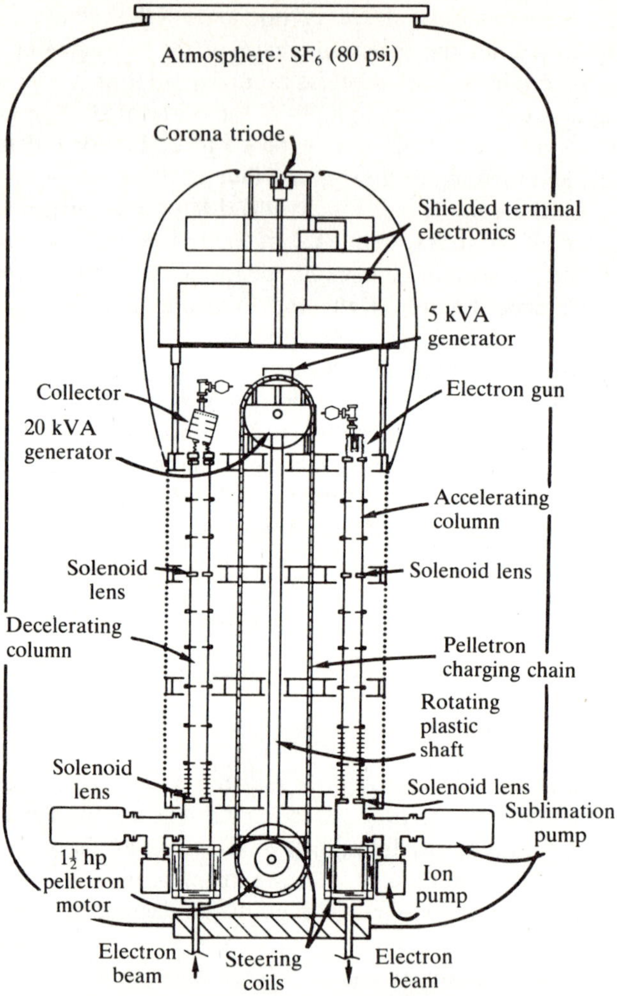

FIG. 11.3. UCSB FEL electron beam system (from Elias and Ramian 1982; © 1982 Addison-Wesley Publishing Co.).

employed 10 bending magnets, 17 quadrupole magnets, and 24 steering coils (Elias and Ramian 1984). Measurements of the frequency and time structure of the FEL output showed a pulsed structure, and a line spectrum which could be explained by the optical cavity's switching from one to another resonance mode following the linear decrease of electron energy caused by the discharge of the high-voltage electrode.

The UCSB FEL has also been operated at the wavelength of 100 μm by raising the accelerator voltage to 6 MV. It has subsequently been dedicated as a user facility where experiments are performed with the emitted radiation.

Table 11.5 Parameters of the UCSB FEL

Undulator parameters
Type of undulator: linear, Sm–Co permanent-magnet in Halbach configuration
Period, $l_u = 3.6$ cm
Length, $L = 5.76$ m
Number of periods, $N = 160$
Peak magnetic field intensity, $b_u = 0.05$ T
Dimensionless vector potential, $A_u = 0.17$

Accelerator parameters
Type of accelerator: electrostatic with e-beam recovery
Electron energy, $\mathscr{E} = 3$ MeV
Recovery efficiency, 90%
Maximum pulse duration, 50 μs (for an energy drift of 1%)
Peak current, 1.25 A
D.c. charging current, 0.2 mA
Emittance, 7.5π mm mrad

Optical resonator parameters
Type of resonator: slab waveguide terminated by cylindrical mirrors
Waveguide width, 1.9 cm
Mirror distance, $D = 7.15$ m
Mirror radii, $R_{1,2} = 5$ m
Output coupling, 11%
Other losses (not including output), < 1.6%

Output characteristics
Wavelength, $\lambda = 400$ μm
Peak power, 10 kW

11.6 Applications of undulators as synchrotron radiation sources

Undulators have been used as emitters of (spontaneous) synchrotron radiation ever since their introduction. During the past decade the interest in synchrotron radiation has been such that several large storage-ring facilities, either previously used for experiments in high-energy physics or built specifically for this purpose, have been dedicated as synchrotron radiation sources. About twenty dedicated facilities exist today or are being built in various laboratories around the world, e.g. Trieste in Italy, Daresbury in the U.K., LURE and ESRF in France, BESSY in West Germany, SSRL, SRC, NSLS, SURF, and ALS in the U.S.A., Kurchatov in the U.S.S.R., and KEK, ISSP, UVSOR, and TERAS in Japan. Other synchrotron-radiation user laboratories are installed as part-time experiments in high-energy-physics laboratories as, e.g., CHESS (U.S.A.), Frascati (Italy), HASYLAB (West Germany) and Novosibirsk (U.S.S.R.). Generally, these facilities are electron storage rings with a design energy between 1 and 2 GeV. A certain number of undulators or wigglers with different characteristics dislocated in the straight sections of the storage ring cover the u.v. and X-ray spectral regions from the near ultraviolet up to photon energies of several keV.

Synchrotron radiation sources are used for all the possible applications of u.v. and X-ray radiation, since they are the brightest, and at some wavelengths the only, sources available in these regions of the spectrum.

The high brightness of synchrotron radiation has allowed investigations of the structure of matter with techniques which were previously unexplored because they involve very small samples. Among these we may quote X-ray powder diffraction, which allows the refinement of complex crystal structures of molecules such as zeolites which form only very small crystals, and the study of crystalline phase transitions under high pressure, where only very small (ten μm) crystals must be used in order to avoid the effect of pressure gradients.

In organic chemistry, the tunability of synchrotron radiation has allowed the study of protein crystal structures by multiwavelength diffraction experiments. The high brightness has made possible time-resolved studies of morphological changes in artificial and biological polymers during chemical reactions.

X-ray microprobe techniques, using a narrow X-ray pencil to show the distribution of trace elements, have been applied to biological and geological specimens; for instance, to determine the distribution of toxic elements in bone, or to study the composition of meteorites.

A very large range of chemical phenomena may be studied by photo-dissociation or photo-ionization experiments, for which synchrotron radiation provides the necessary high-brightness tunable u.v. source. These include photo-electron spectroscopy, photo-ionization mass spectrometry and dispersed photo-ion fluorescence, and have allowed, for instance, high-resolution studies of the intermediate stages of chemical reactions.

The technique of analysis known as EXAFS (Extended X-ray Absorption Fine Structure), which allows a study of the way in which atoms of a particular element are inserted in a compound material, requires a continuous scan of X-ray photon energy from immediately below to hundreds of eV above the electron extraction threshold of the given element, and was very complicated and time-consuming to perform with conventional X-ray sources. Undulators allow an EXAFS scan to be executed in few minutes, or even seconds.

Precision measurements of the electron structure of semiconductors and superconductors, in order to assess the technological potential of new materials, are another application.

Also related to electronic technology is X-ray lithography, which is being investigated at synchrotron radiation facilities in order to develop industrial production processes for ultra-small integrated circuits.

In medicine, the possibility of tuning the X-ray wavelength exactly on the absorption line of the contrast medium (e.g. iodine) allows neat scintigraphs and angiographs to be obtained with a much reduced radiation dose for the patient.

In addition, coherent optical imaging techniques are being moved down to X-ray wavelengths where they can provide new diagnostic means for both medical and technological purposes. X-ray microscopes are also under development.

11.7 Applications of undulators as free-electron lasers

Free-electron lasers provide a high-power tunable coherent light source, the working wavelength of which is not tied to any particular property of matter and may be changed at will. As such, they are the only choice for all those applications of coherent light requiring a wavelength for which no conventional laser is available. This includes the far-infrared interval going from 11 μm up to, say, 1 mm, where conventional sources are sparsely scattered and generally not very powerful, and the ultraviolet wavelengths smaller than approximately 250 nm, for which no coherent source is presently available (Sessler and Vaughan 1987).

Several techniques already perfected with conventional lasers could profitably be extended into new wavelength ranges. In the infrared, the picosecond pulses that characterize most free-electron lasers would permit the study of excitation and relaxation processes of phonons, plasmons, and similar quasi-particles in superconductors and semiconductors, as well as in liquids. Several surface processes are characterized by infrared resonances, including surface-catalysed chemical reactions. It has been suggested that catalysis rates might be greatly improved by the selective vibrational excitation of adsorbed molecules (Shaw and Patel 1980).

All the studies in molecular spectroscopy which have become possible after the introduction of the 10.6 μm CO_2 laser can be repeated with the FEL over a much wider range of wavelengths, and therefore of molecules.

At the other end of the spectrum, many possible uses are awaiting a coherent source below 250 nm. These include all the present uses of synchrotron radiation, which would benefit from an even higher-brightness source. Spectroscopy of electronic levels and microlithography are just two examples. Laser-controlled chemistry is another emerging technology which could profit from the availability of very intense tunable sources in the ultraviolet (Ehrlich 1985).

Finally, all the uses that require a very intense laser source, one such that it would be difficult to achieve a sufficient power density in conventional laser media, seem well-suited to free-electron lasers, since these are intrinsically high-power devices. The first such uses that come to mind are laser fusion and laser acceleration of particles. However, the construction of a free-electron laser specifically designed for such uses has not, so far, been undertaken.

11.8 Related devices

As may be remembered from Section 2.8, the class of devices that are called 'free-electron lasers' is extremely ill-defined. We have intentionally limited the scope of this book to radiation amplifiers that use an ultrarelativistic electron beam streaming through an undulator, the only ones, until today, to have shown the capability of working down to light wavelengths and beyond. Let us conclude by mentioning some devices which, despite their operation being limited to millimetre and submillimetre waves, are sometimes called free-electron lasers (a temptation which we ought to resist, for otherwise the class of free-electron lasers would have to include the majority of the existing microwave amplifiers which, with the exclusion of the maser, all draw power from free electrons). They have in common with undulators (and with travelling-wave tubes and rf particle accelerators) the working mechanism set forth in Section 7.2, which more generally characterizes any device based on synchronous motion between an electron beam and a wave-like force pattern.

The cyclotron-resonance maser, or gyrotron, of Hirshfield and Wachtel (1964), is the most similar of these devices to the undulator, as it employs spiralling electrons to create a quasi-potential out of the beat between the electron motion and the magnetic field of the wave to be amplified, just as a circular-polarization undulator does. The difference is that in a cyclotron maser the spiralling motion is caused by a strong uniform axial magnetic field, rather than by an oscillating transverse field. Very high power in millimetre waves has been obtained from cyclotron masers (cf., e.g., Granatstein 1978). However, they cannot be scaled down to much lower wavelengths, because the frequency of the emitted radiation is just the Larmor rotation frequency of the electrons in the axial magnetic field and, therefore, the minimum working wavelength of a cyclotron maser is determined solely by the maximum axial-field intensity which is technologically feasible (e.g. a field intensity of 4 T yields an output wavelength of 2.78 mm).

The Cherenkov maser is more similar to a travelling-wave tube than to an undulator, in that it uses a slow-wave structure to match the speed of the electromagnetic wave to that of the electron beam. Unlike a travelling-wave tube, however, it simply uses as a slow-wave structure a material medium having an index of refraction higher than a vacuum, so that the radiation emitted when the electron beam travels at the speed of light in the medium, or faster, may be classified as Cherenkov radiation.

The actual physical structure of the device is somewhat more complicated. In order to avoid having the beam go through the dielectric medium, where it would be absorbed in a short distance, destroying both the medium and itself, the dielectric is lined up against the walls of a cylindrical or plane-parallel metallic waveguide, forming a sort of hollow optical fibre. The electron beam is passed through the central bore and interacts with the evanescent field of the wave which travels inside the lining. In this way an axial component of the wave electric field is also created which makes interaction between the parallel electron and radiation beams possible. (No Cherenkov radiation is emitted exactly on axis in a homogeneous medium, even when the phase-matching condition is satisfied, because the electric field and electron velocity vectors happen to be perpendicular).

According to the analysis by Walsh *et al.* (1986), the dispersion curve of a TM mode in a dielectric-lined waveguide begins at low frequency in a region in which the presence of the dielectric is not felt, the phase speed is slightly higher than the speed of light in vacuum, and a large fraction of the power is carried in the central free region, and extends at high frequency into a region in which the phase speed tends to the speed of light in the dielectric and most of the power is carried inside the lining. An optimal interaction point may be found halfway along, depending on the thickness and refraction index of the lining and on the geometric dimensions of the waveguide, a point at which the phase speed of the wave matches the electron speed and the electric field in the central region is still strong enough to allow interaction.

Operation of a Cherenkov maser has been obtained at millimetre and, more recently, submillimetre wavelengths (Garate *et al.* 1987). An experiment with an output wavelength in the range 100–400 μm is in progress (Ciocci *et al.* 1987). Several slab waveguide configurations are under consideration which use a dielectric coating of a few μm of polyethylene.

The main obstacle to scaling down the Cherenkov maser to lower and lower wavelengths is the increasing proximity of the dielectric layer to the electron beam, and its increasing thinness, which make it easily destroyed either by the grazing electron beam or by the radiation power itself. This obstacle is overcome in a device which combines the Cherenkov and undulator effects, proposed by Fauchet *et al.* (1984*a,b*) under the name of gas-loaded FEL. In this device an undulator provides coupling between a transverse electromagnetic wave and the electron beam while a homogeneous gaseous dielectric slows down the wave, thus allowing synchronism at a lower energy for a given wavelength, or at a lower wavelength for a given energy, than would be possible in a simple undulator working in vacuum. As a side effect of the lower energy of operation, a higher gain is also obtained. The most serious problem presented by the operation of a gas-loaded FEL is represented by various beam instabilities which are known to arise in the propagation of an electron beam through a high-pressure gas and which can destroy the beam. However, it has been experimentally shown that no such instabilities develop rapidly enough to become noticeable in the few-picosecond micropulse duration typical of a prebunched rf linac, even at atmospheric or higher pressure. An experiment is presently in progress on the Mark III linac at Stanford (Reid *et al.* 1987).

Similar to both Cherenkov and travelling-wave tubes is the device in which Smith and Purcell (1953) observed emission of light from an electron beam travelling in the proximity of a diffraction grating. In fact, this emission may be ascribed to the synchronism between the e-beam and the slow travelling waves supported by the periodic grating structure (just as by the periodic waveguide structure of a travelling-wave tube). Utilization of the Smith–Purcell effect for a laser amplifier would lead to the same problems stemming from the proximity of the e-beam to the guiding structure which afflict the Cherenkov maser and, as yet, this has not been attempted.

Notes

1. The mechanism cannot be quite the same as that described in Section 7.8 because, as remarked by Warren *et al.* (1986), in the, rather short, Los Alamos undulator at most one synchrotron period fits in the whole undulator length, and there is no possibility of the electrons oscillating in quasi-potential wells. Nevertheless, they advocate a mechanism that is somehow related.

References

Amir, A., Elias, L. R., Gregoire, D. J., Kotthaus, J., Ramian, G. J. and Stern, A. (1986). Spectral characteristics of the UCSB FEL 400 μm experiment. *FEL*85, 35–8.

Billardon, M., Elleaume, P., Lapierre, Y., Ortega, J. M., Bazin, C., Bergher, M., Marilleau, J., and Petroff, Y. (1986). The Orsay storage ring free-electron laser: new results. *FEL*85, 26–34.

Billardon, M., Elleaume, P., Ortega, J. M., Bazin, C., Bergher, M., Velghe, M., Deacon, D. A. G., and Petroff, Y. (1985). Free-electron laser experiment at Orsay: a review. *IEEE J. Quantum Electron.* **21**, 805–23.

Billardon, M., Elleaume, P., Ortega, J. M., Bazin, C., Bergher, M., Couprie, M. E., Lapierre, Y., Petroff, Y., Prazeres, R., and Velghe, M. (1987). Status of the Orsay FEL experiment. *FEL*86, 72–6.

Billardon, M., Elleaume, P., Ortega, J. M., Bazin, C., Bergher, M., Velghe, M., Petroff, Y., Deacon, D. A. G., Robinson, K. E., and Madey, J. M. J. (1983). First operation of a storage-ring free-electron laser. *Phys. Rev. Lett.* **51**, 1652–5.

Bizzarri, U., Ciocci, F., Dattoli, G., de Angelis, A., Fiorentino, E., Gallerano, G. P., Letardi, T., Marino, A., Messina, G., Renieri, A., Sabia, E., and Vignati, A. (1985). Status of the free-electron laser experiment at the ENEA Frascati center. *FEL*84, 213–19.

Bizzarri, U., Ciocci, F., Dattoli, G., de Angelis, A., Gallerano, G. P., Giabbai, I., Giordano, G., Letardi, T., Messina, G., Mola, A., Picardi, L., Renieri, A., Sabia, E., and Vignati, A. (1986). Above threshold operation of the ENEA free-electron laser. *FEL*85, 254–7.

Ciocci, F., Dattoli, G., de Angelis, A., Dipace, A., Doria, A., Fiorentino, E., Gallerano, G. P., Letardi, T., Marino, A., Renieri, A., Sabia, E., Schettini, G., Torre, A., Walsh, J. E., Garate, E., Layman, R., and Shaughnessy, C. (1987). Status of the ENEA-Dartmouth far infrared Cherenkov laser experiment. *FEL*86, 128–32.

Eckstein, J. N., Madey, J. M. J., Robinson, K., Smith, T. I., Benson, S., Deacon, D., and Taber, R. (1982). Additional experimental results from the Stanford 3 micron FEL. *FEL*81, 49–75.

Ehrlich, D. J. (1985). Instruments and methods for fabrication by microcontrolled chemistry. *Nucl. Instr. Meth. Phys. Res.* **A239**, 414.

Elias, L. R. and Ramian, G. J. (1982). Design of the UCSB FEL electron beam system. *FEL*81, 577-602.

Elias, L. R. and Ramian, G. J. (1984). Status report of the UCSB FEL experimental program. *FEL*83, 137-45.

Elias, L. R., Hu, J., and Ramian, G. (1985). The UCSB electrostatic accelerator free-electron laser: first operation. *FEL*84, 203-06.

Elleaume, P., Velghe, M., Billardon, M., and Ortega, J. M. (1985a). Diagnostic techniques and UV-induced degradation of the mirrors used in the Orsay storage-ring free-electron laser. *Appl. Optics* **24**, 2762.

Elleaume, P., Velghe, M., Billardon, M., and Ortega, J. M. (1985b). Diagnostic techniques and VUV-induced degradation of the mirrors used in the Orsay storage-ring free-electron laser. *FEL*84, 263-7.

Fauchet, A. M., Feinstein, J., Gover, A., and Pantell, R. H. (1984*a*). Visible and ultraviolet generation using a gas-loaded free-electron laser. *IEEE J. Quantum Electron.* **20**, 1332-41.

Fauchet, A. M., Feinstein, J., Gover, A., and Pantell, R. H. (1984*b*). The gas-loaded free-electron laser. *FEL*83, 423-9.

Garate, E. P., Walsh, J., Shaughnessy, C., Johnson, B., and Moustaizis, S. (1987). Cherenkov free-electron laser operation from 375 to 1000 μm. *FEL*86, 125-7.

Goldstein, J. C., Newman, B. E., Warren, R. W., and Sheffield, R. L. (1986). Comparison of the results of theoretical calculations with experimental measurements from the Los Alamos free-electron laser oscillator experiment. *FEL*85, 4-11.

Granatstein, V. L. (1978). Gyrotrons and dopplertrons—electromagnetic wave generators based on coherent radiation from magneto-resonant, relativistic electron beams. *FEL*77, 273-300.

Hirshfield, J. L. and Wachtel, J. M. (1964). Electron cyclotron maser. *Phys. Rev. Lett.* **12**, 533.

Reid, M. B., Fischer, A. S., Pantell, R. H., Feinstein, J., Deloney, T. L., and Ho, A. H. (1987). Design of a gas-loaded free-electron laser experiment. *FEL*86, 133-5.

Sessler, A. M. and Vaughan, D. (1987). Free-electron lasers. *Am. Scientist* **75**, 34-43.

Shaw, E. D. and Patel, C. K. N. (1980). Applications in solid-state systems of a tunable far-infrared free-electron laser. *Proc. International Conference on Lasers '80*, 5-8 and 19-22 May 1980, Shangai and Beijing, China.

Smith, S. J. and Purcell, E. M. (1953). Visible light from localized surface charges moving across a grating. *Phys. Rev.* **92**, 1069.

Stein, W. E., Warren, R. W., Winstow, J. G., Fraser, J. S., and Young, L. M. (1985). The accelerator for the Los Alamos free-electron laser-IV. *IEEE J. Quantum Electron.* **21**, 889-94.

Velghe, M. F. and Elleaume, P. (1987). FEL optics coating test device. *FEL*86, 83-7.

Walsh, J. E., Johnson, B., Shaughnessy, C., Ciocci, F., Dattoli, G., de Angelis, A., Dipace, A., Fiorentino, E., Gallerano, G. P., Letardi, T., Renieri, A., Sabia, E., Giabbai, I. and Giordano, G. (1986). A 100 μm Cherenkov laser experiment. *FEL*85, 308-11.

Warren, R. W., Goldstein, J. C., and Newnam, B. E. (1986). Spiking mode operation for a uniform-period wiggler. *FEL*85, 19-25.

APPENDIX A

CONCEPTS OF HAMILTONIAN RELATIVISTIC MECHANICS

The basic structure of classical (i.e. not quantum) relativistic mechanics of a point particle moving in an electromagnetic field can be consistently derived from a variational principle stating that the line integral

$$\int_{O_i}^{O_f} (-mc\,\mathrm{d}s - e\varphi\,\mathrm{d}t + e\boldsymbol{a}\cdot\mathrm{d}\boldsymbol{r}) \tag{A.1}$$

is an extremum with respect to variations of the integration path that leave the initial and final points $O_i(\boldsymbol{r}_i, t_i)$ and $O_f(\boldsymbol{r}_f, t_f)$, representing a given position $\boldsymbol{r}_i$ at a given time t_i and a given position $\boldsymbol{r}_f$ at a time t_f, respectively, unchanged. In eqn (A.1) m is the rest mass of the particle, $\mathrm{d}s = [(c\,\mathrm{d}t)^2 - \mathrm{d}\boldsymbol{r}^2]^{1/2}$ (the positive root is understood), e is the charge, and φ and $\boldsymbol{a}$ are the scalar and vector potentials of the electromagnetic field, considered to be known functions of time and position of the particle. The integral (A.1) is named the action integral and its value is denoted by S. Written explicitly as an integral in time, it becomes

$$S = \int_{t_i}^{t_f} \mathscr{L}\,\mathrm{d}t, \tag{A.2}$$

where

$$\mathscr{L} = -mc[c^2 - (\mathrm{d}\boldsymbol{r}/\mathrm{d}t)^2]^{1/2} - e(\varphi - \boldsymbol{a}\cdot\mathrm{d}\boldsymbol{r}/\mathrm{d}t). \tag{A.3}$$

The quantity $\mathscr{L}$ is named the Lagrangian of the mechanical system.

By enforcing the extremum property of the action integral, the Lagrange equations of motion are obtained which, for a generic Lagrangian $\mathscr{L}(\boldsymbol{r}, \dot{\boldsymbol{r}}, t)$, where $\dot{\boldsymbol{r}}$ is a shorthand for $\mathrm{d}\boldsymbol{r}/\mathrm{d}t$, read

$$\frac{\mathrm{d}}{\mathrm{d}t}\left(\frac{\partial\mathscr{L}}{\partial\dot{\boldsymbol{r}}}\right) = \frac{\partial\mathscr{L}}{\partial\boldsymbol{r}}. \tag{A.4}$$

In particular, for a particle in an electromagnetic field, the Lagrangian of which appears in eqn (A.3), the Lagrange equations are

$$\frac{\mathrm{d}}{\mathrm{d}t}\left[\frac{m\dot{\boldsymbol{r}}}{(1 - \dot{\boldsymbol{r}}^2/c^2)^{1/2}} + e\boldsymbol{a}\right] = -e\frac{\partial\varphi}{\partial\boldsymbol{r}} + e\frac{\partial}{\partial\boldsymbol{r}}\boldsymbol{a}\cdot\dot{\boldsymbol{r}}. \tag{A.5}$$

The total derivative $\mathrm{d}\boldsymbol{a}/\mathrm{d}t$, calculated along the particle path, evaluates to $\partial \boldsymbol{a}/\partial t + \dot{\boldsymbol{r}} \cdot (\partial/\partial \boldsymbol{r})\boldsymbol{a}$. Transporting this term to the r.h.s. of eqn (A.5) gives the usual form of the Lorentz equation of motion,

$$\frac{\mathrm{d}}{\mathrm{d}t}\left[\frac{m\dot{\boldsymbol{r}}}{(1-\dot{\boldsymbol{r}}^2/c^2)^{1/2}}\right] = e\left(-\frac{\partial \varphi}{\partial \boldsymbol{r}} - \frac{\partial \boldsymbol{a}}{\partial t} + \frac{\partial}{\partial \boldsymbol{r}}\boldsymbol{a}\cdot\dot{\boldsymbol{r}} - \dot{\boldsymbol{r}}\cdot\frac{\partial}{\partial \boldsymbol{r}}\boldsymbol{a}\right) = e(\boldsymbol{e} + \dot{\boldsymbol{r}} \wedge \boldsymbol{b}). \quad \text{(A.6)}$$

In fact, the electric field is $\boldsymbol{e} = -\partial\varphi/\partial\boldsymbol{r} - \partial\boldsymbol{a}/\partial t$, and the magnetic field is[1] $\boldsymbol{b} = \text{curl}\,\boldsymbol{a}$. Therefore, the variational principle of least action using the action integral (A.1) is equivalent to the Lorentz equation of motion (A.6).

According to the general principles of Hamiltonian mechanics, the second-order Lagrange equations can be transformed into the first-order Hamilton equations by means of auxiliary variables. These variables form the canonical momentum vector $\boldsymbol{P}$, defined by

$$\boldsymbol{P} = \partial\mathscr{L}/\partial\dot{\boldsymbol{r}} \quad \text{(A.7)}$$

which, for the present relativistic particle with the Lagrangian (A.3), means

$$\boldsymbol{P} = m\dot{\boldsymbol{r}}/(1-\dot{\boldsymbol{r}}^2/c^2)^{1/2} + e\boldsymbol{a}. \quad \text{(A.8)}$$

The presence of the term $e\boldsymbol{a}$ in eqn (A.8) makes canonical momentum a not purely kinematical quantity when electromagnetic fields are present. For this reason a distinction is made between canonical momentum, defined by eqns (A.7) and (A.8), and kinetic momentum, defined as $\boldsymbol{p} = m\dot{\boldsymbol{r}}/(1-\dot{\boldsymbol{r}}^2/c^2)^{1/2}$. (Kinetic momentum is very seldom used in this book, and by writing just 'momentum' we always mean canonical momentum).

First-order differential equations giving the time derivatives of $\boldsymbol{r}$ and $\boldsymbol{P}$ as explicit functions of $\boldsymbol{r}$, $\boldsymbol{P}$, and t can be obtained in the following way. $\dot{\boldsymbol{r}}$ can be expressed as such a function by inverting eqn (A.7). For the Lagrangian (A.3) the result of the inversion of eqn (A.8) is

$$m\frac{\mathrm{d}\boldsymbol{r}}{\mathrm{d}t} = \frac{(\boldsymbol{P} - e\boldsymbol{a})}{[1 + (\boldsymbol{P} - e\boldsymbol{a})^2/m^2c^2]^{1/2}}. \quad \text{(A.9)}$$

On the other hand, the time derivative of $\boldsymbol{P}$ is directly given by eqn (A.4), the r.h.s. of which was evaluated in eqn (A.5), as a function of $\boldsymbol{r}$, $\dot{\boldsymbol{r}}$, and t. Using eqn (A.9) to eliminate $\dot{\boldsymbol{r}}$ from the r.h.s. in favour of $\boldsymbol{P}$ gives

$$\frac{\mathrm{d}\boldsymbol{P}}{\mathrm{d}t} = \frac{\partial\mathscr{L}}{\partial\boldsymbol{r}} = -e\frac{\partial\varphi}{\partial\boldsymbol{r}} + e\frac{\partial}{\partial\boldsymbol{r}}\boldsymbol{a}\cdot\frac{\boldsymbol{P} - e\boldsymbol{a}}{m[1 + (\boldsymbol{P} - e\boldsymbol{a})^2/m^2c^2]^{1/2}}. \quad \text{(A.10)}$$

We can also obtain the first-order equations (A.9) and (A.10) in a different way from a function $\mathscr{H}$ of co-ordinates and canonical momenta, the Hamiltonian of the mechanical system, which is obtained from $\mathscr{L}(\boldsymbol{r}, \dot{\boldsymbol{r}}, t)$ by means of the transformation

$$\mathscr{H}(\boldsymbol{r}, \boldsymbol{P}, t) = \boldsymbol{P}\cdot\dot{\boldsymbol{r}} - \mathscr{L}, \quad \text{(A.11)}$$

where it is implied that both $\dot{\boldsymbol{r}}$ and $\mathscr{L}$ must be expressed as functions of $\boldsymbol{r}$, $\boldsymbol{P}$, and t by inverting eqn (A.7). The change from $\mathscr{L}$ to $\mathscr{H}$ according to eqn (A.11) is mathematically known as a Legendre transformation, and is the same kind of transformation that is used in thermodynamics in order to pass from one thermodynamic potential (internal energy, enthalpy, free energy, etc.) to another. Applying, in particular, eqn (A.11) to the Lagrangian (A.3) of a relativistic particle in an electromagnetic field, and therefore using eqn (A.9) to express $\dot{\boldsymbol{r}}$ as a function of $\boldsymbol{r}$, $\boldsymbol{P}$, and t, one obtains the Hamiltonian

$$\mathscr{H} = e\varphi + [m^2c^4 + c^2(\boldsymbol{P} - e\boldsymbol{a})^2]^{1/2}. \tag{A.12}$$

It can be easily verified that once $\mathscr{H}$ has been calculated eqns (A.9) and (A.10) may be rewritten as

$$\frac{\mathrm{d}\boldsymbol{r}}{\mathrm{d}t} = \frac{\partial \mathscr{H}}{\partial \boldsymbol{P}}, \qquad \frac{\mathrm{d}\boldsymbol{P}}{\mathrm{d}t} = -\frac{\partial \mathscr{H}}{\partial \boldsymbol{r}}. \tag{A.13}$$

The equivalence of eqns (A.13) to eqns (A.4) is indeed general, and applies to any mechanical system. Since eqns (A.13) are sufficient to determine the motion of the particle, i.e. the function $\boldsymbol{r}(t)$, we may say that the Hamiltonian contains as much information about the motion as the Lagrangian. In fact, if only the Hamiltonian is known, as a function of $\boldsymbol{r}$, $\boldsymbol{P}$, and t, the Lagrangian can be calculated from it by rewriting eqn (A.11) as

$$\mathscr{L} = \boldsymbol{P}\cdot\dot{\boldsymbol{r}} - \mathscr{H} \tag{A.14}$$

and expressing $\boldsymbol{P}$ and $\mathscr{H}$ as functions of $\boldsymbol{r}$, $\dot{\boldsymbol{r}}$ and t by inverting the equation $\dot{\boldsymbol{r}} = \partial\mathscr{H}/\partial\boldsymbol{P}$.

The Hamilton equations of motion put in clear evidence the property of canonical momentum, that whenever the Hamiltonian (or, equivalently, the Lagrangian) does not depend on one of the co-ordinates the corresponding component of canonical momentum is a constant of motion. In fact, if the partial derivative with respect to one co-ordinate of the Hamiltonian or of the Lagrangian is zero, the total time derivative of the corresponding momentum results to zero as well. It can be also shown that, whenever the Hamiltonian (or the Lagrangian) does not depend on time, the value taken by the Hamiltonian itself at the position of the system is a constant of motion. This value, whether it be constant or not, is by definition the energy $\mathscr{E}$ of the mechanical system. $\mathscr{E}$ is the total energy, the sum of the kinetic energy and electrostatic potential energy; an analogy may be observed with canonical momentum, which is the sum of the kinetic momentum and a term proportional to the electromagnetic vector potential.

A useful property of the Hamilton equations is that, just like the Lagrange equations, they express the extremality of a certain integral, namely

$$\int_{O_i}^{O_f} (\boldsymbol{P}\cdot\mathrm{d}\boldsymbol{r} - \mathscr{H}\,\mathrm{d}t) = \int_{t_i}^{t_f}\left(\boldsymbol{P}\cdot\frac{\mathrm{d}\boldsymbol{r}}{\mathrm{d}t} - \mathscr{H}\right)\mathrm{d}t \tag{A.15}$$

with respect to independent variations of $\boldsymbol{r}(t)$ and $\boldsymbol{P}(t)$ which leave the initial and final positions (but not necessarily momenta) unchanged. The integral of eqn (A.15) and that of eqn (A.2) can be recognized to be equal when they are evaluated along the true particle path. In fact, eqn (A.14) then holds and

$$\boldsymbol{P} \cdot \mathrm{d}\boldsymbol{r} - \mathscr{H}\,\mathrm{d}t = \mathscr{L}\,\mathrm{d}t. \tag{A.16}$$

A very powerful tool made available by the formulation of a mechanical problem in terms of Hamilton equations is constituted by the canonical transformations, a rather large class of variable transformations which leave the form of the Hamilton equations unchanged. It is in fact possible, in order to solve the problem, to look for a transformation such that the transformed Hamiltonian depends on the least number of co-ordinates, and therefore the greatest number of momenta become constant.

To show what a canonical transformation is, let us introduce two new vectors $\boldsymbol{r}'$ and $\boldsymbol{P}'$ such that the particle position and momentum can be expressed as $\boldsymbol{r} = \boldsymbol{r}(\boldsymbol{r}', \boldsymbol{P}', t)$, $\boldsymbol{P} = \boldsymbol{P}(\boldsymbol{r}', \boldsymbol{P}', t)$. If $\boldsymbol{r}'(t)$ and $\boldsymbol{P}'(t)$ are known, the functions $\boldsymbol{r}(t)$ and $\boldsymbol{P}(t)$ can be uniquely determined from them, and therefore we say that $\boldsymbol{r}'$ and $\boldsymbol{P}'$ represent the motion equally well as $\boldsymbol{r}$ and $\boldsymbol{P}$. In other words, we have performed a transformation of variables from $\boldsymbol{r}$, $\boldsymbol{P}$ to $\boldsymbol{r}'$, $\boldsymbol{P}'$. This transformation is termed canonical if the new and old variables fulfil suitable conditions, a form of which is that a function $F(\boldsymbol{r}, \boldsymbol{r}', t)$ be found such that

$$\boldsymbol{P} = \partial F(\boldsymbol{r}, \boldsymbol{r}', t)/\partial \boldsymbol{r}, \qquad \boldsymbol{P}' = -\partial F(\boldsymbol{r}, \boldsymbol{r}', t)/\partial \boldsymbol{r}'. \tag{A.17}$$

F is then named the generating function of the canonical transformation.

In order to find the evolution equations for the variables $\boldsymbol{r}'$, $\boldsymbol{P}'$, the variational principle connected to the integral (A.15) can be used. First, let us note that the addition to the integrand of a perfect differential does not change the extremality property of the integral. In fact, the integral of a perfect differential does not depend on the integration path and evaluates to a non-essential constant. We can therefore add to eqn (A.15) the integral of $-\,\mathrm{d}F$, and obtain an integral which still has an extremum in correspondence with the same motion. (Actually, since F is a function of $\boldsymbol{r}$, $\boldsymbol{r}'$, and t, this is true only if variations are not allowed which change either $\boldsymbol{r}$ or $\boldsymbol{r}'$ or t at the limits of integration. This is a more restrictive condition than was required in connection with eqn (A.15), as it implies that *both* co-ordinates and momenta are kept constant at the limits of integration, but the added restriction is irrelevant as far as the extremum is concerned.) Since $\mathrm{d}F = (\partial F/\partial \boldsymbol{r} \cdot \mathrm{d}\boldsymbol{r}/\mathrm{d}t + \partial F/\partial \boldsymbol{r}' \cdot \mathrm{d}\boldsymbol{r}'/\mathrm{d}t + \partial F/\partial t)\,\mathrm{d}t$, the resulting integral is

$$\int_{t_i}^{t_f} \left[\left(\boldsymbol{P} - \frac{\partial F}{\partial \boldsymbol{r}}\right) \cdot \frac{\mathrm{d}\boldsymbol{r}}{\mathrm{d}t} - \frac{\partial F}{\partial \boldsymbol{r}'} \cdot \frac{\mathrm{d}\boldsymbol{r}'}{\mathrm{d}t} - \mathscr{H} - \frac{\partial F}{\partial t}\right] \mathrm{d}t.$$

But, if eqns (A.17) hold, this may be transformed into an integral of the same form as eqn (A.15), namely

$$\int_{t_i}^{t_f}\left(\boldsymbol{P}'\cdot\frac{\mathrm{d}\boldsymbol{r}'}{\mathrm{d}t}-\mathscr{H}'\right)\mathrm{d}t$$

by simply defining the new Hamiltonian function $\mathscr{H}'$ as

$$\mathscr{H}'(\boldsymbol{r}',\boldsymbol{P}',t)=\mathscr{H}+\partial F/\partial t. \tag{A.18}$$

Therefore, the equations of motion determining $\boldsymbol{r}'(t)$ and $\boldsymbol{P}'(t)$ are the Hamilton equations obtained from the Hamiltonian function $\mathscr{H}'(\boldsymbol{r}',\boldsymbol{P}',t)$, i.e.

$$\mathrm{d}\boldsymbol{r}'/\mathrm{d}t=\partial\mathscr{H}'/\partial\boldsymbol{P}',\qquad \mathrm{d}\boldsymbol{P}'/\mathrm{d}t=-\partial\mathscr{H}'/\partial\boldsymbol{r}'. \tag{A.19}$$

Notice that $\mathscr{H}'$ is wanted as a function of $\boldsymbol{r}'$, $\boldsymbol{P}'$, and t, and as such is obtained by replacing every occurrence of $\boldsymbol{r}$ and $\boldsymbol{P}$ in the r.h.s. of eqn (A.18) by the functions $\boldsymbol{r}(\boldsymbol{r}',\boldsymbol{P}',t)$, $\boldsymbol{P}(\boldsymbol{r}',\boldsymbol{P}',t)$.

We shall not delve further into the theory of canonical transformations, which would lead us too far from the scope of this book, but we want to recall yet another formulation of the equations of motion, in addition to the variational principle, Lagrange, and Hamilton equations, which turns up in connection with canonical transformations: the Hamilton–Jacobi equation. Therefore, let us go back to the action integral (A.1)–(A.2). When this integral is calculated along the true particle path it assumes a well-defined value which can be thought of as a function of t_i, $\boldsymbol{r}_i$, t_f, and $\boldsymbol{r}_f$. The same value is given also by eqn (A.15). Let us fix t_i and $\boldsymbol{r}_i$ and study the action integral as a function of the final point $S(t_f,\boldsymbol{r}_f)$ (it then takes on the name of action 'function').

The derivatives of the action function can be calculated from eqn (A.15) in the following manner. Let the functions $\boldsymbol{r}(t)$ and $\boldsymbol{P}(t)$ representing the true motion receive an infinitesimal increment $\delta\boldsymbol{r}(t)$ and $\delta\boldsymbol{P}(t)$ because of an infinitesimal change δt_f, $\delta\boldsymbol{r}_f$ in final time and position; the resulting increment in eqn (A.15) is

$$\delta S=\int_{t_i}^{t_f}\left(\delta\boldsymbol{P}\cdot\frac{\mathrm{d}\boldsymbol{r}}{\mathrm{d}t}+\boldsymbol{P}\cdot\frac{\mathrm{d}\delta\boldsymbol{r}}{\mathrm{d}t}-\frac{\partial\mathscr{H}}{\partial\boldsymbol{P}}\cdot\delta\boldsymbol{P}-\frac{\partial\mathscr{H}}{\partial\boldsymbol{r}}\cdot\delta\boldsymbol{r}\right)\mathrm{d}t+\left(\boldsymbol{P}\cdot\frac{\mathrm{d}\boldsymbol{r}}{\mathrm{d}t}-\mathscr{H}\right)\delta t_f. \tag{A.20}$$

Integrating the term $\boldsymbol{P}\cdot(\mathrm{d}\delta\boldsymbol{r}/\mathrm{d}t)$ by parts transforms eqn (A.20) into

$$\delta S=\int_{t_i}^{t_f}\left[\delta\boldsymbol{P}\cdot\left(\frac{\mathrm{d}\boldsymbol{r}}{\mathrm{d}t}-\frac{\partial\mathscr{H}}{\partial\boldsymbol{P}}\right)-\delta\boldsymbol{r}\cdot\left(\frac{\mathrm{d}\boldsymbol{P}}{\mathrm{d}t}+\frac{\partial\mathscr{H}}{\partial\boldsymbol{r}}\right)\right]\mathrm{d}t$$

$$+\boldsymbol{P}\cdot\delta\boldsymbol{r}(t_f)+\left(\boldsymbol{P}\cdot\frac{\mathrm{d}\boldsymbol{r}}{\mathrm{d}t}-\mathscr{H}\right)\delta t_f, \tag{A.21}$$

where $\delta r(t_f)$ is not the same as $\delta \boldsymbol{r}_f$. The total change in final position $\delta \boldsymbol{r}_f$ equals $\boldsymbol{r}(t_f + \delta t_f) + \delta \boldsymbol{r}(t_f + \delta t_f) - \boldsymbol{r}(t_f)$. Expanding this expression with respect to the small variations and retaining only first-order quantities gives

$$\delta \boldsymbol{r}(t_f) = \delta \boldsymbol{r}_f - \frac{\mathrm{d}\boldsymbol{r}}{\mathrm{d}t}\delta t_f. \tag{A.22}$$

On inserting eqns (A.22) and (A.13) into eqn (A.21), the simple relation $\delta S = \boldsymbol{P}\cdot\delta \boldsymbol{r}_f - \mathscr{H}\,\delta t_f$ is obtained. Therefore, the action function has the property that

$$\frac{\partial S}{\partial \boldsymbol{r}} = \boldsymbol{P}, \qquad \frac{\partial S}{\partial t} = -\mathscr{H}. \tag{A.23}$$

By a similar argument, giving an infinitesimal displacement to the lower limit in the action integral, it can also be shown that

$$\frac{\partial S}{\partial \boldsymbol{r}_i} = -\boldsymbol{P}_i. \tag{A.24}$$

Since the Hamiltonian is a given function $\mathscr{H}(\boldsymbol{r}, \boldsymbol{P}, t)$, eqns (A.23) imply a relation between the derivatives of S with respect to time and co-ordinates, namely

$$\frac{\partial S}{\partial t} = -\mathscr{H}\left(\boldsymbol{r}, \frac{\partial S}{\partial \boldsymbol{r}}, t\right). \tag{A.25}$$

Equation (A.25) may be read as a partial differential equation for the action function S, and is known as the Hamilton–Jacobi equation. However, while we have just shown that the action function of a mechanical system, defined by eqn (A.2) as a function of the final point O_f, obeys eqn (A.25), it is not implied that any solution of eqn (A.25) can be also obtained from eqn (A.2). Nevertheless, it is useful to extend the definition of action function so as to include any solution of eqn (A.25). In fact, the motion of a mechanical system can be completely determined from the solution of the Hamilton–Jacobi equation, as is shown below by using the properties of canonical transformations.

Equation (A.25) is a partial differential equation, and as such admits an infinite-dimensional set of solutions which allows the specification of an arbitrary initial function. Let $S(\boldsymbol{r}, t, \boldsymbol{\alpha})$ denote a three-dimensional subset of solutions consisting of a family of functions parameterized by the vector $\boldsymbol{\alpha}$. We can interpret the function $S(\boldsymbol{r}, t, \boldsymbol{\alpha})$ as the generating function F of a canonical transformation leading from the co-ordinates $\boldsymbol{r}$ to the co-ordinates $\boldsymbol{\alpha}$; in fact the conditions (A.17) for the transformation to be canonical can be fulfilled: the first of them is automatically satisfied because of eqns (A.23), and the second can be satisfied by choosing the (otherwise arbitrary) new momenta as $\boldsymbol{P}_\alpha = -\partial S/\partial \boldsymbol{\alpha}$. Since the transformation from $\boldsymbol{r}$, $\boldsymbol{P}$ to $\boldsymbol{\alpha}$, $\boldsymbol{P}_\alpha$ is canonical, eqns (A.18) and (A.19) apply, where one has to read $\boldsymbol{\alpha}$ for $\boldsymbol{r}'$ and $\boldsymbol{P}_\alpha$ for $\boldsymbol{P}'$. On taking into account the last of eqns (A.23), one then finds that the new Hamiltonian $\mathscr{H}'$ is identically zero. As the solution of the Hamilton–Jacobi equation is the generating function of a canonical transformation which leads to a zero Hamiltonian, both $\boldsymbol{\alpha}$ and $\boldsymbol{P}_\alpha$ are constant in time, in accordance with eqns (A.19), and equal their initial values.

Summarizing, once any three-parameter family of solutions of the Hamilton–Jacobi equation has been determined in the form $S(\boldsymbol{r}, t, \boldsymbol{\alpha})$, the curve $\boldsymbol{r}(t)$ representing the motion is implicitly given by the vector equation

$$\partial S/\partial \boldsymbol{\alpha} = -\boldsymbol{P}_\alpha, \tag{A.26}$$

where $\boldsymbol{\alpha}$ and $\boldsymbol{P}_\alpha$ are constant vectors, the values of which must be chosen so as to satisfy the initial conditions. Therefore it can be said that finding a three-parameter family of solutions of the Hamilton–Jacobi equation amounts to a complete determination of the motion of the system.

The usefulness of the Hamilton–Jacobi equation which, being a partial differential equation, seems to be harder to solve than the ordinary Lagrange or Hamilton equations, lies in the fact that exact solutions can sometimes be found for it which are not at all evident in the other formulations. For this reason the Hamilton–Jacobi equation is to be preferred when exact solutions are being sought, but one of the other equivalent formulations of a mechanical problem is better for the purpose of finding approximate analytical or numerical solutions.

The properties of the action function can be also used to show that a transformation using the initial values of co-ordinates and momenta as new variables is canonical. According to the definition of the action function, as the action integral calculated along the true path of motion from a fixed initial position $\boldsymbol{r}_i$ at time t_i to a generic position $\boldsymbol{r}$ at time t, we may consider the action as a function of $\boldsymbol{r}$, t, and $\boldsymbol{r}_i$. The derivatives of this function are given by eqns (A.23) and (A.24). Therefore, if we take $\boldsymbol{r}_i$, $\boldsymbol{P}_i$ to be the new variables $\boldsymbol{r}'$, $\boldsymbol{P}'$, eqns (A.17) are satisfied, with the action function $S(\boldsymbol{r}, \boldsymbol{r}_i, t)$ as generating function F. As a consequence, the transformation leading from the actual co-ordinates and momenta to the initial ones is a canonical transformation, and eqns (A.19) can be applied. Since $\partial S/\partial t = -\mathscr{H}$, the Hamiltonian $\mathscr{H}'$ turns out to be identically zero, according to eqn (A.18), and eqns (A.19) then tell that the initial co-ordinates and momenta are constant, as, of course, they are.

It is fundamental, however, to observe that a transformation's property of being canonical does not depend on the system it is applied to. So, if the Hamiltonian of a mechanical system is represented as $\mathscr{H} = \mathscr{H}_0(\boldsymbol{r}, \boldsymbol{P}, t) + V(\boldsymbol{r}, \boldsymbol{P}, t)$, where $\mathscr{H}_0$ is a simpler Hamiltonian and V is a perturbation, and $\boldsymbol{r}_0(\boldsymbol{r}_i, \boldsymbol{P}_i, t)$, $\boldsymbol{P}_0(\boldsymbol{r}_i, \boldsymbol{P}_i, t)$ represent the solution of the equations of motion obtained from the Hamiltonian $\mathscr{H}_0$, the transformation expressed by $\boldsymbol{r} = \boldsymbol{r}_0(\boldsymbol{r}', \boldsymbol{P}', t)$, $\boldsymbol{P} = \boldsymbol{P}_0(\boldsymbol{r}', \boldsymbol{P}', t)$, obtained by substituting $\boldsymbol{r}'$ for $\boldsymbol{r}_i$ and $\boldsymbol{P}'$ for $\boldsymbol{P}_i$ in the unperturbed motion, is a canonical transformation generated by the action function S_0. When this transformation is applied to the complete Hamiltonian $\mathscr{H}_0 + V$ the transformed Hamiltonian is found, according to eqn (A.18), to be

$$\mathscr{H}' = \mathscr{H} + \partial S_0/\partial t = \mathscr{H} - \mathscr{H}_0 = V. \qquad \text{(A.27)}$$

$\boldsymbol{r}'$ and $\boldsymbol{P}'$ are therefore no longer constant in the presence of the perturbation, but they form a set of slowly varying variables which has many applications in perturbation theory. It may be noticed that this transformation is analogous to the transformation from the Schrödinger to the Heisenberg representation of a system's evolution equation in quantum-mechanical perturbation theory.

Notes

1. The equivalence $(\partial/\partial \boldsymbol{r})\boldsymbol{a} \cdot \dot{\boldsymbol{r}} - \dot{\boldsymbol{r}} \cdot (\partial/\partial \boldsymbol{r})\boldsymbol{a} = \dot{\boldsymbol{r}} \wedge \text{curl}\, \boldsymbol{a}$ can be best seen if it is remembered that the gradient operator $\partial/\partial \boldsymbol{r}$ behaves like a vector in algebraic manipulations if care is taken not to change the function it acts upon. For three vectors $\boldsymbol{v}_1$, $\boldsymbol{v}_2$, and $\boldsymbol{v}_3$, the equivalence $\boldsymbol{v}_2\boldsymbol{v}_3 \cdot \boldsymbol{v}_1 - \boldsymbol{v}_1 \cdot \boldsymbol{v}_2\boldsymbol{v}_3 = \boldsymbol{v}_1 \wedge (\boldsymbol{v}_2 \wedge \boldsymbol{v}_3)$ is well known. If $\boldsymbol{v}_1$ is now identified with $\dot{\boldsymbol{r}}$, $\boldsymbol{v}_3$ with $\boldsymbol{a}$, and $\boldsymbol{v}_2$ with $\partial/\partial \boldsymbol{r}$ (with the understanding that it must act on $\boldsymbol{a}$), the above result follows.

References

Goldstein, H. (1980). *Classical mechanics.* Addison-Wesley, Reading, MA.

Landau, L. D. and Lifschitz, E. M. (1975). *The classical theory of fields.* Pergamon Press, Oxford.

Landau, L. D. and Lifschitz, E. M. (1976). *Mechanics.* Pergamon Press, Oxford.

APPENDIX B

EFFECTIVE MASS

The effective mass approximation affords a simplified description of the motion of an electron in an electromagnetic field differing little from a plane electromagnetic wave (locally plane wave).

This approximation belongs to the general category of two-timing methods, in that the main assumption is the existence of two very different scales of variation, the 'fast' scale of the oscillations of the electron caused by the electromagnetic wave and the 'slow' scale of the drift caused by the slow variation of the wave parameters.

The starting point is the exact solution of the equations of motion of an electron in a pure electromagnetic plane wave. Consider a general electromagnetic wave characterized by the potential four-vector[1]

$$a_j = \mathrm{Re}(A_j e^{i\psi}) \tag{B.1}$$

where

$$\psi = k_j x^j \tag{B.2}$$

and the two conditions $k^j k_j = 0$ and $k^j A_j = 0$ hold. The motion of an electron in this field can be determined in a number of ways. For the present purposes a simple formulation will result from the use of the Hamilton–Jacobi equation (A.25), which in four-dimensional notation reads

$$\left(\frac{\partial S}{\partial x_j} + ea^j\right)\left(\frac{\partial S}{\partial x^j} + ea_j\right) = m_e^2 c^2. \tag{B.3}$$

Since a_j is expressed through the phase ψ, we may try to find a solution of eqn (B.3) in the form

$$S = P_j x^j + S'(\psi) \tag{B.4}$$

P_j being arbitrary constants and S' an unknown function of one only variable. Inserting eqn (B.4) into eqn (B.3) yields

$$(P^j + k^j \, dS'/d\psi + ea^j)(P_j + k_j \, dS'/d\psi + ea_j) = m_e^2 c^2 \tag{B.5}$$

whence, remembering that $k^j k_j = k^j a_j = 0$, one obtains

$$k^j P_j \, dS'/d\psi = m_e^2 c^2 - (P^j + ea^j)(P_j + ea_j). \tag{B.6}$$

Equation (B.6) may be integrated with no difficulty. Since the r.h.s. of eqn (B.6) is an oscillating function of ψ with a generally non-zero average, $S'(\psi)$ will be formed of an oscillating part and a linear part, coming from the integral of this average. We may, without loss of generality, impose the condition that S' be a bounded oscillating function, since the linear part can always be absorbed in the term $P_j x^j$. We shall then obtain a condition that must be fulfilled by the, otherwise arbitrary, constants P_j. Equating the average value of the r.h.s. of eqn (B.6) to zero gives the condition

$$P^j P_j = m_e^2 c^2 - e^2 \langle a^j a_j \rangle, \tag{B.7}$$

angular brackets denoting an average over the values of ψ. (The cross-product $P^j a_j$ gives no contribution to eqn (B.7) because $\langle P^j a_j \rangle = P^j \langle a_j \rangle = 0$).

We have thus shown that a solution of the form (B.4) of the Hamilton–Jacobi equation (B.3), with $S'(\psi)$ a bounded oscillating function, exists for arbitray values of the constants P_j subject to the restriction (B.7). It may be seen that the four-vector P_j then represents the average value of the electron four-momentum. In fact, the electron momentum may be calculated from the action function S as $\partial S/\partial x^j$. Since when S' is bounded the average of its derivative is zero (as explained in Chapter 5 with reference to eqns (5.43) and (5.44)), it follows from eqn (B.4) that $\langle \partial S/\partial x^j \rangle = P_j$. Comparing eqn (B.7) with the relation between four-momentum and rest mass for a free particle of mass M, $P^j P_j = M^2 c^2$, we may conclude that an electron moving in the field of a plane electromagnetic wave behaves as a particle of mass M, where

$$M^2 = m_e^2 - (e^2/c^2)\langle a^j a_j \rangle = m_e^2 + (e^2/c^2)\langle \boldsymbol{a}^2 \rangle - (e^2/c^4)\langle \varphi^2 \rangle. \tag{B.8}$$

Let us now consider the motion of an electron under the action of a four-potential, again given by eqn (B.1), but with A_j and $k_j = \partial\psi/\partial x^j$ slowly varying (with respect to ψ) rather than constant. Building upon the exact solution (B.4) which has been found when A_j and k_j are constant, we may now seek an approximate solution of the Hamilton–Jacobi equation of the form $S = S_{\text{slow}} + S_{\text{fast}}$, where $\partial S_{\text{slow}}/\partial x^j$ is a slowly varying function of space and time and S_{fast} is a bounded rapidly oscillating function of the only variable ψ. With the above position, eqn (B.3) becomes

$$\left(\frac{\partial S_{\text{slow}}}{\partial x_j} + \frac{\mathrm{d}S_{\text{fast}}}{\mathrm{d}\psi} k^j + ea^j\right)\left(\frac{\partial S_{\text{slow}}}{\partial x^j} + \frac{\mathrm{d}S_{\text{fast}}}{\mathrm{d}\psi} k_j + ea_j\right) = m_e^2 c^2. \tag{B.9}$$

For the purpose of determining S_{fast} one can treat $\partial S_{\text{slow}}/\partial x^j$, k_j, and A_j as constants. Equation (B.9) then becomes identical to eqn (B.5), with $\partial S_{\text{slow}}/\partial x^j$ in the place of P_j, and its solution is given by eqn (B.6). The S_{fast} thus obtained is indeed a bounded oscillating function, provided that condition

(B.7) is respected. Condition (B.7) is therefore the equation that determines S_{slow}:

$$\frac{\partial S_{\text{slow}}}{\partial x_j}\frac{\partial S_{\text{slow}}}{\partial x^j} = \left(\frac{1}{c}\frac{\partial S_{\text{slow}}}{\partial t}\right)^2 - \left(\frac{\partial S_{\text{slow}}}{\partial \boldsymbol{r}}\right)^2 = M^2c^2, \tag{B.10}$$

where the effective mass M is given by eqn (B.8).

Since S_{fast} has been forced to be a bounded oscillating function, the average electron momentum is determined by S_{slow}. In fact, repeating the argument applied in the case of a perfect plane wave, the canonical electron four-momentum may be written as $\partial S/\partial x^j = \partial S_{\text{slow}}/\partial x^j + \partial S_{\text{fast}}/\partial x^j$ and $\langle \partial S_{\text{fast}}/\partial x^j \rangle = 0$ when S_{fast} is a bounded oscillating function. By the same token the average kinetic momentum of the electron equals $\partial S_{\text{slow}}/\partial x^j$, as the difference between canonical and kinetic momentum is ea_j, which has zero average, and the average electron velocity is parallel to $\partial S_{\text{slow}}/\partial x^j$. We may conclude that the electron moves, on average, along the trajectory of a particle the Hamilton–Jacobi equation of which is (B.10), performing around this trajectory rapid oscillations similar to those caused by a perfect plane wave approximating locally the actual electromagnetic field. The ideal particle the motion of which is determined by eqn (B.10) takes on the name of guiding centre of the electron trajectory.

The guiding centre equation, eqn (B.10), is the Hamilton–Jacobi equation of a particle having a rest mass M varying in space and time. The Hamiltonian of such a particle may be immediately written by performing the substitutions $\mathscr{H} = -\partial S/\partial t$, $\boldsymbol{P} = \partial S/\partial \boldsymbol{r}$, and is

$$\mathscr{H} = c(\boldsymbol{P}^2 + M^2c^2)^{1/2}. \tag{B.11}$$

The concept of a varying rest mass may look unfamiliar. In order to give it a physical interpretation, let us first consider the non-relativistic limit of eqn (B.11). In order to do so, it is convenient to rewrite the definition of M, eqn (B.8), as $M^2 = m_e^2 + 2m_eU/c^2$, where $U = (e^2/2m_e)\langle a^j a_j \rangle$. Equation (B.11) then becomes $\mathscr{H} = c(m_e^2c^2 + \boldsymbol{P}^2 + 2m_eU)^{1/2}$ which, in the non-relativistic limit ($c \to \infty$), gives

$$\mathscr{H} \simeq m_ec^2 + \boldsymbol{P}^2/2m_e + U. \tag{B.12}$$

Apart for the unimportant constant m_ec^2, eqn (B.12) can be easily recognized as the classical Hamiltonian of a particle moving in the field of forces generated by the potential U. (In fact, the non-relativistic form of the effective-mass approximation defines U as an 'effective potential'.) We may therefore interpret the concept of varying rest mass as the natural relativistic extension of the non-relativistic concept of potential energy, in the same way as inertial mass becomes coincident in relativistic mechanics with total energy.

In conclusion, let us note that the effective mass is, like the usual rest mass, a four-scalar; that is, it is invariant under Lorentz transformations, and so is the form of eqn (B.11). This is an obvious consequence of the fully covariant form of the equations from which effective mass was derived.

Notes

1. In this appendix we use four-dimensional notation, adopting the convention of implied summation on repeated indices. Indices are raised and lowered using a metric with the signature $(+, -, -, -)$. The contravariant co-ordinate four-vector x^j is $(ct, \boldsymbol{r})$, the covariant potential four-vector a_j is $(\varphi/c, -\boldsymbol{a})$.

References

Kibble, T. W. (1966). Mutual reflection of electrons and photons. *Phys. Rev.* **150**, 1060–9.

Luchini, P. and Solimeno, S. (1982). Effective mass of an electron in an electromagnetic field. *Phys. Lett.* **91A**, 438–40.

APPENDIX C

UNDULATOR AND FEL FORMULARY

This appendix collects the most frequently needed formulae of undulator and FEL theory. Cross-reference to the main text is provided by chapter and equation numbers. SI units are used throughout.

General symbol definitions

l_u Undulator period
L Undulator length
N $= L/l_u$, Number of periods
λ Radiation wavelength
k $= 2\pi/\lambda$, Free-space radiation wavenumber
β Radiation propagation constant (where different from k)
I_b Electron beam current
j_b Electron beam current density
G $= P_{out}/P_{in}$, Power gain (output power/input power)
g $= (P_{out} - P_{in})/P_{in}$, Incremental gain

Electron motion in undulator

Dimensionless undulator vector potential:

$$\boldsymbol{a}_\perp = \mathrm{Re}(A_u e^{ik_u z}) \quad (0 \leqslant z \leqslant L), \qquad a_z = \varphi = 0, \tag{3.21}$$

where $k_u = 2\pi/l_u$, and A_u is non-dimensionalized with respect to $m_e c/e$. With b_{peak} (peak magnetic field) in teslas and l_u in metres, $A_u = 93.37 f b_{peak} l_u$, where $f = 1$ for linear polarization and $f = \sqrt{2}$ for circular polarization. The dimensionless number A_u (or sometimes $A_u/\sqrt{2}$) is also called the deflection parameter K.

Effective mass:

$$M = (1 + A_u^2/2)^{1/2}. \tag{3.26}$$

Peak amplitude of transverse electron oscillation:

$$\Delta x_u = A_u/\gamma k_u. \tag{3.29}$$

Characteristic lengths

Betatron oscillations period:

$$l_{\text{bet, typ}} = 2\gamma l_u/A_u. \tag{3.51}$$

Gain length:

$$L_g = \gamma(I_A/\pi A_u^2 k_u j_b)^{1/3}. \tag{5.26}$$

Synchrotron oscillation period:

$$L_{s,\min} = (M^2/2A_u A_w)^{1/2} l_u, \tag{7.48}$$

where A_w is the dimensionless vector potential of the laser electromagnetic wave. With P_{in} in W/m^2 and λ in metres, $A_w^2 = 2\zeta e^2 P_{\text{in}}/k^2 m_e^2 c^4 = 7.31 \times 10^{-11}\lambda^2 P_{\text{in}}$.

Spontaneous radiation

Fundamental resonance wavenumber:

$$k_r = 2k_u/(M^2/\gamma^2 + \vartheta_{we}^2), \tag{4.42}$$

where ϑ_{we} is the angle between direction of wave propagation and electron trajectory.

Relative linewidth of emission near nth harmonic of k_r:

$$\delta\lambda_{fwhm}/\lambda \simeq \delta k_{fwhm}/k = 0.89/nN. \tag{4.43}$$

Angular half-aperture of the central-frequency cone:

$$\delta\vartheta_{we,\,fwhm} = 0.94(nN)^{-1/2}M/\gamma = (1.78\lambda/L)^{1/2}. \qquad \text{(4.44a, b)}.$$

Effective undulator vector potential for linear polarization and emission on axis on the $(2p + 1)$th harmonic:

$$A_{ueff,\,2p+1} = \left\{\mathrm{J}_p\left[\frac{(2p+1)A_u^2}{4 + 2A_u^2}\right] - \mathrm{J}_{p+1}\left[\frac{(2p+1)A_u^2}{4 + 2A_u^2}\right]\right\}A_u. \tag{4.51}$$

Detuning parameter:

$$\nu_n = \pi N(k - nk_r)/k_r. \tag{4.41}$$

Power per unit frequency and solid angle:

$$S_\omega = \frac{k^2 L^2 \zeta e I_b}{32\pi^2\gamma^2} A_{ueff,\,n}^2 \frac{\sin^2\nu_n}{\nu_n^2}. \tag{4.46}$$

Small-signal, small-gain regime

Conditions of validity:

$$l_u \ll L \ll L_g, \quad L \ll L_{s,\min} \quad (A_w \ll M^2/A_u N^2).$$

Basic FEL gain formula:

$$g_0 = N^3 A_u^2 \gamma^{-3} \frac{\pi^2 l_u^2 j_b}{I_A} \frac{\cos 2v - 1 + v \sin 2v}{v^3}, \tag{5.23}$$

where I_A (Alfvén current) = 17.03 kA.

Maximum gain (for $v = -1.303$):

$$g_{0\max} = 0.54 N^3 A_u^2 \gamma^{-3} \pi^2 l_u^2 j_b / I_A = 0.27 (L/L_g)^3. \tag{5.24–28}$$

Gain for other-than-circular polarization and/or higher harmonics:

$$g = \frac{|A_{ueff,n} \cdot \hat{E}_w^*|^2}{A_u^2} g_0, \tag{5.64}$$

where $\hat{E}_w$ is the polarization vector of the input electromagnetic wave.

Maximum gain for a Gaussian radiation beam with the optimal Rayleigh range $z_R = 0.25\,L$ and an infinitely thin electron beam:

$$g_{\max} = 0.93 \frac{I_b}{I_A} \frac{(2\pi n N)^2 A_{ueff,n}^2}{\gamma M^2}. \tag{5.82}$$

Maximum admissible angular spread:

$$\delta\vartheta_{i,fwhm} = (1.78\lambda/L)^{1/2}. \tag{4.77}$$

Maximum admissible relative energy spread:

$$\delta\gamma_i/\gamma_i \simeq 1/2nN. \tag{4.80}$$

Maximum admissible beam emittance:

$$\varepsilon \ll \lambda. \tag{5.84}$$

Quantum regime:

$$\lambda \lesssim N\lambda_C/\gamma, \tag{4.89}$$

where λ_C (Compton wavelength of the electron) = 2.4×10^{-12} m.

Small-signal, large-gain regime

Conditions of validity:

$$l_u \ll L_g \ll L \ll L_{s,\mathrm{min}}, \quad A_u^2 L_g \gg M^2 l_u.$$

Dispersion equation:

$$p(p - \nu)^2 + 1 = 0, \tag{6.29}$$

where $p = (\beta - k)L_g$, $\nu = k_u L_g(k - nk_r)/k_r$.

Gain:

$$G = K\mathrm{e}^{2\mathrm{Im}(p)L/L_g}, \tag{6.30}$$

where $K = |(p - \nu)/(3p - \nu)|$.

Maximum gain (for $\nu = 0$):

$$G_{\mathrm{max}} = \frac{1}{9}\mathrm{e}^{1.732L/L_g}. \tag{6.31}$$

Linewidth, admissible energy spread: replace L by $2L_g$ in the small-gain formulae.

Raman large-gain regime

Conditions of validity: $l_u \ll L_{g(\mathrm{Raman})} \ll L, \quad A_u^2 L_g \ll M^2 l_u$.

Gain length:

$$L_{g(\mathrm{Raman})} = (4M^2 l_u/\pi A_u^2 L_g)^{1/4} L_g = (8\gamma^3 I_A/\pi A_u^4 k_u^2 j_b)^{1/4}. \tag{6.49}$$

Gain:

$$G_{\mathrm{max}} = K\mathrm{e}^{2L/L_{g(\mathrm{Raman})}}. \tag{6.50}$$

Backward-wave instability

Suppressed as long as

$$L/L_g < (\pi^3 L_g/2l_u)^{1/2}. \tag{6.99}$$

Large-signal, small-gain regime

Conditions of validity:

$$l_u \ll L_{s,\mathrm{min}} \ll L \ll L_g.$$

Relation between gain and energy extraction:

$$g = (8\pi\gamma j_b/k^2 A_w^2 I_A)\eta. \tag{7.4}$$

Detuning parameter:

$$\nu = (M^2/2A_uA_w)^{1/2}(k - k_r)/2k_r. \tag{7.54}$$

Trapped-electron fraction:

$$f_t = (2/\pi)\arccos \nu. \tag{7.55}$$

Approximate formula for energy extraction at complete saturation:

$$\eta_{\text{sat}} = -\left(\frac{2A_uA_w}{M^2}\right)^{1/2} \frac{\nu \arccos \nu}{\pi/2}. \tag{7.58}$$

Maximum energy extraction (attained for $\nu \simeq -0.62$ and $L/L_{s,\min} \simeq 0.68$):

$$\eta_{\max} = 0.44/N. \tag{7.66}$$

Gain at maximum energy extraction: one fourth of small signal gain (7.67).

Optical klystron

Small-signal gain:

$$g = \frac{\nu_d}{2\pi}\frac{L_1^2}{l_u^2} A_{ueff,n}^2 \gamma^{-3} \frac{\pi^2 l_u^2 j_b}{I_A} \frac{\sin^2 \nu_n}{\nu_n^2} \frac{\partial}{\partial \nu_d}(4\cos^2 \nu_d), \tag{5.97}$$

where

$$\nu_d = k[(1 + \langle a_1^2\rangle)L_1 + (1 + \langle a_d^2\rangle)L_d]/2\gamma^2. \tag{5.94}$$

Saturation parameter:

$$q = \frac{4\pi^2\langle a_d^2\rangle L_d L_1}{M^4 l_u^2} A_u A_w. \tag{7.79}$$

Intensity of nth harmonic in frequency multiplication:

$$|F_{2,n}| = |\mathrm{J}_n(nq)| \tag{7.83}$$

Tapered undulator

Maximum wave amplitude:

$$A_{w,\max}^2 = 8\pi\gamma j_b/k^2 I_A. \tag{8.49}$$

Nominal energy yield:

$$\eta_{\text{nom}} = \pi^2 N^2 A_{w,\max}^2 (A_u/M^2)^2. \tag{8.53}$$

Condition for near-unity yield:

$$A_{w,\max} \simeq M^2/NA_u. \tag{8.54}$$

Nominal gain:

$$G_{\text{nom}} = (1 + \pi N A_{w,\max}^2 A_u / M^2 A_{w,\text{in}})^2. \tag{8.55}$$

Gain limitation:

$$G_{\text{nom}} \ll N^{2/3}. \tag{8.59}$$

Storage-ring FEL operation

Renieri limit:

$$P_{\text{out}} \simeq P_{s,\text{tot}}/2N. \tag{9.15}$$

AUTHOR INDEX

SUBJECT INDEX